Plumbing and Mechanical Services:

a textbook

Plumbing and Mechanical Services:
a textbook

A. H. Masterman and R. M. Boyce

Stanley Thornes (Publishers) Ltd

First published in 1984 by:
Hutchinson Education

Reprinted in 1990 by:
Stanley Thornes (Publishers) Ltd
Delta Place
27 Bath Road
CHELTENHAM
GL53 7TH
United Kingdom

04 05 06 / 15 14 13 12 11 10

A catalogue record for this book is available from the British Library

ISBN 0 7487 0368 3

Printed in Croatia by Zrinski

Contents

Preface

Plumbing and Mechanical Services provides a learning resource for students of plumbing, heating, gas and allied industries studying for qualification at City & Guilds Craft Certificate Level.

The text closely follows the requirements of the syllabus in craft technology, working processes and associated subjects, the primary aim being to assist the student towards that qualification.

Although mainly written for those commencing a career in the mechanical services industry this book will contain much of interest for the mature craftsman and those who wish to keep abreast with the changing techniques and advancing technology in this section of the construction industry. In addition to those mentioned above, this book should prove invaluable for technician students and for students attending Links, Foundation or Youth Training Scheme programmes.

Acknowledgements

The authors and publisher are grateful to the following for permission to reproduce textual material and illustrations:

The Swiss Society of Engineers and Architects for Figures 7, 8, 9, 10, 11, 21, 22, 23, 31, 32, 40, reproduced from *Berufskunde für Spengler*; HMSO for Table 10, reproduced from Building Research Establishment Digest 69: *Durability and application of plastics*; BCIRA for Figures 54 and 55; Tubela Engineering Co. Ltd for Figures 111 and 112; Bartols Plastics for Figures 208 and 209; Barking-Grohe for Figures 221–3; Armitage Shanks Ltd for Figures 279, 280, 282, 284–9, 291, 292, 294–9, 308; UBM Building Supplies Ltd for Figures 290, 294, 305, 306, Phetco (England) Ltd for Figure 301; Twyfords Bathrooms for Figure 310; The Lead Development Association for Figures 232–5, 327, 328, 330(b), 331–7; The Copper Development Association for Figures 351–6 and material from *Technical Note TN6* used in Chapter 8; British Alcan Aluminium Ltd for material from *Aluminium Roof Weatherings* used in Chapter 8; Table 18 from BS 1566: Part 1: 1972 is reproduced by kind permission of the British Standards Institution, 2 Park Street, London W1A 2BS from whom complete copies of the standard can be obtained.

Every effort has been made to reach copyright holders, but the publisher would be grateful to hear from any source whose copyright they may unwittingly have infringed.

1 Safety

After reading this chapter you should be able to:

1. State the basic requirements of the Health and Safety at Work etc. Act 1974.
2. List the main hazards involved in handling materials and equipment, including scaffolding.
3. State the correct use of and maintenance procedure for tools.
4. State the important points to observe when lifting.
5. Explain the possible hazards of falling items, fragile roofs and welding.
6. State the requirements to observe when working from ladders.
7. State the safe working procedures for electrical tools.
8. List the problems associated with untidy sites.
9. Demonstrate general first aid including artificial respiration.
10. Name the correct equipment for fire fighting.
11. State the correct procedure in the event of an accident.

Introduction: safety is everybody's business

There are far too many accidents in the construction industry, many of which could be avoided with thought and common sense. The Government publishes a guide entitled Health And Safety At Work etc. Act 1974. If the procedures in this were adhered to by both employers and employees, then the number of accidents would be very greatly reduced.

Accidents are generally caused by people disregarding the recommended procedures. They may feel that accidents only happen to other people and that, in any case, they have done that operation many times before without any problem.

Regardless of how good an Act is, it will only succeed if all participants, both employers and workers, are awake to their obligations and respond accordingly.

The two basic principles of accident prevention are:

1. Implement safe methods of working to reduce the chance of a mistake.
2. Implement precautions to reduce the chance of injury even if somebody does make a mistake.

This chapter looks briefly at matters of general safety and highlights many of the common causes of accidents in the construction industry. The procedure and the correct type of extinguishers to be used in the case of fire are listed. Basic first aid treatment and the methods of resuscitation in case of shock are described, and, finally, the reporting procedure of accidents is dealt with.

Health and Safety at Work etc. Act 1974

This Act has wide implications. Its purpose is to provide the legislative framework to promote, stimulate and encourage high standards of health and safety and welfare of all personnel at work as well as the health and safety of the public as affected by work activities. Its broad aim concentrates on the promotion of safety awareness and effective safety organization and performance channelled through schemes designed to suit the particular industry or organization. The Act is an enabling measure superimposed over existing health and safety legislation and consists of four main parts: Part I relates to health, safety and welfare at work; Part II relates to the Employment Medical Advisory Service; Part III relates to the building regulations; and Part IV relates to a number of miscellaneous and general provisions.

The Act established a new Health and Safety Commission and brought together a number of government inspectorates into one body called the Health and Safety Executive (HSE). Under the Act, the Health and Safety Commission is responsible for the work of the Executive who have the power to enforce statutory requirements on employers to provide and maintain a safe and healthy place of work. If an inspector discovers a contravention in any Act or regulation he or she can issue a prohibition notice or an improvement notice. He or she can also prosecute any person contravening a relevant statutory provision. He or she can also seize, render harmless or destroy any substance or article considered to be the cause of imminent danger or serious personal injury.

The scope of the Act includes all 'persons at work', whether employers, employees or self-employed persons but not domestic servants in a private household. It also covers the keeping and use of dangerous substances and their unlawful acquisition, possession and use. Requirements can be made imposing controls over dangerous substances in all circumstances, including all airborne emissions of obnoxious or offensive substances which are not a danger to health but which would cause a nuisance or damage the environment.

In 1975, the Act came into full operation and HSE set up a Construction Central Operations Unit to undertake a detailed study to produce a successful organization for safety in the industry.

Tools

Most accidents with tools are caused by workers striking themselves or other workers; through the use of defective tools, or through misuse of tools. Figures 1–4 show some typical examples of defective tools. They must be repaired immediately or taken out of service. Figures 5–8

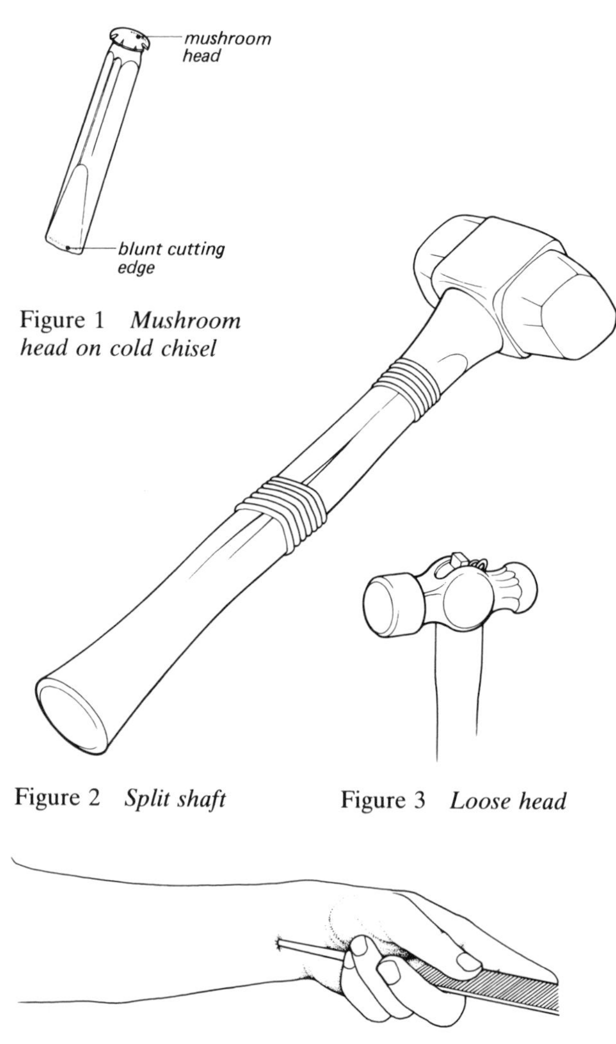

Figure 1 *Mushroom head on cold chisel*

Figure 2 *Split shaft*

Figure 3 *Loose head*

Figure 4 *Defective handle on file*

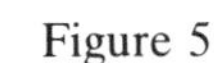

Figure 5

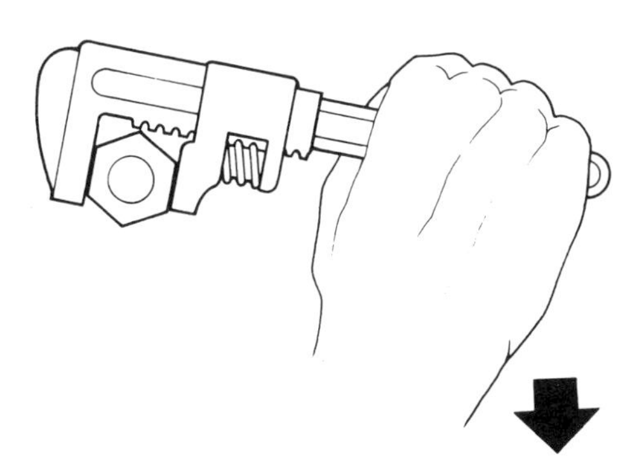

Figure 6

show a few typical examples of the correct application of tools. The basic principles to follow are:

Make sure you use the right tool for the job – do not make do with the wrong one.

Wear unbreakable goggles when:
- chipping welds;
- de-scaling boilers;
- cutting concrete, brick etc.

When using stillsons and wrenches make sure the pull forces the jaws together, otherwise the tool might slip away.

Never leave a defective tool about for others to use.

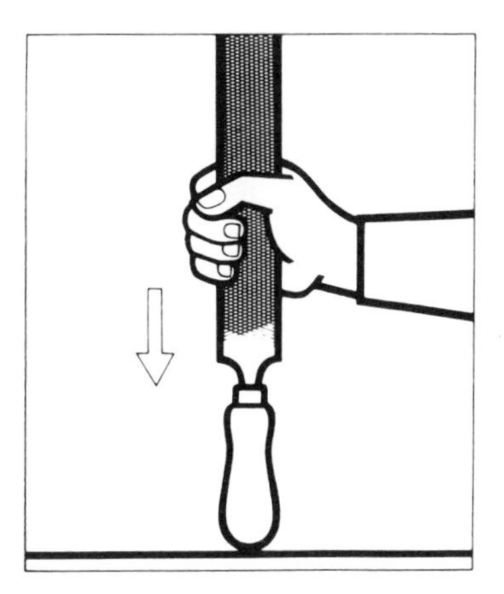

Figure 7

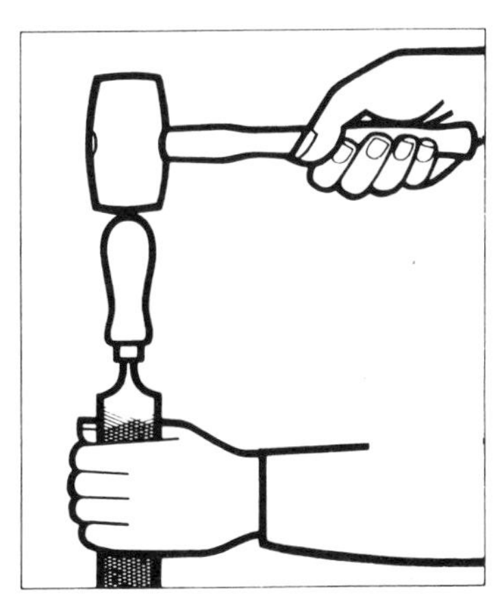

Figure 8

Lifting

The six major points to remember when lifting are:

1 Back straight.
2 Chin in.
3 Arms close to body.
4 Feet slightly apart.
5 Bend knees and lift by straightening the legs.
6 Grip with palm of hands, not just fingers.

The correct and incorrect methods of lifting are illustrated in Figures 9 and 10. The following checklist details the procedure to be observed:

Size the job up: look out for splinters and jagged edges on the object to be lifted.
If you are going to carry an object make sure you have an unobstructed path.
Beware of slippery surfaces
Feet 200–300 mm apart, one foot in advance of the other, pointing in the direction you intend to go – feet together can cause a rupture.
Chin in – avoid dropping head forwards or backwards
Bend the knees to a crouch position – back straight but not necessarily vertical.
Arms close to the body as possible so that the body takes the weight (instead of the fingers, wrists, arm and shoulder muscles).
Get a firm grip with the palms of the hands and the roots of the fingers – using just the finger tips means more effort and more chance of dropping the object.
Lift with the thigh muscles by straightening the legs.
Lift by easy stages – from floor to knee, from knee to carrying position.
Make sure you can see over your load when carrying.
Do not change grip while carrying – rest the load on some firm support, then change.
Reverse the lifting procedure to set the object down.
Wear gloves when handling sharp or slippery objects.

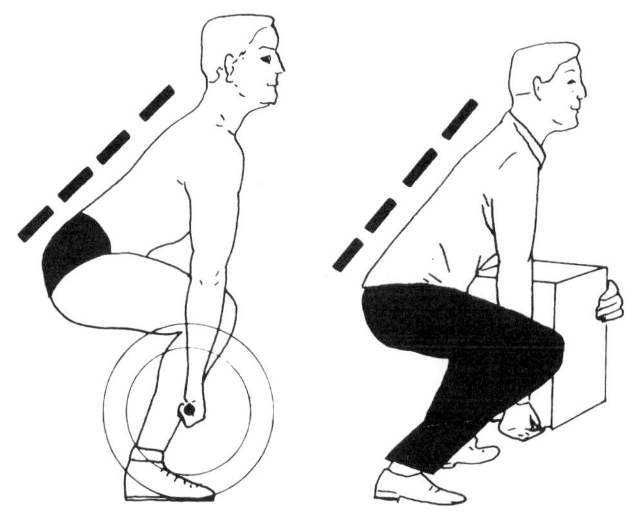

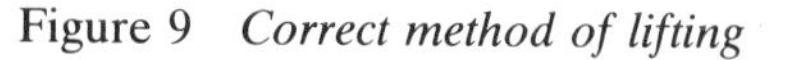

Figure 9 *Correct method of lifting*

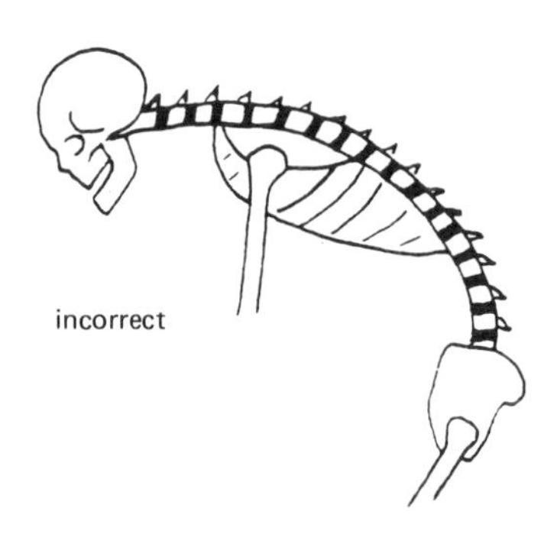

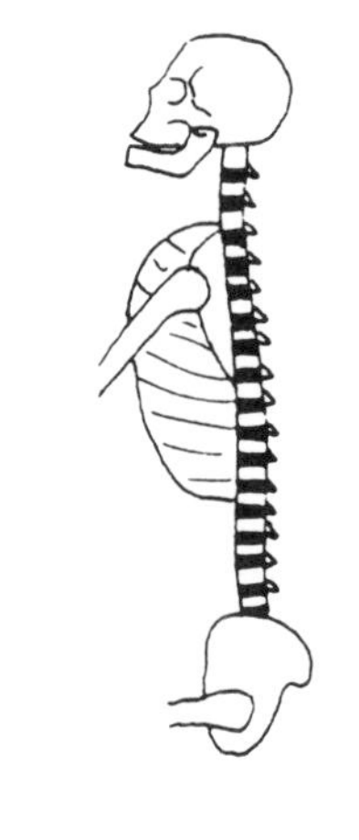

Figure 10

Team lifting

The same basic principles apply when two or more men are lifting the same object (see Figures 11 and 12). In addition, remember the following procedure:

Lifting gangs must work as a team.
Everyone in a lifting team should be roughly the same height.
The operation should be planned from 'lifting' to 'setting down' – route to be taken, signals etc.
Make sure everyone in the team knows precisely what to do.
Appoint one person as team leader.

Figure 11

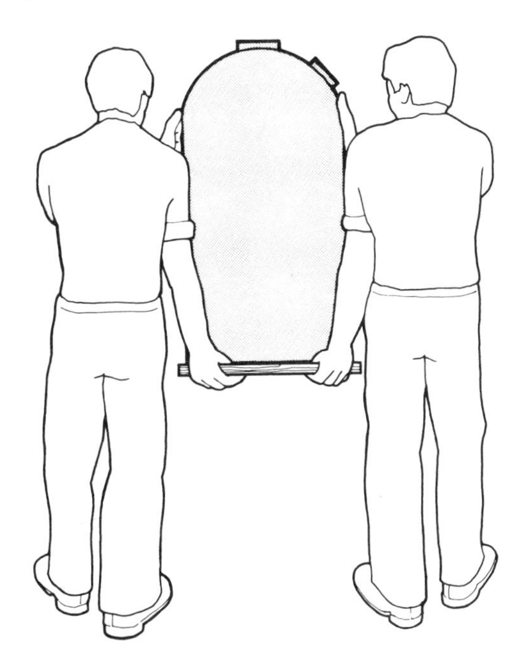

Figure 12 *Team lifting*

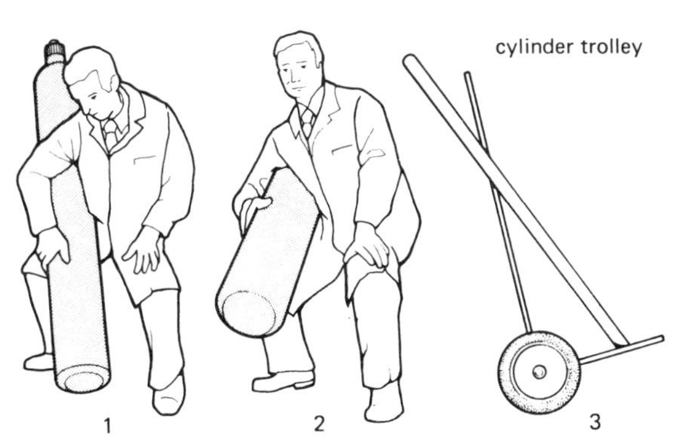

Figure 13 *Lifting a cylinder*

Lifting gas cylinders

Special care should be taken when moving gas cylinders. Use a trolley wherever possible. To lift a cylinder on to a trolley, first lift to the vertical position with a straight back and bent knees and then use your thighs as shown in Figure 13.

Falling materials

It is well to remember that an object gathers tremendous energy when falling. For example, a 19 mm nut falling and striking a person's unprotected head from 20 m high can kill him or her. It is therefore essential that great care must be taken not to place tools, tins of jointing paste

Figure 14 *Lacerated scalp*

etc. on pipes or ledges, where they could topple off. Always observe the following safety rules:

Do not place materials where they are likely to fall or get knocked off.
If possible position yourself and other workers where there is no danger from falling objects.
When working under scaffold make sure it has a toe-board.
Lower materials – never throw things down from scaffolds or ladders or throw things to other workers on the ground.
Make sure materials are securely stacked and withdrawn in the correct order.
Always wear safety helmets and safety boots.
Provide tool boxes for tools.

Stacking materials

Location
Materials should be readily accessible and as close as possible to the point of use, but:

Not in quantities so great as to limit working space unnecessarily.
Not where they will cause obstruction.
Not close to edges of excavations.
Not close to moving machinery or overhead lines.
Not where they will interfere with new deliveries.

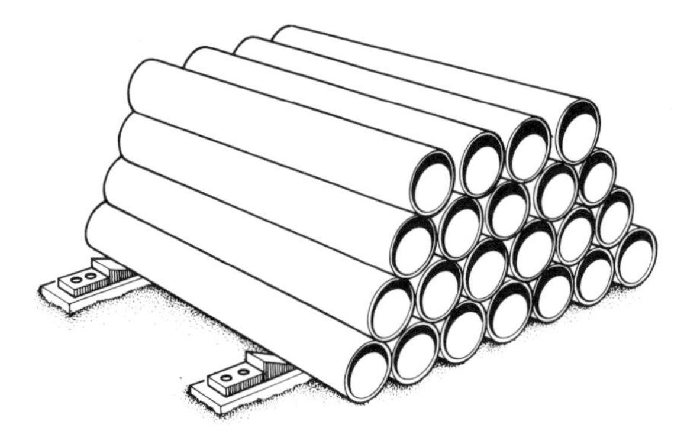

Figure 15 *Acceptable stacking of materials*

Foundations and supports
A firm, even base is essential. The foundation – for example, the floor – must be strong enough to support the total weight, which may be considerable. Stacking materials against a wall may be dangerous, as the wall may not be designed to take the sideways thrust.

Size
Stacks should not normally be higher than a man, to permit easy withdrawal. The shorter base should be about one-third the height.

Structure
Batter (i.e., step back) every few tiers. Chock or stake rolling objects (for example, drums, pipes) with sound material. Bond to prevent the stack collapsing. Avoid unnecessary protrusions – protrusions which cannot be avoided should have a distinctive marker tied to them. Oxygen cylinders may be stacked horizontally. Acetylene cylinders *must* be stored and used vertically.

Use
Withdraw materials from top of stack – never from bottom or sides. Do not climb on to stacks – use a ladder.

Excavations

Accidents can arise when people slip into trenches, sometimes while trying to jump across them or while climbing out of them or when supports give way. Other accidents are due to people falling into uncovered manholes while walking across a site. Always observe the following safety rules (see Figures 16 and 17):

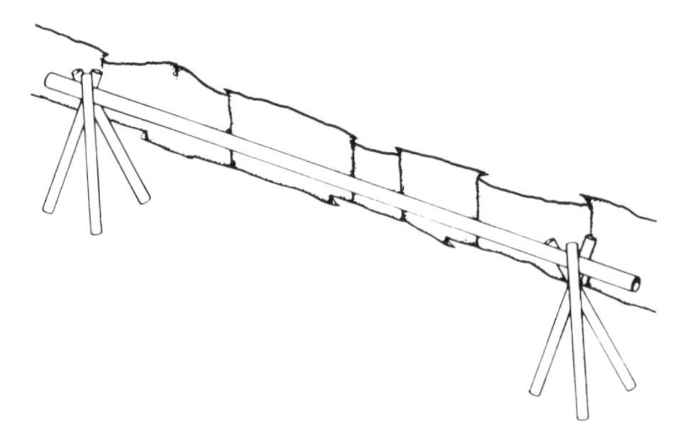

Figure 16 *Trench fenced off with scaffold tubes*

Figure 17 *Mark covers over holes*

Warn people to look where they are going.
Provide proper walkways across trenches.
Erect barriers round excavations where necessary.
Do not leave manholes uncovered or unfenced.
Take care when doing manhole work
Fence openings or cover them with heavy material appropriately marked.
Never walk along pipes.

Ladders

Used incorrectly or in bad condition a ladder becomes a hazard. Working from a ladder is inherently dangerous. Where possible always provide a working platform. Most falls from ladders are the result of a person simply slipping or falling from the ladder, but movement of the ladder also causes a considerable number of accidents – the ladder either slips outwards at the bottom or sideways at the top. Accidents are also caused by missing or broken rungs or by the ladder itself breaking. Always observe the following points:

Use of ladders
Stand on a firm, even base. Trussed side underneath if reinforced.
Never wedge one side up if ground is uneven. Either level ground or bury foot of ladder.
Set at correct angle of repose – four up to one out. The ladder should project at least 1 m above any landing place.
Beware of wet, icy or greasy rungs. Clean any mud or grease from boots before climbing.
Watch out for live overhead cables, particularly when using metal ladders.

Typical examples of the incorrect use of ladders which results in accidents
Figures 18–23 show six examples of the dangerous uses of ladders.

Never use a ladder which is too short or stand it on something, for example a drum or dustbin, to get extra height (Figure 18).
Do not over-reach sideways from a ladder – move it (Figure 19).

Figure 18 *Asking for trouble*

Figure 19 *Never over-reach*

Never support a ladder on its rungs (Figure 20).
Never work on a ladder set at the incorrect angle (75° is recommended) (Figures 21 and 22).
Never overload a ladder (use a hoist) (Figure 23).

Correct method of climbing and descending a ladder

1 Be aware of your limitations.
2 Check ladder for security
3 Clean mud etc. from your footwear.
4 Face ladder squarely. Using both hands either
   (a) grasp stiles, or
   (b) grasp rungs (fireman fashion).

Figure 20 *Never do this*

5 Keep feet placed well into the rungs.
6 Eyes should be directed at working level above – do not look down
7 Don't carry anything in your hands. Tools and materials may fall when carried up ladders, even if carried in pockets. Where possible provide a hoist line. Alternatively, use a shoulder bag.

Securing ladders

Too much importance cannot be placed on this part of your work. In addition to the obvious lashing of the top of the ladder, the foot of the ladder should also be suitably anchored. When it is not possible to tie the top then side guys should be used. These should be secured to the stiles (never the rungs) and should form an angle with the horizontal of approximately 45° (see Figures 24 and 25).

Platforms

A hook-on foot is useful in providing something more comfortable than a rung to stand on (see Figure 26). Foot platforms should be capable of being easily fitted and removed and should give a level surface with the ladder placed at 75°. Platforms are designed to protect either in front of or behind the ladder. One disadvantage is they can be difficult to climb past. Various hook-on tray attachments are available to enable tools and components to be readily accessible.

Figure 21 *Too steep*

Figure 22 *Not steep enough*

Figure 23 *Never overload a ladder – use a hoist*

Figure 24 *Ladder staked and guyed*

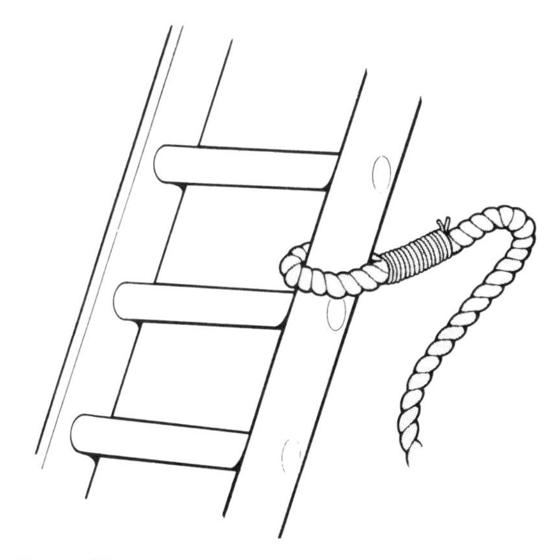

Figure 25

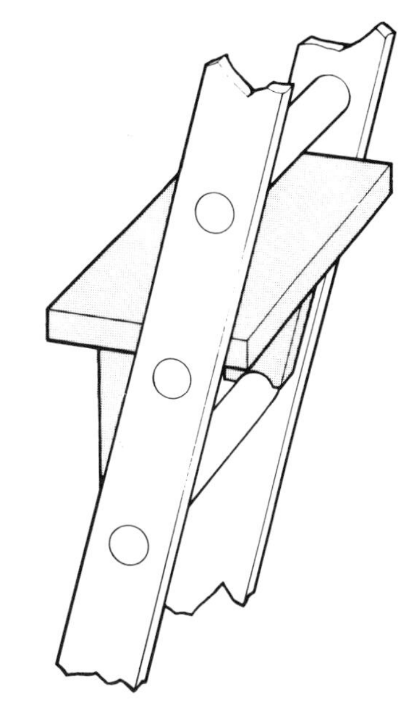

Figure 26 *Platform*

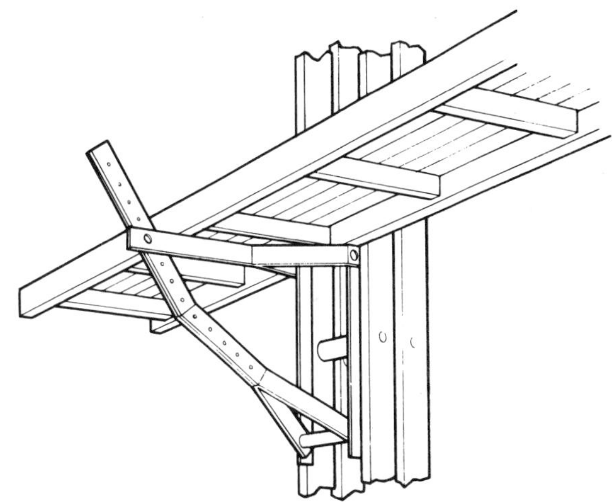

Figure 27 *Cripples*

Cripples (ladder brackets)

A pair of ladder cripples enables a light working platform to be erected between two ladders (see Figure 27). Some are adjustable for angle while others have a fixed angle to suit the ladder at 75° rake.

Standoff

These fitments hold the top of the ladder off from the wall (see Figure 28). They are particularly

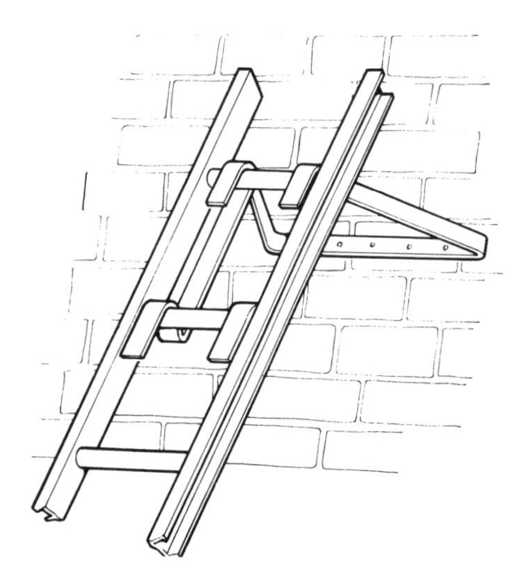

Figure 28 *Standoff*

useful when working on gutters etc. as they overcome the need to lean outwards. It is advisable to secure them to the ladder to prevent side-to-side movement.

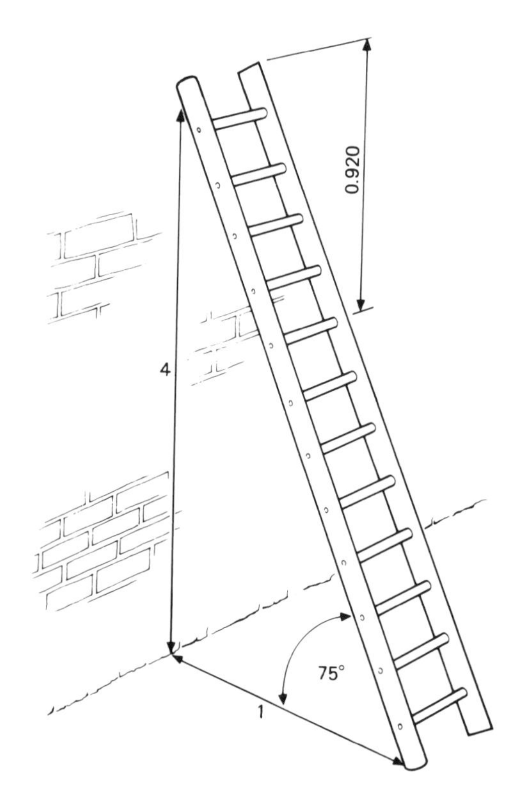

Figure 29

Correct positioning of a ladder

The recommended ideal working angle of a ladder is 75° to the horizontal or one unit out to four units up (see Figure 29). It is also recommended that the top of the ladder should extend 1 m past the working platform. The lift of a single ladder should not be more than 8–10 m.

Raising and carrying a ladder

Figure 30 illustrates the correct method of raising a ladder.

Ladders should be carried vertically or with the front end elevated (see Figures 31 and 32). It takes two people to move a tall ladder.

Storage of ladders

Do not leave ladders on wet ground or leave them exposed to the weather. Store at normal temperatures under cover to prevent warping. Support at intermediate points and not just the ends (see Figure 33).

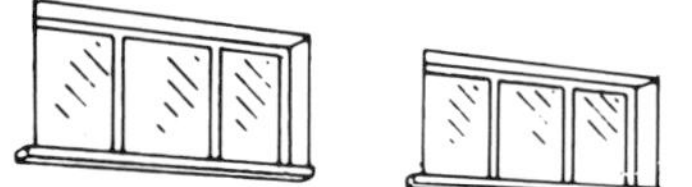

Figure 30 *Correct method of raising a ladder*

Figure 31 *Correct method of carrying ladders*

Figure 32

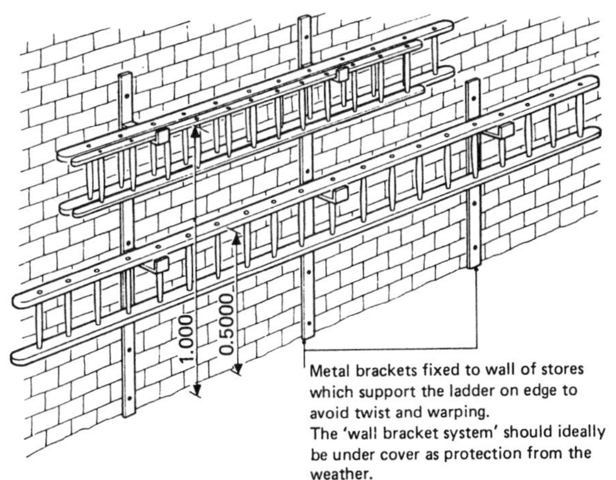

Figure 33 *Correct method of storage*

Inspection and maintenance

Always inspect a ladder before use. Ladders should *not* be subjected to a severe test. Always check:

The rungs, particularly at the point where they enter the stiles.
The wedges, which should be properly in position.
The stiles for warping, cracking or splintering.
The condition of the feet.
Any ropes or metal attachments.
To test the rungs tap each rung with a mallet – a dull sound indicates a defective rung.

Take defective ladders out of service immediately, mark them defective and do not use them again until they are repaired. Destroy unfit ladders which cannot be repaired.

Treat new ladders with clear wood preservative, particularly round the end grain of the rungs and coat with clear varnish. Painting a ladder is not recommended as the paint may hide defects.

Electrical work

The following are the most common causes of faults in electrical apparatus:

1 Improvised junction boxes with, for example, wires jammed in sockets with matchsticks or nails.
2 Insulation damage through flexing – cables should be protected by heavy rubber sleeves at the point where they enter the tool plug.
3 Earth wires pulled out of terminals and touching live conductors. (When making connections the earth wire should have some slack so that, if the cable tends to pull out, the earth wire will be the last to fail.)
4 Powered tools run off lamp sockets so that they cannot be earthed. There is also the danger that the earth wire, if tucked into the lamp adaptor, may make contact with a live conductor.
5 Wrong connections in plugs or joints, usually caused by unauthorized persons tampering with electrical apparatus or confusion of British and Continental colour codes.

6 Plugs forced into the wrong socket.
7 Use of wrong fuses. Fuses should be rated as closely as possible to the normal working current of the tool. Using a 30-A fuse for a 3-A load will result in greater injury should there be an accident.
8 Cables lying around where they can get damaged or wet.
9 Cables hung up on nails, which is not a secure method of fastening and can damage the insulation. Proper cleats should be used.

Remember:
Be especially careful of overhead power lines when using mobile scaffolding.
Do not work near trailing cables – get them suspended.
Inexperienced people should not tamper with electrical connections.
In case of failure in breathing owing to electrical shock, artificial respiration must be started immediately.
Always use 110 V supply, centre tapped to earth, for portable equipment

Powered tools

In inexperienced hands, powered tools can become a source of danger.

Electrically operated tools

Look out for:
faulty leads;
trailing leads;
faulty plugs;
unearthed equipment.
Check that all tools are properly earthed and insulated before use.
Make sure that leads are suspended and not trailing in oil or water.
Be especially careful with unguarded threading and cutting machines.
If a machine has a guard, make sure it is fitted.

Step-by-step method of connecting a plug

See Figure 34:
1 Prepare the cable. The outer sheath should then be stripped back 32 mm and the insulation on the three cores should be removed to expose 15 mm of conductor.
2 Prepare the plug. Remove the cover, take out the fuse and remove the terminal fixing nuts of the MK Safetyplug, or loosen the terminal screw in other makes of plug, to enable the conductor to be inserted. (The patented terminal nut used in the MK Safetyplug ensures a greater area of contact.)
3 Secure the cable in the cord grip. The outer sheath should pass through the cord grip. It is essential that the cable is retained firmly in the cord grip, to prevent conductors being pulled out in service; the patented cord grip in the MK Safetyplug ensures a secure grip.
4 Connect the conductors: green/yellow to the earth (centre) terminal; brown to the live (right hand) terminal; and blue to the neutral (left hand) terminal. Twist the conductors to ensure that there are no loose strands and wind them around the terminal pillar of the MK Safetyplug in a clockwise direction so that they are tightened as the fixing nut is screwed down.
5 Have a final check at the end to ensure that all the fixing nuts are securely tightened, and that there are no loose strands of wire.
6 Replace the (ASTA certified) fuse. For small appliances up to 700 W, for example table lamp, hair dryer, mixer/blender, sewing machine, hi-fi and television (black and white only), it is preferable to replace the brown 13 A fuse by a red 3 A fuse. Larger appliances that will need the 13 A fuse include deep freezer, refrigerator, electric fire, washing machine and colour TV.
7 Replace the cover, making sure that the conductors are within the appropriate wiring channels in the plug.
8 Tighten the captive cover fixing screw. (Part insulation of the pins on the MK Safetyplug prevents risk of the plug pins being touched whilst the plug is being inserted or removed. They also prevent contact from knives and other implements inserted between the plug and socket-outlet by children.)

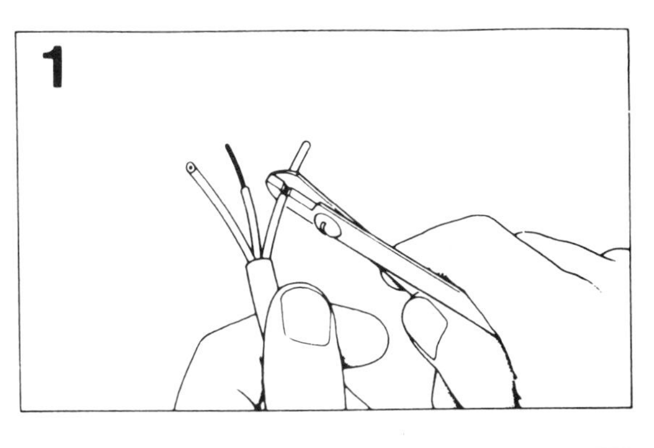

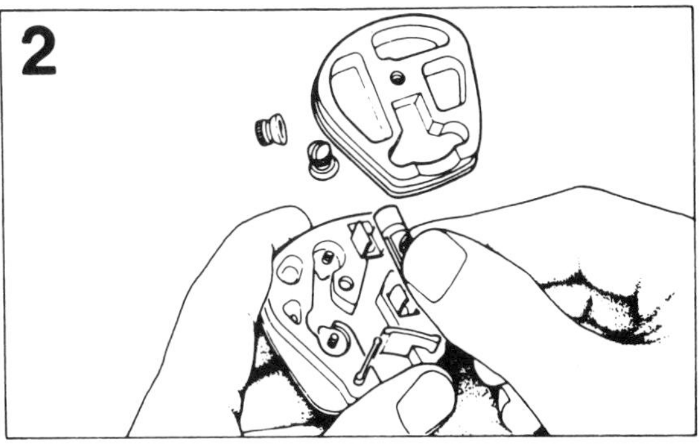

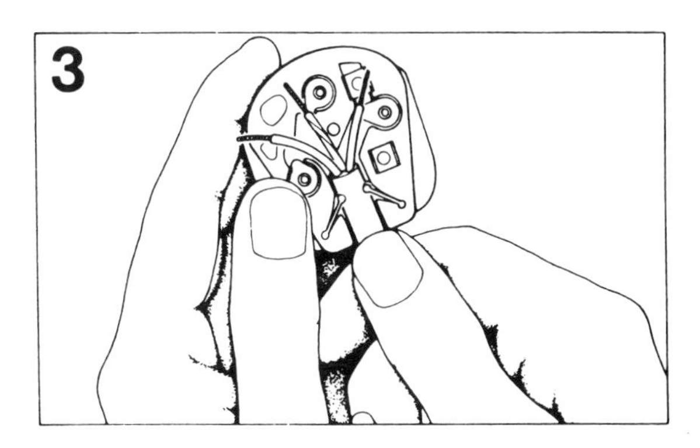

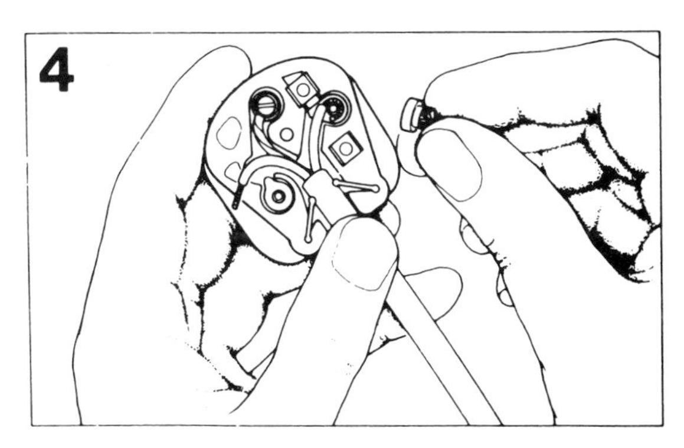

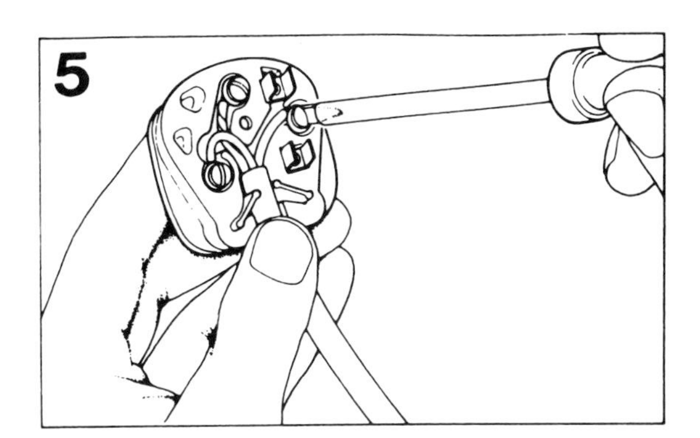

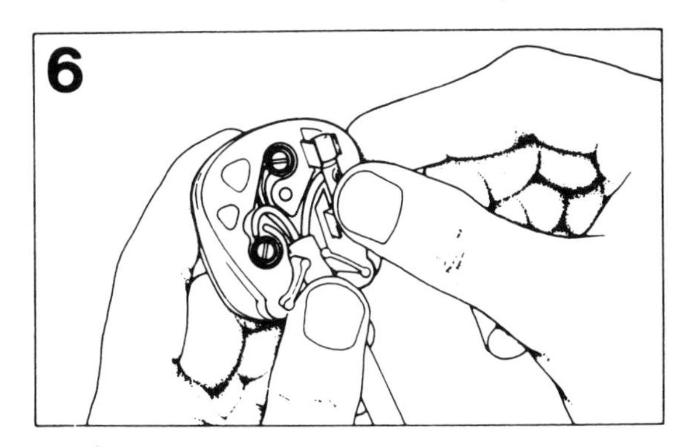

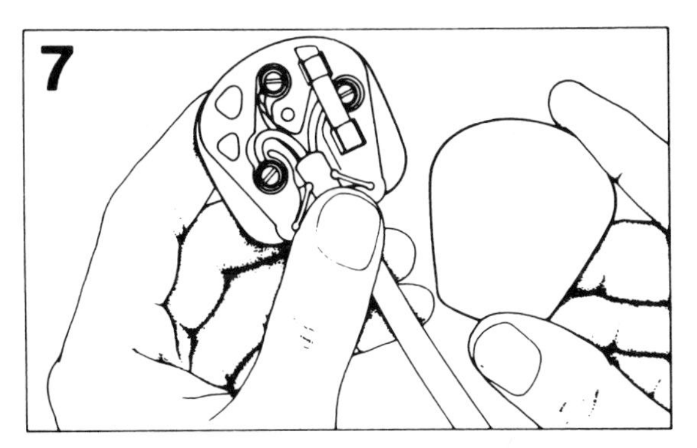

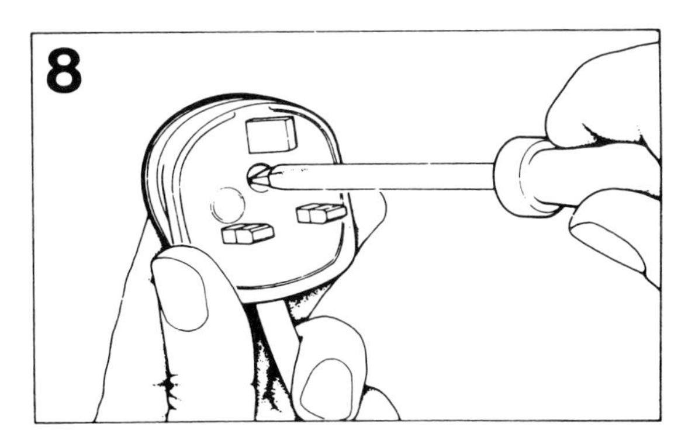

Figure 34 *Connecting a 13 A plug (UK only)*

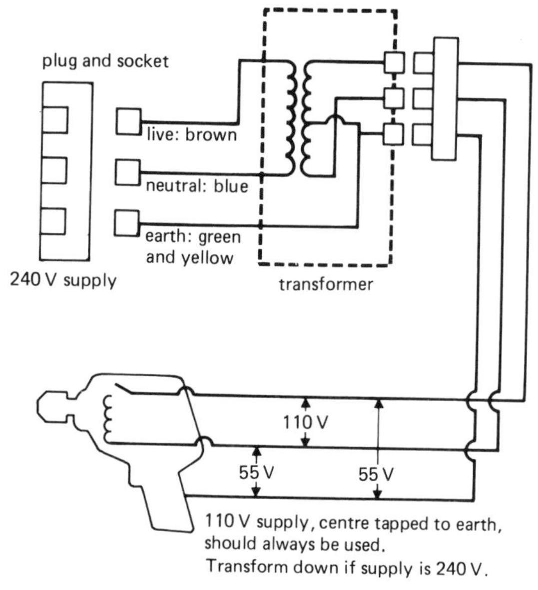

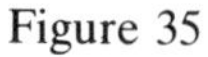
Figure 35

Figure 36 *Goggles – wear them over your eyes*

Figure 35 shows the method of connecting the power (240 V) via the plug and transformer to the local power tool.

Cartridge operated tools

Anyone using these tools must be fully instructed in their use.
Unbreakable goggles must be worn.
Always check the material into which the fixing stud is to be fired – make sure there is no danger of the stud going right through the material.
Make sure protective guard is fitted.
Never leave unexploded cartridges lying about.
Always unload when not in use.

Goggles

Make sure goggles are of the unbreakable sort.
Always wear goggles when:
grinding;
cutting concrete;
using a cartridge operated tool;
drilling metal;
Wear goggles over your eyes – they are no good on top of your head (see Figure 36).

Scaffolding

Falls from scaffolds are usually serious, sometimes fatal. The main causes are:

1 Badly constructed scaffolds.
2 Alterations.
3 Obstructed walkways.
4 Absence of guard-rails and toe-boards.

Defects to look for

Something removed and not replaced.
No toe-boards.
No guard-rails.
Split or knotted boards.
Loose boards.
Overlapping or protruding boards.
Gaps between boards.
Obstructed walkways.
Uneven foundations.
Bent or rusty poles.
No bracing.
No tie-ins.
No base plates.
Wrong couplers used.
Worn or rusty couplers.
Ledgers protruding.

Use of scaffolds

Erection, dismantling and alteration of scaffolding should be done by experienced workers.

Always inspect scaffolding before starting work on it and also after wet or frosty weather.

If you have to alter scaffolding to get a particular job done get it altered by a scaffolder and make sure the alteration is made good after the work has been done.

Never overload a scaffold – if in doubt, find out what the safe load is.

Make sure stacked materials cannot fall off – get wire mesh frames between the guard-rail and toe-board.

Keep scaffolds tidy.

Provide unobstructed passageways for workers and materials.

Use the access ladders – never jump from or climb up and down a scaffold.

Hoist materials up and down – don't throw them.

When using suspended scaffolds or cradles, make sure that the rope, pulley and hoisting gear are in good condition and that the cradle or scaffold is adequately counterbalanced for the weight it is taking.

When using a trestle scaffold make sure it is firm and that the platform is properly supported.

Fragile roof coverings

The most common roof covering in this category is either plain or corrugated. It can look and feel deceptively strong and safe but is liable to break suddenly under a concentrated load, i.e. a person standing or walking on it.

Safety rules

1 Warning notices should be clearly displayed on all asbestos and other fragile roofs (see Figure 37).
2 No person to work directly off the roof.
3 Loads to be distributed over as large an area as possible by means of battens and/or crawling boards (roof ladders) (see Figures 38 and 39).

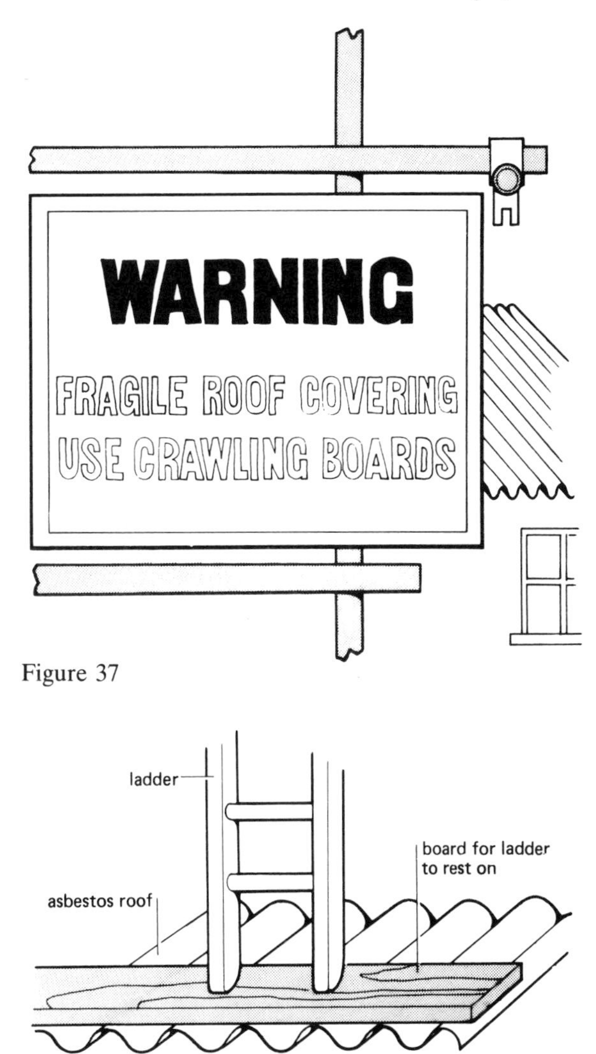

Figure 37

Figure 38 *Method of support when working on a fragile roof*

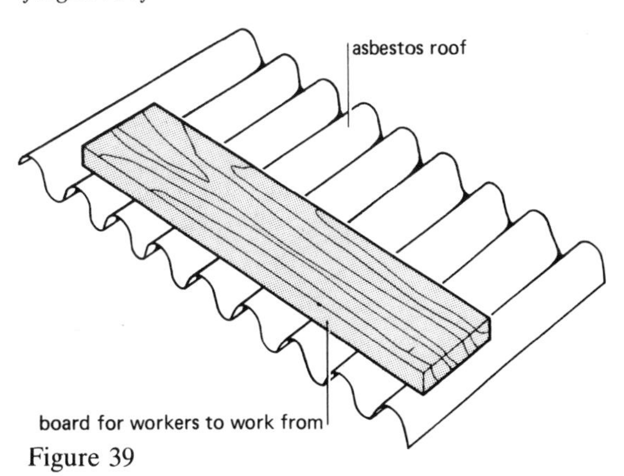

Figure 39

Welding

Eye protection and burns

Great care must be exercised by welders to avoid the possibility of workers being burned or fires or explosions. With the increased use of arc welding it is necessary to look at some of the possible safety points.

Electric arc welding

Face shields for protection against arc-eye must be worn by operators and by anyone watching (see Figure 40).

If necessary, to protect passers-by, erect barriers around the work area and display warning notices.

Check plugs and leads regularly.

All equipment must be properly earthed and insulated.

Remove all combustible materials from the surrounding area.

Use asbestos blankets to cover combustible materials which cannot be moved.

Always have a fire extinguisher nearby.

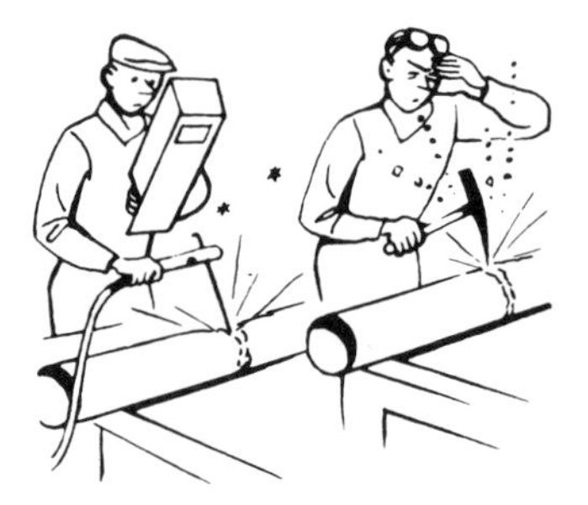

incorrect

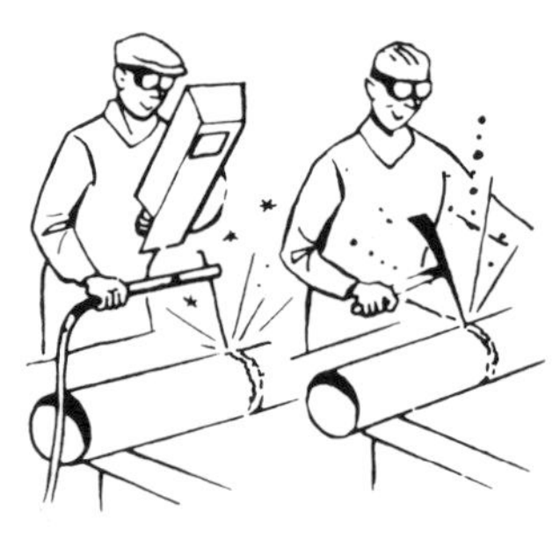

correct

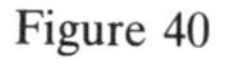

Figure 40

Gas cylinders

Keep all gas cylinders, especially the regulators, clear of oil, grease and dirt; do not handle them with greasy hands; spontaneous combustion may occur if gas and oily vapours mix.

Do not expose gas cylinders to excessive heat.

Gas cylinders should not be loaded loosely on lorries.

Store oxygen cylinders separate from cylinders containing combustible gases – acetylene, propane, butane, hydrogen, coal gas.

When storing or stacking oxygen cylinders make sure they are secured – do not lean them up against a wall. They must not be stacked more that four high.

Acetylene cylinders and liquefied petroleum gas cylinders must be kept upright, in storage and in use.

Propane should be stored above ground in a place with adequate ventilation and away from excavations. Propane is heavier than air, and if stored below ground level or near excavations any leakage will collect in the lower level.

Precautions to be taken during gas welding

Wear protective clothing – goggles and gloves.

Wear a leather apron when sparks are flying.

Make sure the surrounding area is free of combustible materials and that the cylinders are clear of falling sparks.

Use asbestos blankets to cover materials which cannot be moved.

Keep hoses clear of walkways.

Purge hoses before using equipment – the explosion of mixed gases in hoses causes the majority of welding accidents.

Check all equipment before use, especially hoses and regulators for leaks.

Use soapy water to check for leaks *not* a naked flame.

See that the nozzel of the blow pipe is free from obstruction.

Mark completed work *'hot'*.

If welding in a confined space, see that adequate ventilation is provided.

Welding or cutting tanks or vessels

Tanks or vessels which have contained inflammable or explosive materials should always be cleaned out thoroughly before welding or cutting.

Clean the container by steaming or with boiling water. Resteam daily if work is to continue on following days.

Never blow out the container with oxygen.

Tidy site and workshop

It may suprise you to learn that untidiness is a major contributor to accidents involving people falling. Tidiness is everybody's business.

Safety rules

1 A basic safety rule is 'keep the site tidy' (not like Figure 41).
2 Keep walkways free of obstructions.
3 Watch where you are going.
4 Remove any hazard you come across, even those left by others.
5 Nails should be removed or knocked down flat in discarded timber (not left as in Figure 42).

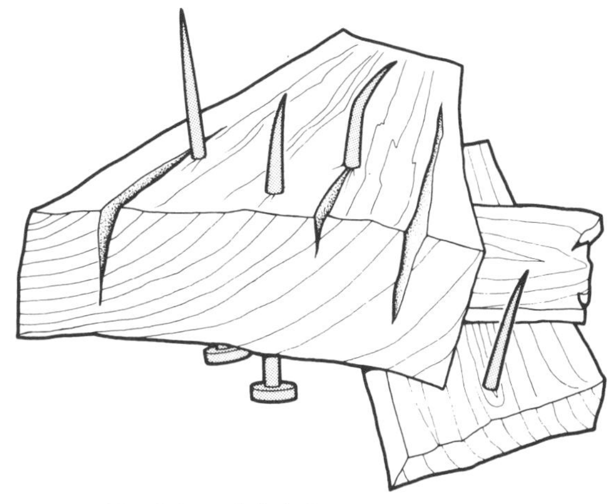

Figure 42 *Discarded timbers*

Figure 41 *Unsafe and inefficient*

Fire

Section 51 of the Factories Act 1961 requires that in every factory there shall be provided and appropriately maintained means for fighting fire, which shall be so placed as to be readily available for use. For fire fighting in rooms where there are no exceptional risks to life, portable fire extinguishers are normally sufficient, provided that the right type is used. For fire fighting in certain buildings, such as factories and large commercial premises, it is necessary to install a fixed system because of the risk of fire breaking out after working hours. Premises which are deemed to have a fire certificate fall within: Section 40 of the Factories Act 1961; Section 29 of the Offices, Shops and Railway Premises Act 1963; Hotels and Boarding Houses (Fire Precautions) Act 1971; and all premises specified under Section 1 (2) of the Fire Precautions Act 1971 as and when designated by a Commercial Order.

Sources of fire danger

Heating devices and stoves.
Electric wiring and equipment.
Flammable liquids, materials and pastes such as fuels, lubricants, paint, timber, and oily rags.
Welding operations, particularly below overhead welding.
Beneath and near hot roofing.
All areas exposed to sparks and heat if refuse burning takes place
Compressors, engine generators and all internal combustion engines and their fuel supplies.

Fire precautions

Maintain free access from street and hydrant.
Provide hosing where possible near water taps.
Provide and maintain fire extinguishers.

Fire extinguishers

Provide the proper kind of extinguisher for the risk (see Table 1 and Figure 43)
Extinguishers should be inspected regularly and recharged immediately after use.
Protect from freezing in cold weather by enclosing them.
Instruct workmen on use of extinguishers.

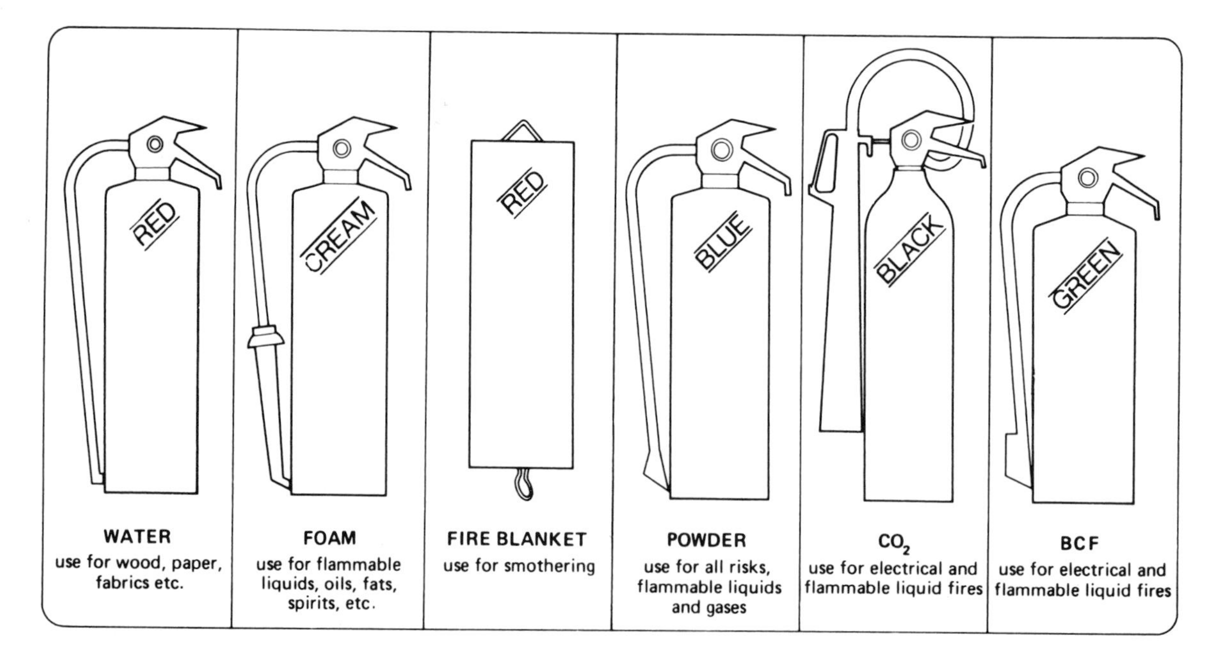

Figure 43 *Modern fire extinguishers, their colour codes and uses*

Table 1 *Choice of portable extinguisher extinguishing agents*

Class of fire risk	*Remarks*	*Water*	*Dry powder e.g. sodium bicarbonate*	*Carbon dioxide*	*Foam*	*Chlorobromomethane*
Wood, cloth, paper or similar combustible materials	Water best agent	Most suitable	Not recommended except for small surface fires	Not recommended except for small surface fires	Not recommended except for small surface fires	Unsuitable – dangerous fumes given off
Flammable liquids – petrols, oils, greases, fats	Smothering effect required	Unsuitable	Most suitable for general use	Most suitable where contamination by deposits must be avoided	Most suitable where reignition risk is high*	Effective for small fires but dangerous fumes given off
Live electrical plant but not including electric wiring or individual electric motors	Smothering required by non-conductor of electricity	Unsuitable – dangerous	Suitable	Suitable	Unsuitable – dangerous	Effective but dangerous fumes given off

*Special foams required for liquids which mix with water.

General first aid

This advice is concerned only with first aid. It is not a substitute for attention by a doctor or a trained nurse. If medical aid is going to be needed urgently, send for a doctor or ambulance immediately.

General

If the casualty has stopped breathing from whatever cause, artificial respiration must be started at once before any other treatment is given and should be continued until breathing is restored (see page 31). Where there is shock, keep the casualty lying down and comfortable. Cover with a light blanket or clothing, but do not apply hot water bottles. Do not give drink or anything by mouth if there sems to be an internal injury. Wash your hands before treating wounds, burns or eye injuries.

Minor wounds and scratches

All wounds and scratches, even minor ones, should receive attention immediately. Delay increases the risk of infection. Cover the wound as soon as possible with a sterilized dressing* or adhesive wound dressing†. If it is necessary to clean the skin round the wound, avoid washing the actual wound because this can wash germs into it. Warn the casualty that this is the first dressing and that further attention may be needed; if an injury becomes inflamed, hurts or festers, he or she should get medical attention.

Serious injuries

Bleeding

Stop the bleeding at once and send promptly for a doctor or an ambulance. To control bleeding by direct pressure, apply a pad of sterilized dressing(s), bandage firmly, adding, if need be, sterilized cotton wool; finally apply a triangular bandage. It will sometimes be possible to stop arterial bleeding by pressing the artery with the finger or thumb against the underlying bone.

**Sterilized dressing*: an unmedicated complete dressing with bandage, sterilized and put up in an individual sealed packet.
†*Adhesive wound dressing*: special type of dressing approved by certificate of HM Chief Inspector of Factories.

If bleeding cannot be controlled by direct pressure, a rubber bandage or pressure bandage may be applied to a limb between the wound and the heart for no longer than 15 minutes at a time, pending medical attention. It is most important that this time limit is not exceeded.

Fractures
Do not attempt to move a casualty with broken bones or injured joints until the injured parts have been secured with triangular bandages so that they cannot move. An injured leg may be tied to the uninjured one, and the injured arm tied to the body, padding between with cotton wool.

Electric shock
Switch off the current. If this is impossible, free the person using something made of rubber, cloth or wood or a folded newspaper; use the casualty's own clothing *if dry*. Do not touch the skin before the current is switched off. If breathing is failing or has stopped, give artificial respiration and continue for some hours if necessary. Get help and send for a doctor.

Gassing
Carry the casualty into fresh air; do not let him or her walk. If breathing has stopped, give artificial respiration (see page 31), get help and send for a doctor or an ambulance.

Mild cases should be kept resting and, after recovery, sent home by car.

Special injuries

Burns and scalds
If serious, send promptly for a doctor or ambulance.

Put a sterilized dressing on the burn or scald. Never use an adhesive wound dressing. If extensive, cover with clean towels and secure loosely. Do not burst blisters or remove clothing sticking to the burn or scald.

Chemical burns
Remove all contaminated clothing and flush the burn with plenty of cold water. Apply a sterilized dressing.

Eye injuries

Something in the eye
If the object cannot be removed readily with sterilized cotton wool moistened with water, or if the eye hurts after removal of the object, cover the eye with an eye-pad‡ and bandage firmly so as to keep the eye shut and still. Send the casualty to a doctor or hospital quickly. If there is likely to be considerable delay in getting medical attention first insert eye ointment.§

Injury from a blow
Cover the eye with an eye-pad and send the casualty at once for medical attention. Do not apply an eye ointment.

Chemical in the eye or chemical burn
Flush the open eye at once with clean cold water and continue washing the eye for at least 15 minutes. (A good method is to get the casualty to put his or her face under water and blink his or her eyes.)

Then cover with an eye-pad. Do not apply eye ointment. Send the casualty to a doctor or hospital quickly.

Heat burns
Cover the eye with eye-pad and send the casualty immediately to a doctor or hospital. Do not apply eye ointment.

Bandaging for eye injuries
The eye-pad is kept in place by the covering bandage running under the ear next to the injured eye and above the other ear.

Arc-eye
Apply cold compresses to the eyes and bathe them with an astringent lotion, obtainable from a chemist. Do not use any eye-drops unless prescribed by a doctor.

‡*Eye-pad*: a pack containing a sterilized pad with a long bandage attached.

§*Eye ointment*: an ointment approved by certificate of HM Chief Inspector of Factories.

First aid box

An adequately stocked first aid box should always be available (see Figure 44). All items used should be replaced immediately. It is also good practice to have names, addresses and telephone numbers of qualified helpers, i.e. local first aiders; doctor; hospital permanently displayed inside the lid of the box.

Where more than fifty people are working on the site, a trained first aider must be in charge of the box.

Local branches of the Red Cross and St John Ambulance run evening courses which qualify for Trained First Aider Certificate.

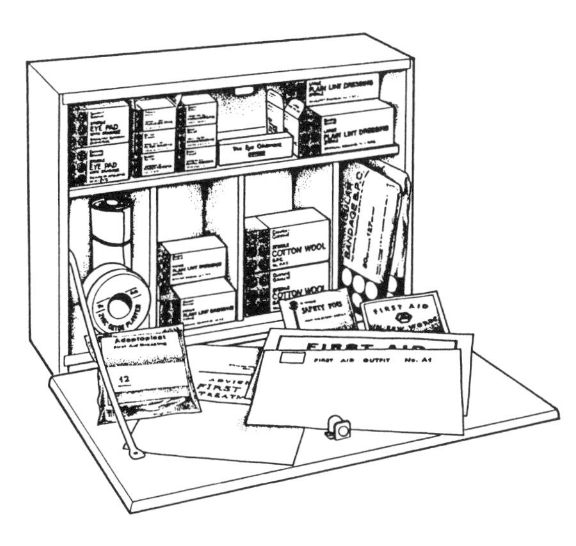

Figure 44 *First aid box*

Artificial respiration

Electric shock, gassing, drowning or choking may cause breathing to stop. In any of these cases *artificial respiration must be started without delay*. Do not find help if you are alone – only go for help when the patient is breathing.

Mouth-to-nose respiration is by far the most effective method of artificial respiration – the mouth-to-mouth should only be used if the mouth-to-nose is impossible.

Mouth-to-nose respiration

1. Lay the patient on his or her back, and, if on a slope, have the stomach slightly lower than the chest.
2. Make a brief inspection of the mouth and throat to ensure that they are clear of obvious obstruction.
3. Give the patient's head a backwards tilt so that the chin is prominent, the mouth closed and the neck stretched to give a clear airway (see Figure 45).
4. Open your mouth wide, make an airtight seal over the nose of the patient and blow. The hand supporting the chin should be used to seal the patient's lips (see Figure 46).

Figure 45 *Tilt head back*

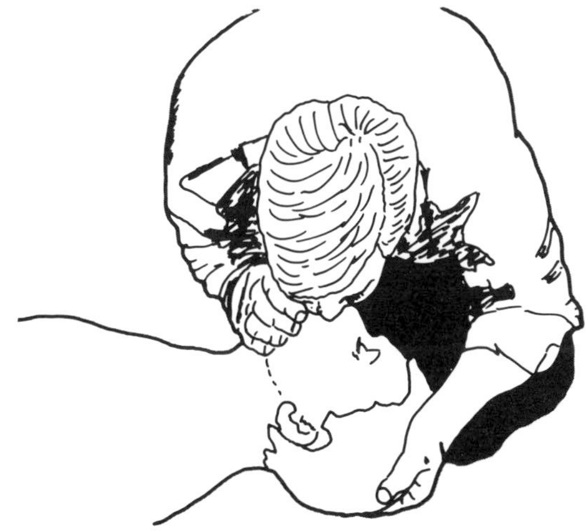

Figure 46 *Blow down nose*

5 After blowing, turn your head to watch for chest movement, while inhaling deeply in readiness for blowing again.
6 If the chest does not rise, check that the patient's mouth and throat are free from obstruction and that the head is tilted back far enough.
7 Blow again.
8 If air enters the patient's stomach through blowing too hard, press the stomach gently with the head of the patient turned to one side. If at any time the patient vomits, turn the head to one side so that he or she cannot inhale the vomit.
9 Commence resuscitation with four quick inflations of the patient's chest to give rapid build up of oxygen in the patient's blood and then slow down to twelve to fifteen respirations per minute or blow again each time the patient's chest has deflated.

NOTICE OF ACCIDENT OR DANGEROUS OCCURRENCE

1. OCCUPIER OF PREMISES
Name
Address
Nature of Business

2. EMPLOYER OF INJURED PERSON (if different from above)
Name
Address

3. INJURED PERSON
Surname
Christian Names
Date of Birth
Address
Resident/Staff
Widow/Widower
Married/Single
Occupation
Name and address of parent or guardian

4. PLACE WHERE INCIDENT OCCURRED
Address
Exact Location (e.g. staircase to office, canteen storeroom, classroom
Name of Person supervising

5. INJURIES AND DISABLEMENT
Fatal or non Fatal
Nature and extent of injury (e.g. fracture of leg, laceration of arm, scalded foot, scratch on hand followed by sepsis)

6. ACCIDENT OR DANGEROUS OCCURRENCE
Date
Time
Full details of how the incident occurred and what the injured person was doing.
If a fall of person or materials, plant, etc. state height of fall.

Name and address of any witness.

If due to machinery, state name and type of machine
What part of the machine caused the accident?
Was the machine in motion by mechanical power at the time?

7. ACTION FOLLOWING THE ACCIDENT
What happened?
When was the doctor informed?
When did he attend?
Name of Doctor (address and telephone)
If taken to hospital, say when and where
Names and addresses of friends or relatives who have been notified of the accident:

When and how were they informed?

Signature of injured person or person completing this form:
Date:
If the form is completed by some person acting on the injured person's behalf, the address and occupation of such person should be entered.

Figure 47 *Sample accident form*

10 Continue until the patient is breathing normally.

Mouth-to-mouth respiration
Proceed in the same way for mouth-to-nose respiration but close the nostrils by pinching patient's nose between your fingers and seal your mouth around the patient's mouth. Remember – only use the mouth-to-mouth method if the mouth-to-nose is impossible.

Get medical attention as soon as the patient is breathing normally.

Accident forms

All accidents should be fully reported and recorded. Figures 47 and 48 show sample accident forms.

ACCIDENT INVESTIGATION REPORT

PERSONNEL

Date .... Day .... Time ....
Name Of Injured Person(s) .... Age .....
.... Age .....
Name Of Any Other Person Involved .... Age .....
Instructor In Charge .... Location ....
Witnesses ....

INJURIES

Detail Of Injury ....
....
Actual Cause Of Injury ....
Treatment

First Aid	Visit To		Detained At		Operation	
	Doctor	Hospital	Home	Hospital	Temporary	Permanent

ACCIDENT

Machine Equipment or Tool Involved ....
Other Than Above Involved ....

Shop Lights	On	Off	Machine Light	On	Off
Machine Switch	On	Off	Were Machinery Parts Stationary	On	Off

Possible Distraction At Time Of Accident ....
Any Other Factor ....

CAUSES

Known Contributory Causes ....
....
....
....
....

PREVENTION

Action Required To Prevent Re-occurrences ....
....
....
....
....
....
....
....

Notified
Safety Committee ☐ ☐

Figure 48 *Sample accident form*

Useful contacts for information on safety

Her Majesty's Stationery Office (HMSO),
49 High Holborn,
London WC1

HM Factory Inspectorate,
Baynards House,
1–13 Chepstow Place,
Westbourne Grove,
London W2

Institution of Industrial Safety Officers,
23 Queen Street,
London WC1

Royal Society for the Prevention of Accidents (RSPA),
Head Office,
Cannon House,
Priory Queensway,
Birmingham

Fire Protection Association,
Aldermary House,
Queen Street,
London EC4

British Standards Institution,
2 Park Street,
London W1

British Research Station,
Buckmalls Lane,
Garston,
Watford,
Herts

The National Federation of Building Trade Employers,
Cavendish Street,
London N1

The Federation of Civil Engineering Contractors,
6 Portugal Street,
London WC2

Central Electricity Generating Board,
15 Newgate Street,
London EC1

Self-assessment questions

1 (1) The Health and Safety at Work etc. Act covers all people at work.
(2) The Health and Safety at Work etc. Act excludes domestic servants in private households.
(a) Both statements (1) and (2) are correct
(b) Both statements (1) and (2) are incorrect
(c) Only statement (1) is correct
(d) Only statement (2) is correct

2 The colour coding for a 3-core flex is:

	live	*neutral*	*earth*
(a)	red	black	green
(b)	brown	blue	green
(c)	red	blue	green and white
(d)	brown	blue	green and yellow

3 Which fire extinguisher should *not* be used on electrical fires?
(a) dry powder
(b) carbon dioxide
(c) foam
(d) vapourizing liquid

4 Tilting the head backwards when giving mouth-to-mouth resuscitation ensures:
(a) an effective breathing in position for the rescuer
(b) a clear airway into the lungs of the victim
(c) a good supply of blood to the victim's brain
(d) the victim's chest will rise and fall automatically

5 When treating an unconscious person for electric shock, a number of steps need to be taken immediately:
(1) switch off the supply
(2) seek medical help
(3) treat the burns
(4) carry out artificial respiration
(5) keep the patient warm

What is the correct sequence of events?
(a) 2, 1, 3, 5, 4
(b) 1, 2, 3, 4, 5
(c) 3, 5, 1, 2, 4
(d) 1, 4, 2, 5, 3

6 A person lifting a load from the ground by hand should use:
(a) bent knees and straight back
(b) straight legs and arched back
(c) straight legs and curved back
(d) bent knees and curved back

7 The fuse used in a 13 A plug is intended to:
(a) avoid the use of an earth wire
(b) maintain a steady voltage
(c) allow the use of double insulated tools
(d) fail as soon as the system is overloaded

8 Which is the correct procedure to be adopted for severe external bleeding if no dressing is available?
(a) wash the wound thoroughly
(b) lie flat and keep warm
(c) hold the sides of the wound firmly together
(d) apply an antiseptic lotion

9 By law an industrial accident must be reported without delay to the factory inspectorate if it involves:
(a) hospital treatment
(b) two or more persons
(c) site visitors
(d) absence from work longer than three days

10 Mobile scaffolding should be fitted with:
(a) locking wheels
(b) lifting handles
(c) wheels at one end only
(d) an independent ladder

The answers to the self-assessment questions are gathered together on pages 242–3.

2 Materials

After reading this chapter you should be able to:

1 Identify and name materials used for pipes, fittings and components.

2 State what is meant by the properties of materials.

3 Define the terms 'ferrous' and 'non-ferrous'.

4 State the advantages and disadvantages of various metals.

5 Select a suitable material for a particular location or application.

6 Describe the chemical and physical properties of plastics material.

Introduction

The installation of pipework systems for the supply of cold and hot water, heat and gas, together with the systems for the removal of surface, waste and foul water, form the major part of day-to-day work for plumbers. Included in the operations must be the manufacture and fitting of weathering components to roofs and the outsides of buildings. This means that it is essential for the craftsman plumber to have a comprehensive understanding of the characteristics, properties and performance of materials in current use.

British Standards

The British Standards Institution was founded in 1901 to standardize industrial activities such as design, installation and manufacturing practice. The British Standards Specifications refer to standards of manufacture, for example low carbon steel tube is made to conform with the specification contained in BS 1387.

British Standard Codes of Practice are recommendations related to methods of good practice in installation work, for example CP 5572: 1978 deals with sanitary pipework. By having standards which have been agreed by manufacturers and industrial experts, the process of obtaining the right materials and correct design and installation techniques or procedure is made easier, and leads to a better quality of completed

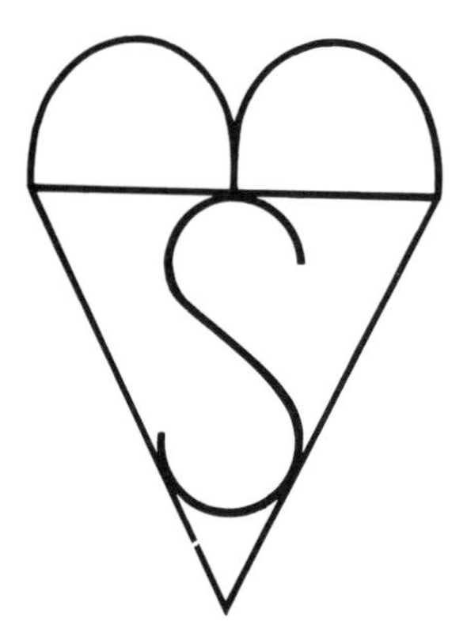

Figure 49 *British Standard kitemark*

job. Architects and others involved with design work need only to specify that material must conform to the relevant British Standard and that work is to be undertaken and installed in accordance with the relevant British Standard Code of Practice, to ensure that materials used and work undertaken is of a satisfactory and acceptable standard. Figure 49 shows the British Standard kitemark.

Properties of materials

Strength

One of the main properties of a material is its strength, i.e. its ability to withstand force or resist stress. Stress can be applied to a material in many ways (see Figure 50), for instance:

(a) Tensile, or stretching.
(b) Compressive, crushing, or squeezing.
(c) Bending, which is both (a) and (b) on either side of an axis.
(d) Shear, or cutting.
(e) Torsion, or twisting.

$$Stress = \frac{load}{area}$$

and this means that the stress can be increased by either increasing the *load* or by decreasing the *area*. So that when a piece of material is cut out of a support it has the effect of increasing the *load* acting upon it. Therefore care must be taken when it becomes necessary to cut out structural support such as a joist or beam.

In addition to strength, materials have other properties which may affect their selection or use. These are:

Brittleness

This means that the material is easily fractured or broken. Brittleness in metals is usually associated with hardness, and it is often necessary to reduce the hardness of a material in order to give it greater strength.

Malleability

This is the property which means that the material is capable of being formed or shaped by the use of hand tools or machines (see Figure 51).

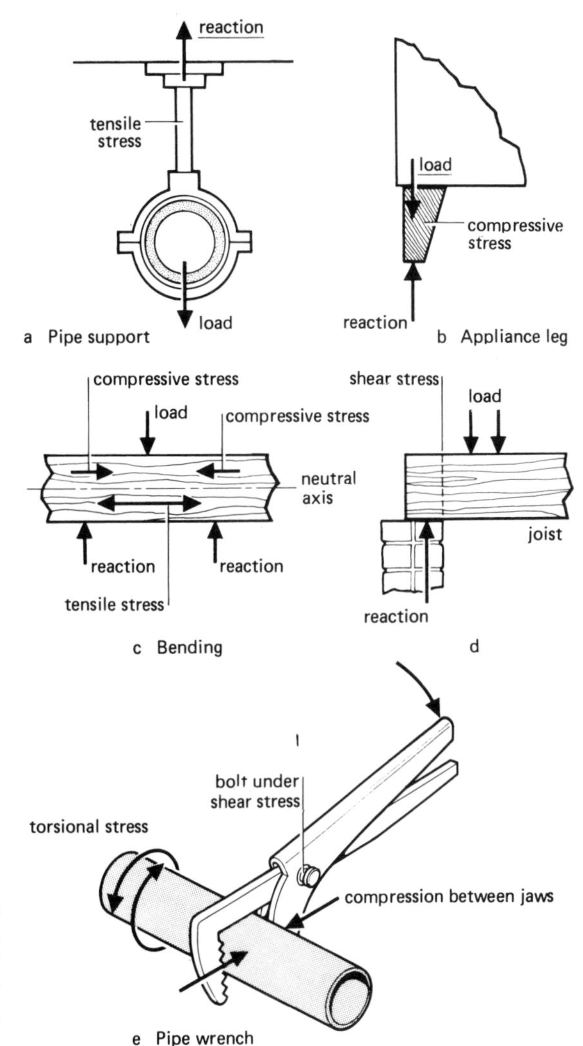

Figure 50 *Types of stress: (a) tensile; (b) compressive; (c) bending; (d) shear; (e) torsion*

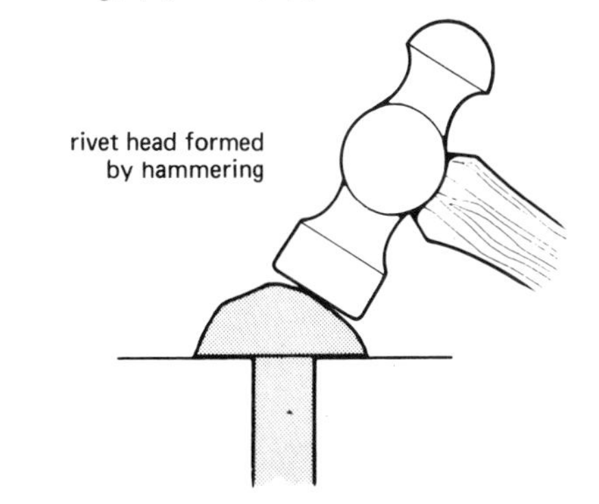

Figure 51 *Malleability*

Ductility

This is the property of many metals and enables them to be drawn out into a slender wire or thread without breaking (see Figure 52). Ductile materials are easily bent.

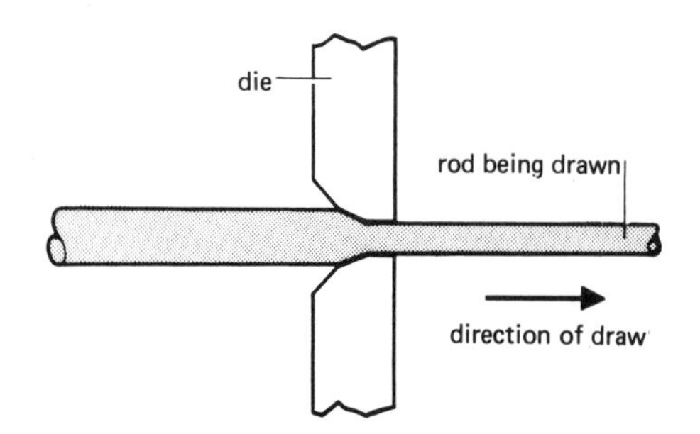

Figure 52 *Ductility*

Elasticity

The ability of a material to return to its original shape, or length after a stress (stretching) has been removed (see Figure 53). The 'elastic limit' is the greatest strain that a material can take without becoming permanently distorted.

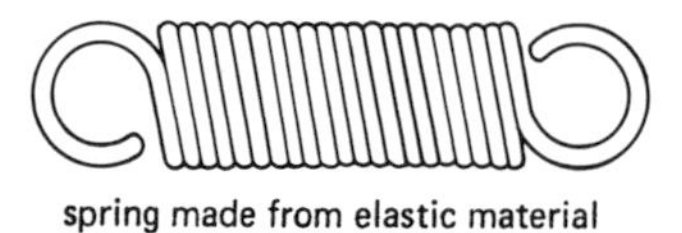

Figure 53 *Elasticity*

Hardness

The property of a material to resist wear or penetration. This is an essential requirement for the cutting edge of a tool. The hardest naturally occurring substance is diamond.

Tensile strength

The tensile strength of a material is its ability to resist being torn apart. Samples are subjected to increasing loads until breakage occurs. The loading which causes breakage being the ultimate tensile strength.

Temper

The temper of a metal is the degree or level of hardness within the metal and can vary from dead soft to dead hard. Copper is a good example related to temper, being in a dead soft condition following annealing its temper changes as it is hammered or worked into a dead hard, cold-worked condition.

Work hardening

Describes the increase in hardness caused by hammering, bossing or other cold working techniques. These working techniques cause the grains or crystals which form the metal to become mis-shaped. This deformation of the metal structure reduces malleability and ductility which prevents further cold working.

Annealing

This is the description of the heat treatment applied to soften temper and to relieve the condition of work hardening. The metal is heated to a specified temperature, during which the deformed crystalline structure returns to its normal condition. This treatment softens the metal, thereby reducing internal stresses and allows further working processes to be carried out.

Corrosion resistance

Corrosion is a chemical or electrochemical action which causes the 'eating away' of some metals. Different metals react in varying ways to the different forms of corrosive attack, some having strong resistance, others need protection where attack may occur.

Table 2 *Properties of materials*

Material	*Chemical symbol*	*Density (kg/m^3)*	*Coeff. linear exp. (°C)*	*Melting point (°C)*
Aluminium	Al	2705	0.0000234	660
Copper	Cu	8900	0.0000160	1083
Iron (cast)	Fe	7200	0.0000117	1526
Iron (wrought)	Fe	7700	0.0000120	2200
Lead (milled)	Pb	11300	0.0000293	327
Non-metallic sheet	—	1021	0.0000188	not applicable
Tin	Sn	7300	0.0000210	215
Zinc	Zn	7200	0.0000290	416

Note: Properties vary according to condition, that is, whether the material is cold-worked, hot-worked, hard or annealed.

Categories of materials

Materials can be divided into different categories in a variety of ways. The easiest division is into: *metals* and *non-metals*.

Metals differ from non-metals in chemical properties as well as in the more obvious physical properties. There is not a rigid dividing line between the two categories. Some metals possess characteristics of both groups, while some non-metal materials behave like metal in some respects, for instance the conducting of electricity.

Table 3 gives examples of some of the differences between the two categories.

Table 3

Metals	*Non-metals*
Characteristic appearance, 'metallic' sheen or gloss	No particular characteristic appearance
Good conductor of heat	Usually poor conductor of heat, may be insulator
Good conductor of electricity	Usually poor conductor of electricity, may be insulator (except carbon, silicon)
Electrical resistance usually increases as temperature rises	Electrical resistance usually decreases as temperature rises
Density usually high	Density usually low
Melting point and boiling point usually high	Melting point and boiling point usually low

Metals can be subdivided into *element metals* and *alloys*.

The element or pure metals are those which do not have any other metal or materials mixed with them.

Alloys are produced when two or more metals, or a metal and a non-metal, are blended together, usually by melting (see Table 4). Many alloys have useful properties which their parent metals do not possess. For example, a much lower melting point, as in the case of solders.

Table 4 *Composition of common alloys*

Alloy	*Main elements*
Brass	Copper and zinc
Bronze	Copper and tin
Gunmetal	Copper, tin and zinc
Pewter	Tin and lead
Steel	Iron and carbon
Steel (stainless)	Iron, chromium and nickel
Soft solder	Lead and tin

Table 5 lists many metals which are used by the plumber.

Metals may be further classified as:

1 Ferrous metals, that is those which contain iron.
2 Non-ferrous metals, those that do not contain iron.

The basic difference between the two categories is that, with the exception of stainless steels, ferrous metals will rust when in contact with water and oxygen. Also, ferrous metals may be magnetized.

Non-ferrous metals do not rust, although they may corrode under certain circumstances, and they are non-magnetic.

Table 5

Pure metal	*Alloy*
Aluminium	Brass
Chromium	Bronze
Copper	Chrome-vanadium steel
Gold	Duralumin
Iron	Gunmetal
Lead	Invar
Magnesium	Nickel silver
Mercury	Pewter
Nickel	Rose's alloy
Platinum	Solder
Silver	Stainless Steel
Sodium	White-metal (bearing metal)
Tin	
Tungsten	
Zinc	

Ferrous metals

Grey cast iron

This is a mixture of iron with 3.5 to 4.5 per cent carbon and very small amounts of silicon, manganese, phosphorous and sulphur. The carbon is in the form of 'flake graphite' and it is these flakes which make the material very brittle (see Figure 54). The iron fractures easily along the lines of the flakes.

Cast iron is hard and has a much higher resistance to rusting than steel. It is used for drainage and discharge system pipework, appliance components, manhole covers and frames. Its melting point is 1200 °C.

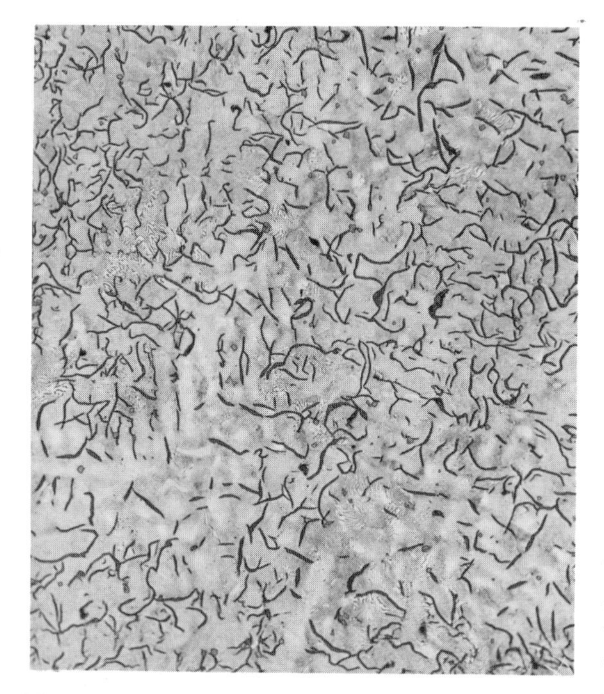

Figure 54 *Microsection of grey cast iron showing flake graphite*

Ductile cast iron

This has approximately the same carbon content as grey cast iron. The difference is in the shape, size and distribution of the carbon particles which are changed from flakes into ball-like nodules or 'spheroids' (see Figure 55). The change is due to alloying the iron and carbon with magnesium compounds before casting. Subsequent heat treatment at 750 °C changes the brittle iron carbide into softer, ductile ferrite.

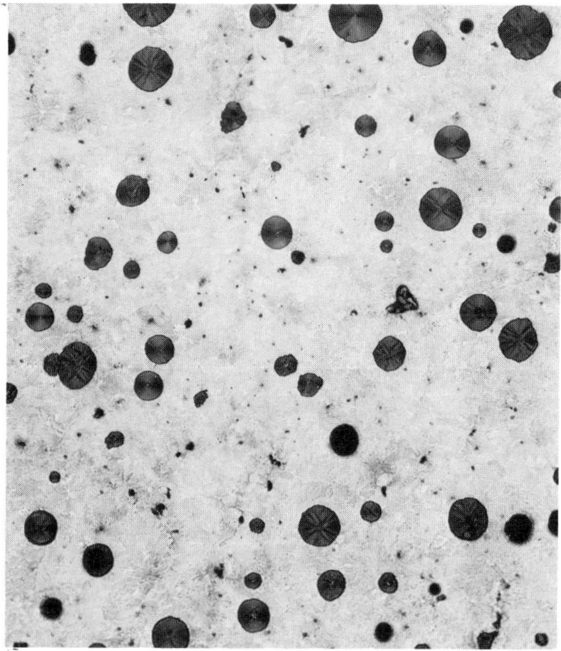

Figure 55 *Microsection of pearlitic nodular cast iron*

Used to make pipes for gas mains and drainage pipelines, ductile cast iron has the strength and ductility of lower grade steel and the high corrosion resistance of cast iron.

Wrought iron

This is the purest commercial form of iron containing very little carbon. Made from heat-treated cast iron it is rolled or hammered to give it the required grain structure. It is still used for chains, horse-shoes, gates and ornamental iron-work. Because of its high cost it is no longer used for pipes or fittings.

Malleable iron (malleable cast iron)

Malleable iron is white cast iron in which the carbon has combined with the iron to form iron carbide. This results in a very hard and brittle material which is annealed.

Most of the ferrous pipe fittings used in conjunction with low carbon steel pipe in internal installations are malleable iron.

Low carbon steel (LCS)
This is another mixture of iron and carbon. Its carbon content is between 0.15 and 0.25 per cent. The material has a wide use and is commonly known as mild steel, although its correct title is low carbon steel (LCS). In building, it is used for general structural work, pipes, sheets and bars. It can be pressed, drawn, forged and welded easily.

The LCS tube used by plumbers is manufactured to BS 1387 in three grades of weight: light, medium and heavy.

The outside diameter of each grade of tube is similar, the difference being the tube wall thickness. The grades are identified by 50 mm wide colour band coding, brown, red and blue, to correspond with light, medium and heavy grades. Low carbon steel tubes used for domestic water supplies must be protected with a zinc coating; this is called 'galvanized'.

Tubes are generally available in 3 m and 6 m lengths, with bore size ranging from 6 mm to 150 mm.

Stainless steel
This material is mainly used in plumbing in tube form for domestic water and gas services or for the manufacture of sanitary appliances such as sink units, urinals, WC pans etc.

The metal is an alloy and has a composition of chromium (18 per cent), nickel (10 per cent), manganese (1.25 per cent), silicon (0.6 per cent), a maximum carbon content of 0.08 per cent, the remainder being iron with small amounts of sulphur and phosphorous. The chromium and nickel provide the material with its shiny appearance and its resistance to corrosion, this being due to a microscopic film of chromium oxide, which forms on the tube surface and prevents further oxidation.

Stainless steel tube is manufactured to BS 4127 and is available in nominal bore sizes ranging from 6 mm to 35 mm in one grade only (light gauge) with an average wall thickness of 0.7 mm. The outside diameter of the tube is comparable to copper tube manufactured to BS 2871 Table X so that most types of capillary and compression fittings can be used. The mechanical strength of stainless steel is much higher than that of copper giving it greater resistance to damage and a better ability to support itself and therefore will require less fixing positions during installation. Stainless steel tubes of 15 mm to 35 mm diameter are supplied in 3 m and 6 m lengths.

Non-ferrous metals

Aluminium
This is useful metal because it is so light. It is much lighter than steel, and some of its alloys are as strong as steel. It is also a good conductor of heat and electricity. Aluminium and its alloys are therefore widely used where lightness and strength are needed. The material does not rust as ferrous metals do when exposed to the atmosphere. Although the surface does oxidize, this oxide film acts as protection against corrosive attack. Aluminium is primarily alloyed with copper, magnesium, and silicon.

Aluminium is the most common metal in the earth's crust. It is found in common clay, although removal is a difficult and expensive process. The only suitable ore is bauxite, which contains alumina, or aluminium oxide. Producing aluminium from bauxite is very different from the usual methods of obtaining metals from their ores. Firstly the bauxite is crushed and heated with caustic soda. The alumina dissolves and leaves behind the impurities. The pure alumina obtained is then dissolved in molten cryolite, and aluminium is produced when electricity is passed through this solution.

Aluminium in a soft condition is malleable and ductile, thus giving it good working properties, although it will work harden if cold worked. Annealing is easily completed, but because of its low melting point, care must be taken to avoid overheating.

BS 1470 specifies five grades of aluminium, but only two (S1 and S1B) are recommended for roofwork. Grade S1 is known as super purity aluminium.

The two grades of aluminium (S1–S1B) used for roofworks are obtainable in standard rolls of 8 m length, with widths of 150 mm, 300 mm, 600 mm, and 900 mm. Two thicknesses are available, these being 0.7 mm (standard) generally used for flashings, and 0.9 mm (heavy duty) which is suitable for bays and situations where foot traffic may occur.

Antimony

This is a hard, grey, crystalline metal which expands slightly when it solidifies. It is principally used in alloys of lead and tin (solder) to produce a degree of hardness and making the alloy more resistant to corrosion.

Bronze

True bronzes are alloys containing copper and tin as the basic metals, but the name bronze has been so loosely applied that certain alloys containing no tin at all are sold as bronze.

The useful bronzes contain not less than 65 per cent of copper. As the tin content is increased the alloys become progressively harder, although bronzes containing above 25 per cent of tin become weak and brittle. On the other hand, if too much copper is present, the cast metal becomes porous. The properties of many bronzes may be greatly improved by the addition of a third element – lead. When zinc is present in this alloy the material is known as 'gunmetal'.

Phosphorous is added to some copper/tin alloys to deoxidize the metal prior to casting, these alloys are termed 'phosphor bronze' although the percentage of phosphorous added may be very low.

Chromium

This metal is plated on other metals and materials to protect them. It is hard and does not lose its shine through corrosion. Steel is made stainless by the addition of chromium and nickel. Nickel and chromium together form important alloys much used for plating and the manufacture of tools. Chromium is found as chromite, or chrome iron ore.

Copper

Copper is obtained from the ore copper pyrites. The USA is a very large producer, although most of the copper used in Great Britain comes from Canada, Chile and Zambia.

This metal is probably the most important of the non-ferrous metals and is used extensively in sheet, strip, rod, wire, tube and other fabricated forms. It is also employed in the making of a wide range of alloys, of which brass and bronze are best known.

Although copper is a comparatively inexpensive metal, physically and chemically it is closely allied to silver and gold, and there are many similarities in the properties of the three. The outstanding features of copper are its high malleability (particularly when in a soft or annealed state), its electrical and thermal conductivities and its good resistance to corrosion.

Copper exposed to air forms on its surface a natural protective skin or 'patina' which effectively prevents corrosive attack under most conditions. On copper exposed to the atmosphere the patina takes the form of a green covering of copper salts (mainly sulphate or carbonate). The formation of a somewhat similar protective film also takes place on the interior surface of copper water pipes, this film is insoluble in water, therefore making copper an ideal material for carrying water.

The first British Standard for copper tube was issued in 1936 to standardize the various sizes of tube on the basis of their outside diameter. Prior to this, copper tube had been used since the turn of the century, when jointing was achieved by screwed and socketed fittings which required a tube wall thick enough to thread. As time progressed and methods of jointing which did not require threading of the tube developed, the wall thicknesses of the tube were reduced, and tubes were identified as 'light gauge'.

Copper tube for use in the construction industry is manufactured to BS 2871. Part 1 defines three specifications of tube, which are denoted as Tables X, Y, and Z. Table X specifies the requirements for light gauge tube of half-hard temper. Table Z refers to tubes of hard temper, whose increased hardness allows for reductions in wall thicknesses, but makes the tube unsuitable for bending. Tubes complying with Tables X and Z are not intended for below ground use. Table Y covers tube of either half-hard or fully annealed temper, this is tube that can be laid underground for the conveyance of water or gas.

Table 6

Specification	*Mass (kg/m)*	*Wall thickness (mm)*
Table X	0.2796	0.7
Table Y	0.3923	1.0
Table Z	0.2031	0.5

The basic differences between these tubes is their temper and wall thicknesses, the wall thickness factor affecting the mass. A comparison of 15 mm diameter tube manufactured to the three different specifications is given in Table 6.

Copper tubes to BS 2871 Part 1 Table X and Z (half-hard and hard tempered respectively) are available in standard lengths of up to 6 m. Tubes to Table X can be bent with the aid of a bending machine or spring or by sand loading. Tubes to Table Z are hard drawn and thin walled and should not be bent – changes of direction must be achieved with the use of fittings.

Tubes to BS 2871 Part 1 Table Y (annealed condition) can be obtained in standard coil lengths from approximately 10 m to 30 m (depending on diameter size). This tube in its soft condition can be bent by hand and is most commonly used for below ground service pipes, it can be obtained with a protective plastic covering fitted during manufacture for use in soils which have a corrosive nature.

Sheet copper is malleable and highly ductile so it can be easily worked particularly when in a soft or fully annealed state. Work hardening, however, occurs with cold working so as little manipulation as possible should be done.

As soon as the material becomes hard it should be annealed. This is carried out by heating to a dull red colour. The metal can then be allowed to cool naturally or may be doused with water to assist the cooling process.

Copper for weathering is available in the form of rectangular sheets or in strip form.

Strip copper is usually supplied in rolls or coils of varying length and width. The standard strip widths are 114 mm, 228 mm, 380 mm, 475 mm, 533 mm, and 610 mm. The length is determined by the thickness and mass of the roll which is usually limited to 25 kg or 50 kg to assist safe handling.

Sheet copper is obtainable in a variety of sizes, although standard sizes of 1.83 m × 0.91 m and 1.83 m × 0.61 m are normally specified. All copper for weathering purposes is sold by mass.

The techniques and methods used for laying and securing copper weatherings are fully described in Chapter 8.

Lead

Lead is obtained from the ore galena and is mined and processed in many different parts of the world, the main producing countries being Australia, the USA, Canada, Mexico and the USSR.

The name 'plumber' derives from the Latin name for lead which is plumbum. Lead is one of the six metals known to humanity from the early days of history, and the development of the use of lead in building work for sanitary, roofing and weathering purposes was also the development of the plumber's trade.

The ore galena is a compound of lead containing, in the pure form, 86.6 per cent lead and 13.4 per cent sulphur. Galena has a cubic crystalline structure, is very lustrous and of a dark grey colour. The lead produced from the smelting process usually contains small quantities of other metals, for example antimony, tin, copper, gold and silver, and these are removed as required by refining processes. The lead which is obtained possesses a very high degree of purity – a good commercial lead being over 99.9 per cent pure.

Lead is the softest of the common metals and has a very high ductility, malleability and corrosion resistance. It is capable of being shaped easily at normal temperatures without the need of periodic softening (lead does not appreciably work harden). Lead sheet and pipe are therefore easily worked with hand tools, and can readily be manipulated into the most complicated shapes. Lead is very seldom corroded by electrolysis attack when in contact with other metals.

Lead pipe, because it is soft and heavy, requires frequent or even continuous support along its length. It also requires a high degree of skilled workmanship to install, and this, coupled with high initial cost and public fears about lead poisoning, means that lead pipe is not widely used nowadays. However, it is still specified for certain uses and in specialist situations, and thousands of buildings still contain lead pipes which from time to time need repair or alteration, so it is necessary for plumbers to have knowledge and experience of the material.

Lead pipe for other than chemical purposes is manufactured to BS 602 and BS 1085, the essential difference being the composition of the lead, which is in fact a lead alloy. Pipe manufactured to BS 602 for water supplies must contain not less than 99.8 per cent lead. Lead pipe up to 25 mm bore is usually sold in coils of 20 m length or 50 kg mass, pipes from 25 mm to 50 mm bore are obtainable in coils of 12 m length or 50 kg mass.

Lead sheet is produced by either casting or milling. Cast lead is still made as a craft operation by the traditional method of running molten lead over a bed of prepared sand. A comparatively small amount is produced by specialist leadworking companies, mainly for their own use, in particular for replacing old cast lead roofs and for ornamental leadwork. There is no British Standard for this material. The available size of sheet is determined by the casting table bed size.

Milled sheet lead is manufactured on rolling mills. The process involves passing a slab of refined lead about 125 mm thick backwards and forwards through the mill until it is reduced to the required thickness.

Milled sheet lead is manufactured to BS 1178 and is supplied by the manufacturer cut to dimensions as required or as large sheets 2.4 m wide and up to 12 m length. Lead strip is defined as material ready cut into widths from 75 mm up to 600 mm. Supplied in coils, this is a very convenient form of lead sheet for most flashing and weathering applications. The thickness of lead is designated by a BS specification code number, and by an identifying colour. These are shown in Table 7 which also includes the relative thickness.

The thickness of sheet chosen for a particular situation or application depends upon several factors, for example roof design, type of building, shape and location of component, money available. Cost is of course an important consideration and so the thinnest lead to suit the particular fixing position will usually be used.

The thickness of lead sheet for various situations as recommended by the Lead Development Association are set out in Table 8.

Table 7

BS code no.	*Thickness (mm)*	*Colour*	*Weight (kg/m^2)*
3	1.25	Green	14.18
4	1.80	Blue	20.41
5	2.24	Red	25.40
6	2.50	Black	28.36
7	3.15	White	35.72
8	3.55	Orange	40.26

Table 8

Fixing situation	*BS code no.*
Damp proof courses	3, 4 or 5
Pipe weatherings, chimney flashings	4 or 5
Soakers	3 or 4
Cornice weatherings	4, 5 or 6
Small flats with no foot traffic	4 or 5
Large flats with or without foot traffic	5, 6 or 7
Gutters, valley, box, parapet etc.	5 or 6
Dormer cheeks and roofs	4 or 5
Hip, ridge and cover flashings	4 or 5
Vertical claddings	4 or 5

Zinc

This metal is obtained from several ores, the most common being zinc blende, sphalerite and calamine. Commercial sheet zinc has low ductility and is the least malleable of common roofing metals. This means that manipulation is difficult when compared with materials such as lead or aluminium, and jointing or shaping is achieved by folding, cutting or soldering. These disadvantages have led to the development of zinc alloys.

Zinc alloys have good ductility, and a linear

expansion rate of less than two thirds that of commercial quality zinc sheet (overcoming the problems of creep often associated with zinc sheet). Two of these alloys are currently available, namely zinc/lead and zinc/titanium, these being produced under the respective trade names of Metiflash and Metizinc.

Metiflash is very malleable and is most suitable for flashings and small weathering details such as bay window tops, dormers and canopies. Metizinc is most suitable for covering large roof areas.

Metiflash (zinc/lead) is obtainable in 10 m length rolls, in widths of 150 mm, 240 mm, 300 mm, 480 mm and 600 mm and with a thickness of 0.6 mm. Rolls of 6 m length with a width of 900 mm are also available.

Metizinc (zinc/titanium) can be obtained in sheets up to 3 m long by 1 m wide, although the standard sheet is 2.438 m × 0.914 m. Various thicknesses are available ranging from 0.2 mm to 2 mm, the usual thickness being the same as those for commercial zinc.

Commercial zinc is obtainable in sheets 2.438 m × 0.914 m with a recommended thickness of 0.6 mm for flashings and 0.8 mm for other roof areas. Zinc sheet has a grain which runs through the length of the sheet, and although this occurs in all rolled sheet metals the effect is of more significance in zinc, making it more difficult to obtain a sharp fold or turn along the length of a sheet than across its width. For this reason it is advisable to arrange weathering details so that the majority of folds or turns are made across the sheet, i.e. across the grain. In cold weather zinc becomes brittle and should be warmed slightly before folding is attempted, otherwise cracking or fracture may occur. Such folds should not be too sharp – a rounded fold with a radius at least twice the thickness of the metal will be satisfactory.

An electrolytic action is set up when zinc is in contact with copper in the presence of moisture, and for this reason zinc and copper should never be allowed to touch each other in roofing or as water service piping.

Non-metallic sheet

One of the most popular non-metallic sheet roofing materials is Nuralite. The material is an asphalt bonded asbestos sheet of laminated (layer) construction.

Nuralite is intended primarily to replace zinc, copper, aluminium and lead as a fully supported roofing material, and is particularly suitable for flashings and other types of weathering accessories.

The material was introduced in the mid-1940s by British Uralite Limited, 'Nuralite' being its trade name. The material is semi-rigid which means it has a degree of natural flexibility, although any attempt to bend or form it at normal outside temperatures will result in fracture or tearing.

Shaping and forming is usually achieved by the application of a gentle heat (optimum moulding temperature 182 °C). The material has a low tensile strength and will tear if attempts are made to stretch it. Compared with metals Nuralite has a low density, this lightness coupled with its low creep characteristics makes it satisfactory for vertical or steeply pitched surfaces.

Nuralite is available in standard sheets 2.4 m × 0.9 m, and is approximately 2 mm thick. Its mass is 2.44 kg/m^2.

Solders

A better understanding of the nature of the solders, and how to select one for a specific application, can be obtained by observing the melting characteristics of metals and alloys. Melting of pure metals is easy to describe as they transform from solid to liquid state at one temperature. The melting of alloys is more complicated as they may melt over a temperature range.

Solders of the tin/lead alloys constitute the largest portion of all solders in use. They are used for joining most metals. Impurities in tin/lead solders can result from carelessness in the refining and alloying operations, but can also be added inadvertently during normal usage. The soldering properties of tin/lead solders are affected by small traces of aluminium and zinc. As little as 0.005 per cent of either of these metals may cause lack of adhesion and grittiness. Above 0.02 per cent of iron in a tin/lead solder is harmful and will cause hardness and grittiness. The presence of above

0.5 per cent of copper will have the same harmful effects.

Antimony can play a dual role in tin/lead solders. Depending on the purpose for which the solder is to be used, it can be considered as either an impurity or as a substitute for some of the tin in the solder. When the amount of antimony is not more than 6 per cent of the tin content of the solder, it can be completely carried in solid solution by the tin. If the antimony content is more than the tin can carry in solid solution, tin/antimony components of high melting point crystallize out, making the solder gritty, brittle and sluggish.

Antimony content of up to 6 per cent of the tin content increases the mechanical properties of the solder but with slight impairment to the soldering characteristics. The use of lead/antimony/tin solders is not recommended on zinc or zinc-coated metals, such as galvanized iron. Solders containing antimony, when used on zinc or alloys of zinc, form an intermetallic compound, causing the solder to become brittle.

Tin/zinc solders are used for joining aluminium. Corrosion of soldered joints in aluminium (electrogalvanic) is minimized if the metals in the joint are close to each other in the electrochemical series. The addition of silver to lead results in alloys which will wet steel and copper. Flow characteristics, however, are very poor. The addition of 1 per cent tin to a lead/silver solder increases the wetting and flow properties and, in addition, reduces the possibility of humid atmospheric corrosion.

All solders are alloys, and as mentioned previously the melting point of an alloy varies with the different percentages of element metals comprising the alloy (see Table 9). The graph in Figure 56 illustrates the various melting points of tin/lead solders.

A solder with about 36 per cent lead and 64 per cent tin has the lowest melting point and changes from solid to liquid between 183 °C and 185 °C. This is known as the 'eutectic', which is the composition at which the alloy behaves as a pure metal. Adding either more lead or more tin raises the final melting point, although the alloy still begins to melt at about 185 °C. In its intermediate stage it becomes 'plastics' and may be 'wiped' or formed into its required shape or position before it becomes completely molten.

Table 9 *Solders BS 219: 1977*

BS grade	*Lead (%)*	*Tin (%)*	*Antimony (%)*	*Temperature range (°C)*
Grade A (electrical)	35	65	0	183–185
Grade B (tinning)	47	50	3	183–212
Grade C	58	40	2	183–234
Grade D (wiping)	68.3	30	1.7	185–248
Grade F (tinning	50	50	0	183–212
Grade G (wiping)	60	40	0	183–234
Grade J (wiping)	70	30	0	185–248

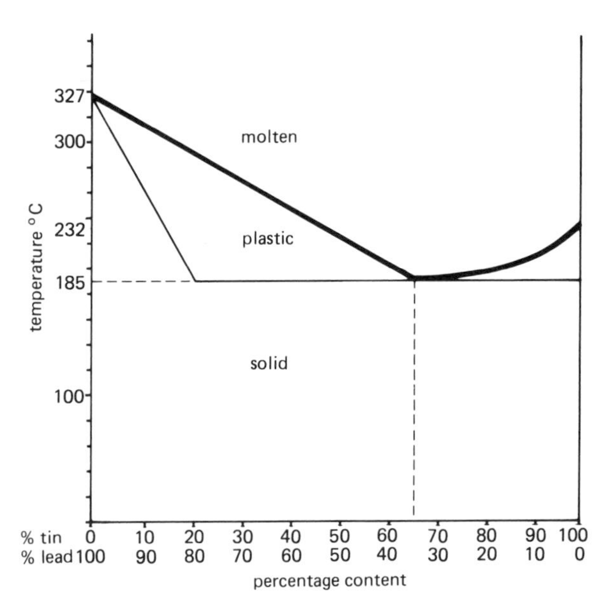

Figure 56 *Tin/lead diagram*

Plastics

The word 'plastics' came into being as a general description for many materials of a similar nature, such as celluloid, casein and bakelite. The name merely signifies that the material is capable of being moulded into shape. In this sense the word is not a suitable description for the many materials that have since been developed.

The term 'plastics' describes a group of man-made organic chemical compounds which can broadly be divided into two types – *thermoplastics* and *thermosetting plastics*.

Thermoplastics

Thermoplastics soften when heated and harden on cooling. They can be softened again afterwards provided that the heat applied is not sufficient to cause them to decompose. Among the thermoplastic materials used in plumbing work are polythene, polyvinyl chloride (PVC), polystyrene, polypropylene, acrylics (perspex) and nylon.

Thermosetting plastics

Thermosetting plastics are those which soften when first heated for moulding, and which then harden or set into a permanent shape which cannot afterwards be altered by the application of further heat. These are also called thermosetting resins, the most important of these are formaldehyde, phenol formaldehyde and polyester resins.

The advantages of plastics materials are their light weight, and the simple methods used in the jointing processes. Plastics materials used in plumbing have many common physical characteristics. Those used for pipes are invariably thermoplastics, all of which have a high resistance to corrosion and acid attack. They have a low specific heat, which implies that they do not absorb the same heat quantity as metals. They are also poor conductors of heat and electricity.

One of the disadvantages in the use of plastics is their high rate of linear expansion, although manufacturers make provision for this in the design of their components. Another disadvantage is that resistance to damage by fire attack is low. Plastics are not such stable materials as metals and are all, to some extent, affected by the ultra-violet rays of the sunlight. This causes long term degradation or embrittlement of the material.

Most plastics in their natural state are clear and colourless, and to reduce the effects of degradation, manufacturers add a darkening substance or agent to give the material colour and body.

It is important for a plumber to be able to identify the type of plastics he or she is using, as the method of jointing suitable for one type may be unsuitable for another. Basic tests will be described which will enable him or her to correctly identify a particular material. Table 10 sets out typical properties of plastics used in building.

Polythene

Polythene pipes up to 150 mm nominal bore are manufactured to BS 3284 and are classified as 'low' or 'high' density. The low density pipe is comparatively flexible. High density polythene is more rigid and has a slightly higher melting temperature. Both types are resistant to chemical attack and are much used as a material for laboratory and chemical waste installations. Polythene is sufficiently elastic not to fracture if the water in the pipe should freeze, although normal frost precautions are recommended to prevent freezing and loss of supply.

Polypropylene

This material is in the same family group as polythene. It is tough, having both surface hardness and rigidity. It is able to withstand relatively high temperatures, and is in this respect superior to polythene, ABS or PVC.

Polypropylene can withstand boiling water temperature for short periods of time making it suitable material for the manufacture of traps. Methods of jointing are similar to those used for polythene, e.g. compression joints or 'O' ring couplings. Solvent welded joints are not suitable.

Polypropylene and polythene belong to the family of synthetic plastics known as polyofins. They have a waxy touch and appearance, and if ignited burn with a flame similar to a paraffin wax candle.

Table 10 *Typical properties of plastics used in building*

Material	*Density kg/m*	*Linear expansion per °C Coefficient*	*mm/m*	*Max. temperature recommended for continuous operation °C*	*Behaviour in fire*
Polythene* low density	910	20×10^{-5}	0.2	80	Melts and burns like paraffin wax
high density	945	14×10^{-5}	0.14	104	
Polypropylene	900	11×10^{-5}	0.11	120	Melts and burns like paraffin wax
Polymethyl methacrylate (acrylic)	1185	7×10^{-5}	0.07	80	Melts and burns readily
Rigid PVC (UPVC)	1395	5×10^{-5}	0.05	65	Melts but burns only with great difficulty
Post-chlorinated PVC (CPVC)	1300–1500	7×10^{-5}	0.07	100	Melts but burns only with great difficulty
Plasticized PVC	1280	7×10^{-5}	0.07	40–65	Melts, may burn, depending on plasticizer used
Acetal resin	1410	8×10^{-5}	0.08	80	Softens and burns fairly readily
ABS	1060	7×10^{-5}	0.07	90	Melts and burns readily
Nylon	1120	8×10^{-5}	0.08	80–120	Melts, burns with difficulty
Polycarbonate	1200	7×10^{-5}	0.07	110	Melts, burns with difficulty
Phenolic laminates	1410	3×10^{-5}	0.03	120	Highly resistant to ignition
GRP laminates	1600	2×10^{-5}	0.02	90–150	Usually inflammable. Relatively flame-retardant grades are available

Key: UPVC = unplasticized polyvinyl chloride GRP = glass–reinforced polyester PVC = polyvinyl chloride ABS = acrylonitrile/butadiene/styrene copolymer

*High density and low density polythene differ in their basic physical properties, the former being harder and more rigid than the latter. No distinction is drawn between them in terms of chemical properties or durability. The values shown are for typical materials but may vary considerably, depending on composition and method of manufacture.

Polyvinyl chloride

Generally abbreviated to PVC, this is possibly the most common plastics material used for drainage and discharge pipe systems. The material is a thermoplastic produced on the basic reaction of acetylene with hydrochloric acid in gas form in the presence of a catalyst. The material is rigid, smooth, light and resistant to corrosion.

PVC is often confused with polythene and polypropylene. One method of identifying the material is to drop a small piece of it into water, and if it sinks it is PVC which is heavier than water, whereas the other two materials are lighter and therefore float. PVC unlike many other plastics materials will not burn easily, another fact which can be used for its identification.

Unplasticized polyvinyl chloride (UPVC) is the basic material without softening additives. Plasticized polyvinyl chloride is produced by adding a small amount of rubber plasticizer to the basic material during the manufacturing process. The result is a slightly more flexible material which is more resistant to impact damage than UPVC. All UPVC pipes for cold water supply should comply with BS 3505.

Polyvinyl chloride for sanitary pipework should conform to BS 4514. This standard requires that the material should not soften below 70 °C for fittings and 81 °C for pipes and it should be capable of receiving discharge water at a higher temperature than these for short periods of time.

Jointing methods used include solvent welding, rubber 'O' ring joints and compression type couplings.

Acrylonitrile butadiene styrene

Known as ABS, this is a material used mainly for small diameter waste and discharge pipes or overflows. The material itself is a toughened polystyrene which can be extruded or moulded. It can withstand higher water temperatures for a longer period of time than PVC and for this reason some manufacturers produce full ABS waste systems. It also retains its strength against impact at very low temperatures thus providing greater resistance to physical damage. ABS has a duller matt appearance than PVC, and if ignited burns with a bright white flame. The material is slightly more dense than water and will therefore not float.

Acrylic (perspex)

This thermoplastic is tough and durable with good resistance to abrasion. The material can be transparent or opaque, it is easily machined and parts can be joined together by cementing. It is used mainly in plumbing for bathroom accessories.

Polystyrene

It is a white thermoplastic produced by the polymerization of styrene (vinyl benzine). This material is very light and brittle, and is mainly used for thermal insulation in granule, sheet or foam form.

Poly tetra flora ethylene (PTFE)

This material can be used at temperatures up to 300 °C, and because it is chemically inert it is used for lining pipes and components where chemical resistance is necessary. Used by the plumber in tape or paste form as a sealant or jointing material.

Nylon

This is a thermoplastic material which is produced from phenol or benzine. It is rot proof and strong. Nylon is widely used in the form of a solid plastic, often as valve seatings, taps, gears and bearings. Moving nylon parts need no oiling because they slide easily over each other. The word nylon was made up by its inventors Du Pont.

Synthetic rubber

There are many forms of synthetic rubber, some of which contain a proportion of natural rubber. The most common synthetic rubber used in plumbing is called Neoprene which is generally used for the manufacture of 'O' ring seals for various jointing techniques. Neoprene is a trade name. The substance resists attack from oil, grease and heat, and is more stable against oxidation than natural rubber.

Asbestos

This is a mineral compound of calcium, magnesium and silica. It occurs in various forms in different parts of the world and is used extensively for its heat and acid resisting properties.

Asbestos is a fibrous material which can be made into rope or cloth or mixed with cement to form sheets, pipes and cisterns. The plumbing industry has used asbestos rope for flue pipe jointing and asbestos cloth as protection blankets when using a flame close to a combustible material or surface.

Asbestos cement compounds are generally used in the manufacture of flue pipes and fittings, and cold water storage cisterns and their covers. With the recent discovery that crocidolite or blue asbestos is detrimental to health alternatives to asbestos have been sought, although there is no evidence that grey asbestos constitutes a hazard if proper precautions are taken.

These basic precautions are:

1 Avoid making a dust if possible, and avoid inhaling it at all costs when cutting asbestos cement sheets, pipes or cisterns.
2 Do the cutting outside.
3 Dampen the material to arrest dust.
4 Use a sharp, medium toothed saw or cutter.
5 Dispose of the swarf safely.
6 Wash your hands after completion.

Ceramics

Sanitary equipment may be made from one of the three ceramic materials – fireclay, earthenware or vitreous china. These materials have different qualities and characteristics. Each is, therefore, suitable for different uses which may be divided broadly into public, industrial and private use.

Where rough or heavy usage is expected, strength is a most important factor, and this is the outstanding quality of both fireclay and vitreous china. For private purposes, where good appearance is important, vitreous china is used.

Fireclay

This ware is made from fireclay which can be fired at very high temperatures, resulting in an article both heavy and strong. Large articles can be made in fireclay with the minimum of distortion, which due to the shrinkage which occurs during firing is inevitable with all ceramic materials. Fireclay is used in the manufacture of large sanitary appliances such as sinks, urinal ranges, laboratory sinks and special hospital fittings. Fireclay is heavy in weight and has a buff coloured porous body protected by the hard glaze which covers it. It has the highest initial cost of all three ceramic materials.

Earthenware

Ball and china clays are the most important constituents of earthenware, which is considerably lighter in weight than fireclay. This material has a pleasing appearance due to the clean lines and sharp definition which characterize articles manufactured from this material. Earthenware has a white porous body protected by a hard impervious glaze. This material is cheaper than either fireclay or vitreous china.

Vitreous china

This material is made from the same clays as earthenware but feldspar is also included. This gives two additional qualities, great strength and a vitrified body impervious to water. It is lighter in weight and less costly than fireclay, and has the same clean lines as earthenware. Apart from industrial use it is particularly good for hotel and household use since it combines pleasing appearance and strength.

Since the body is vitrified, vitreous china does not rely upon its skin of glaze for its sanitary properties, and is glazed only to give it a smooth, glossy finish and to allow easy cleaning. Should the glaze become damaged, the appliance remains impervious to water, and, therefore, completely sanitary. For this reason, and because of its strength and moderate cost, vitreous china is the most economic pottery material, both from the point of view of initial cost and maintenance costs.

During firing, vitreous china tends to distort more readily than either fireclay or earthenware, and for this reason great care has to be taken to ensure that the finished articles are of good shape. For this reason some of the larger sanitary appliances such as sinks, urinals and hospital equipment are not manufactured in vitreous china.

Vitreous china lends itself to the manufacture of articles of curved and rounded design, which minimizes the production problem of good shape and results in articles that are practical in use and conform to contemporary ideas of good practice in design.

Weight for weight, vitreous china is the strongest of the three ceramic materials used in the manufacture of sanitary appliances.

Self-assessment questions

1. Name *four* non-ferrous metals used by plumbers.
2. Define the term 'alloy'.
3. State the main constituents of non-metallic sheet.
4. Name *six* common alloys.
5. Name the ores from which aluminium, copper and lead are extracted.
6. State the type of copper tube which is manufactured for below-ground service pipes.
7. Name the two methods of manufacturing sheet lead.
8. Name the two groups into which plastics material can be subdivided.
9. State two disadvantages related to plastics materials.
10. State three advantages that plastics materials may possess when compared with metal.

3 Tools and equipment

After reading this chapter you should be able to:

1 Name the various hand tools used by a plumber.

2 Recognize different types of hand tools.

3 Select the correct tools for a particular operation.

4 Distinguish between the cutting action of various tools.

5 State the use of given tools and equipment.

6 Understand and state the purpose of different tools and items of workshop or site equipment.

Introduction

Good tools are indispensable to the craftsman, and buying them can be an expensive business. Some tools can only be used for one specialized task, whilst others, like hammers and pliers, may be used for a variety of jobs.

Most apprentices and plumbers will buy their tools as experience grows, and as they need them for the job they are doing. A kit of good quality tools, built up in this manner, is a sound investment. Some employers will buy tools for their employees, which helps to reduce the cost, payment often being made by an agreed weekly deduction from the employee's wages. The cost of the tool kit will, of course, depend upon the quality and quantity of tools bought.

Manufacturers produce such a wide variety of tools that there is no limit to the possible contents of a kit. Each tool has its own particular advantages and disadvantages and everyone has his or her own preferences and prejudices – many plumbers also make or adapt tools to suit their own special needs. The following list has been agreed between the Joint Industry Board for Plumbing Mechanical Engineering Services in England and Wales and the Electrical, Electronic, Telecommunication and Plumbing Union to be a full kit of tools which should enable a plumber to complete a reasonable job.

Recommended tool list

Adjustable spanner (300 mm long)
Bent pin or bolt
Brace
Gas blow torch and nozzle
Boxwood dresser (large or small)
Boxwood bending dresser
Boxwood mallet (large or small)
Bradawl
Compass saw (padsaw)
Fixing clamps or points
Footprint wrench (225 mm long)
Flat chisel for wood (225 mm × 55 mm)

Floor board cutter
Gimlet for lead pipe (pipe opener)
*Glass cutter and putty knife
*Hacking knife
Hacksaw frame
Hammers (small and large, 1 kg maximum)
Junior hacksaw
Knife, large pocket type
Lavatory union key
Pliers (two holes, gas)
Rasp (250 mm long)
Rule (metric 3 m tape)
Screwdrivers (large and small)
Shave hook
Snips (250 mm Tinmans)
Spirit level (225 mm long)
Springs for bending 15 mm and 22 mm light gauge copper tube
Steel chisels for brickwork (up to 500 mm long)
Stillson or similar pipe wrench (up to 300 mm long)
Tan pin
Tank cutter
Tool bag
Trowel (small)
Tube cutters suitable for light gauge copper tubes
Wiping cloths (3)

Plumbers able to provide tools from the list when they are needed for a job are eligible for a tool maintenance payment for every day which plumbing work is done. This money is used to maintain the kit in good working order, and to replace items as they wear, or are lost.

The list does not include items such as pipe cutters, pipe vices, welding apparatus, stocks and dies, and bending machines – these and other larger and more expensive items of equipment are provided by the employer. Neither does the list contain various tools for specialized or unusual tasks such as may be encountered when working on sheet lead, copper, aluminium, zinc or non-metallic roof weatherings. Many of these tools are made by the plumber, and will be described as the need for them arises.

A plumber is responsible for his or her tools and is dependent on them for his or her livelihood... he or she should:

Select, from the tool kit, the correct tool for the job.
Ensure that it is in good working order.
Use it correctly.
Transport it safely.

Failure to do this can result in:

Use of more physical effort than is necessary.
Waste of time.
Damage to fittings, appliances or client's property.
Injury to the user or others.

The maintenance and care of tools is important from both the practical and safety point of view. This applies to all tools, whether they are your own or provided by your employer for your use. Damaged, blunt or worn tools will not produce good work, and they could prove dangerous to the user and others working nearby.

The rest of this chapter is an introduction to the tools commonly used by a plumber.

Metal saws

The hacksaw has a pistol grip and the frame is adjustable to take various lengths of blade, usually 250 mm or 300 mm (see Figure 57).

Two types of blade are commonly used, one having 22 teeth per 25 mm of blade length and the other 32 teeth per 25 mm. The coarser blade is

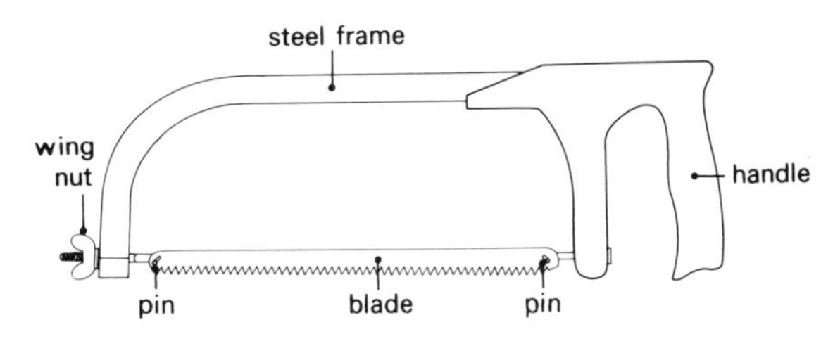

Figure 57 *Hacksaw*

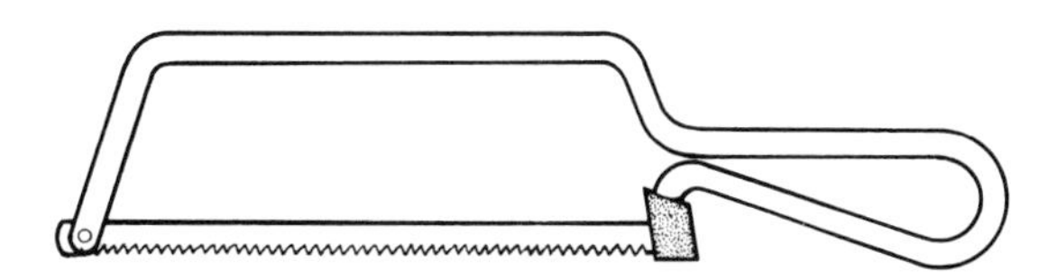

Figure 58 *Junior hacksaw*

*If glazing is usually done by plumbers in the district.

used for steel pipe and the finer blade for copper tube, although many plumbers use a junior hacksaw for cutting the smaller sizes of copper tube (see Figure 58).

The 'shetack' saw is particularly useful when cutting heavy gauge metal or corrugated materials. Its design enables the blade to cut material of an unlimited length or width, something which is not possible with an ordinary hacksaw, as the length of cut is limited by the depth of its frame.

Wood saws

There are many types of wood saw available; one of the most useful is the tenon saw. This can be used for cutting floorboards. These saws are generally 250–350 mm long and usually have about 12 teeth per 25 mm of blade. The blade teeth are 'set', that is folded outwards on alternate sides to provide clearance for the blade in the cut. A plumbers' saw has a double edged blade and is suitable for wood or lead. The blade length is usually 400 mm.

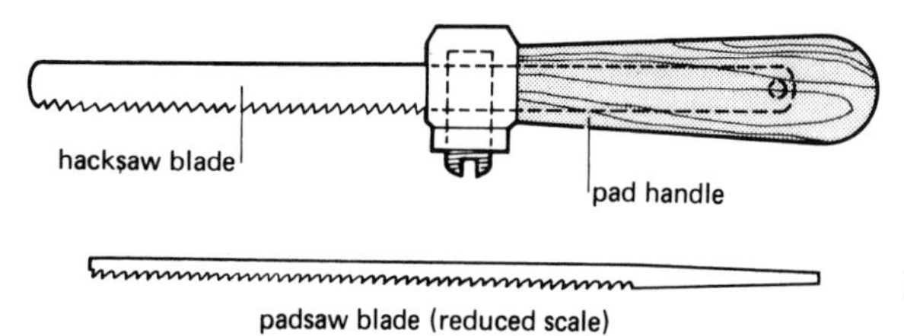

Figure 59 *Padsaw*

The padsaw has a plastic or metal handle into which is fitted a blade (see Figure 59). The blade can be a tapered one with about 10 teeth per 25 mm of blade and used for cutting wood. Alternatively a hacksaw blade can be fitted for cutting metal. The padsaw is most useful in awkward corners or when cutting floorboards in position.

Drilling tools

A ratchet brace can be used to hold a variety of drills or 'bits'. The ratchet allows the brace to be used close to a wall or corner. The chuck usually has two jaws and is most suitable for holding drills with a squared tapered shank. The jaws are called alligator jaws (see Figure 60).

The hand brace can be used in more confined spaces than a ratchet brace. The hand brace has a three-jaw chuck and holds round shanked drills or bits. Its gearing enables it to turn much quicker than a ratchet brace and is most suitable for drilling holes up to about 7 mm in diameter in metal, wood or masonry. The hand and breast drill is a larger version of the hand brace (see Figure 61).

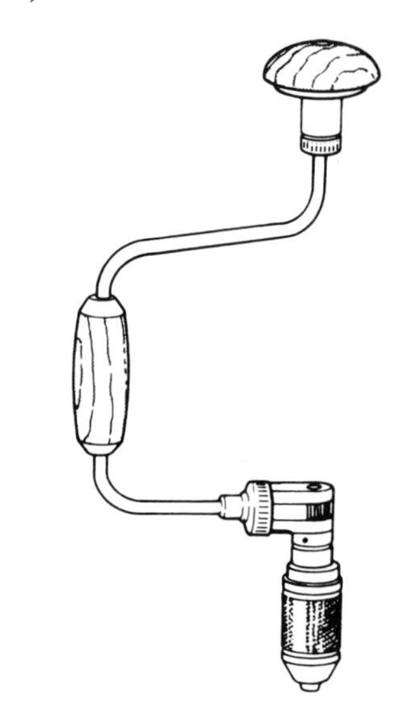

Figure 60 *Ratchet brace*

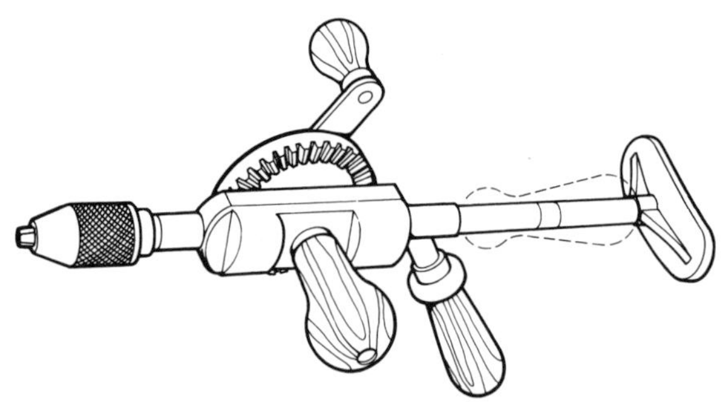

Figure 61 *Hand and breast drill*

Hammers

There are many different types of hammer and they are identified by their weight and head pattern. The 'pein' or 'pane' is the end of the head

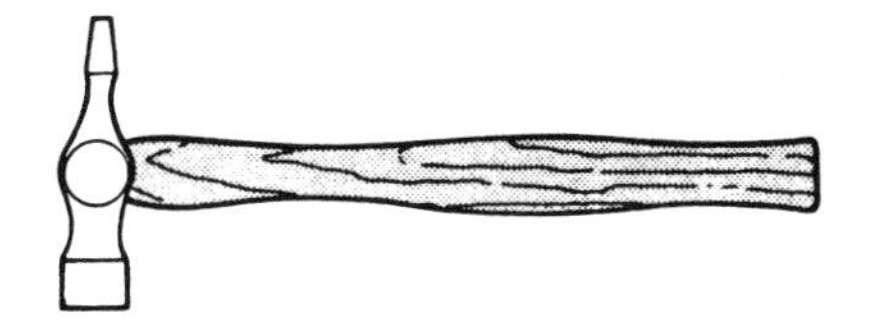

Figure 62 *Cross pane hammer*

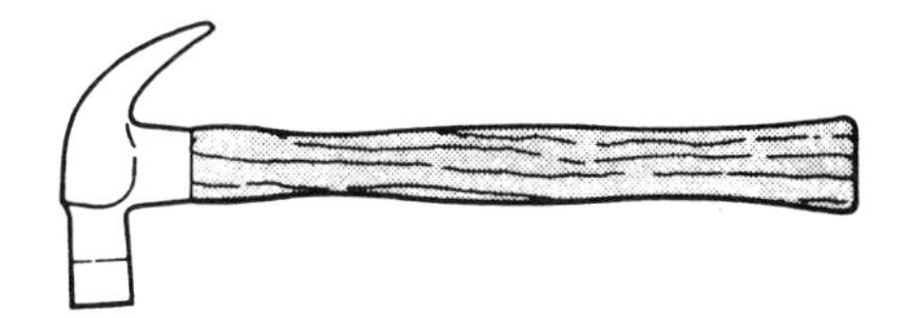

Figure 63 *Claw hammer*

opposite to the face of the hammer. Hammer heads are usually made of cast steel and the handles or shafts are ash or hickory. A claw hammer is useful for removing nails from roof timbers or floorboards. The range of hammers is extensive and the choice will depend upon the particular work operation, personal opinion and experience (see Figures 62 and 63).

Chisels

Wood chisels are made in a variety of types and sizes. In the past these chisels had a steel blade and wooden handle, but they are now available made completely of steel (see Figure 64). These are most useful to a plumber who will usually have a hammer in his tool kit but not always a wooden mallet to strike the chisel. Plugging chisels are used for cutting out slots or joints between bricks (see Figure 65).

Cold chisels are so called because they can cut mild steel when it is cold, although a plumber will also use this tool for cutting brick and concrete (see Figure 66). As with wood chisels, cold chisels are available in a variety of types, shapes and sizes, and for normal domestic work a selection of chisels ranging from 150 mm to 450 mm in length and 12 mm to 20 mm in diameter will be most useful. Chisels for lifting floorboards are available (see Figure 67). These have a parallel blade about

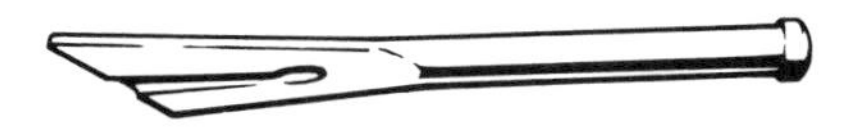

Figure 65 *Plugging chisel*

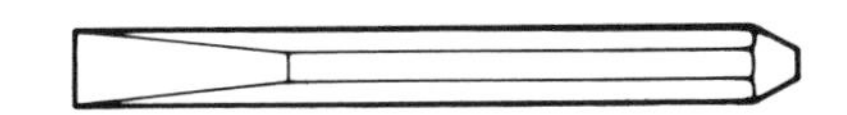

Figure 66 *Flat cold chisel*

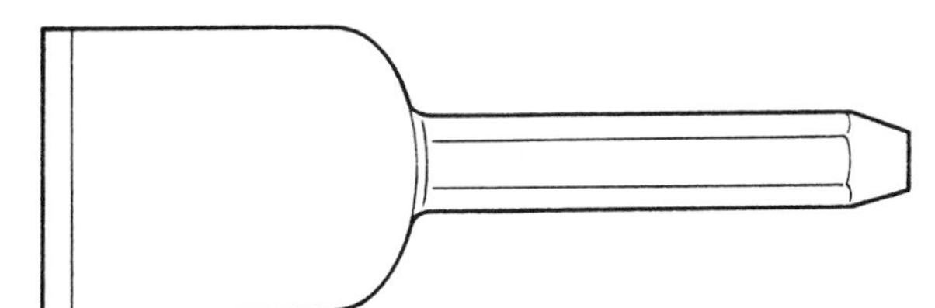

Figure 67 *Floorboard chisel*

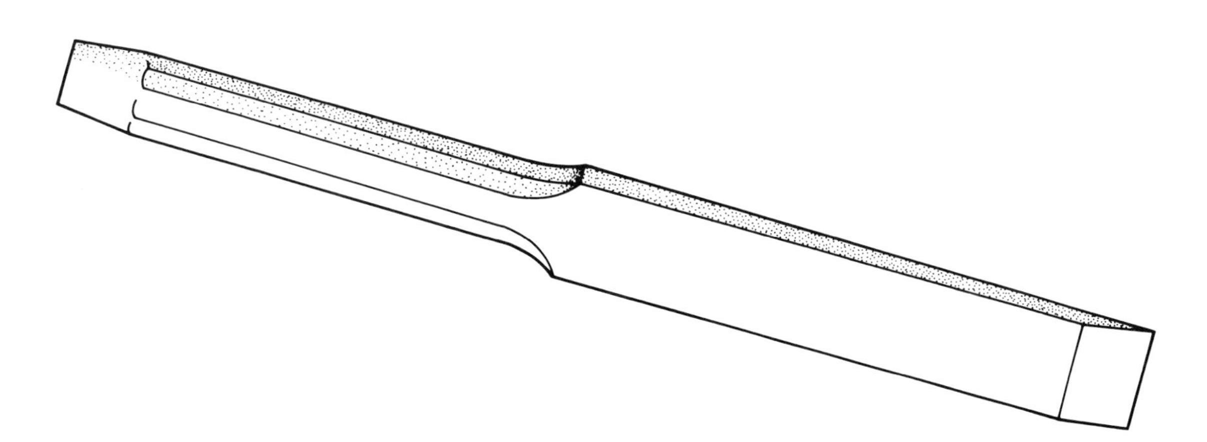

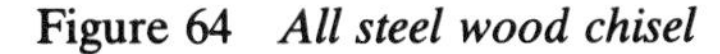

Figure 64 *All steel wood chisel*

75 mm wide. This blade is driven between the floorboards which are then raised by leverage via the chisel. This type of chisel is also used for cutting bricks.

Pliers

There are many types of pliers in common use, and most plumbers have several of these to enable them to perform different tasks.

Engineers' or combination pliers are available in several sizes, 150 mm to 200 mm are generally the most useful length (see Figure 68). Models are available with insulated grips for use on electrical circuits.

Gas pliers are an essential item in any plumber's kit. Their circular jaws make them most useful for holding pipes or bulky components (see Figure 69).

Long snipe pliers are useful when riveting sheet metal and on domestic servicing work, seaming pliers are mainly used when working on sheet (aluminium, zinc or copper) weatherings to assist with folding and welting (see Figure 70). A variation of these have the jaws in line with the handle and formed into a 'V'. These are called dog earing pliers and are used for that operation.

Gland nut pliers can be obtained in sizes from 100 mm to 350 mm and more than one size may be included in a tool kit (see Figure 71). They are extremely useful for many jobs, but like most pliers with serrated jaws can damage brass or chromium surfaces or fittings.

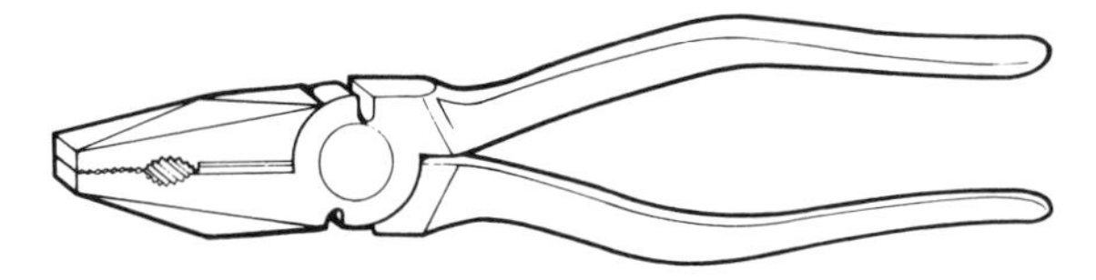

Figure 68 *Engineer's pliers*

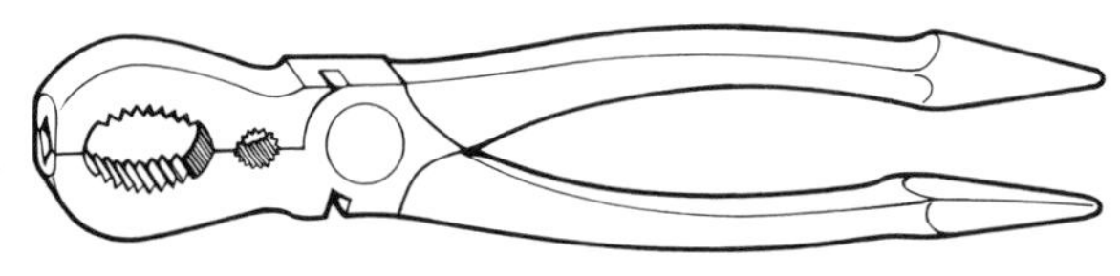

Figure 69 *Gas pliers*

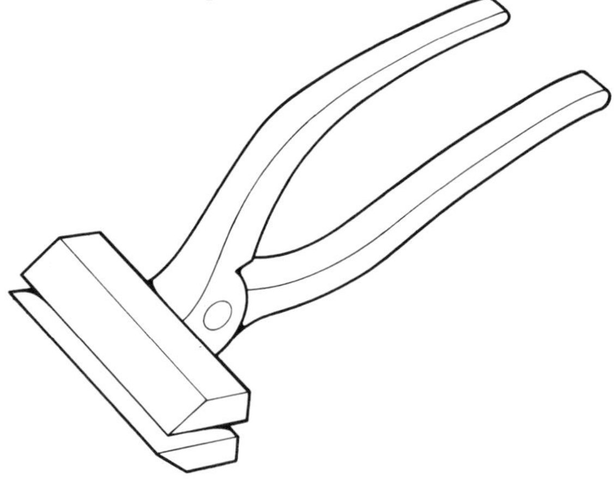

Figure 70 *Seaming pliers*

Files and rasps

Files and rasps are made of cast steel. One end is formed into a tang on to which fits a wooden handle (see Figures 72 and 73).

Some files and rasps are available with the tang formed into a handle. Files and rasps are identified by their:

Length 100 mm to 350 mm, in 50 mm steps.
Shape and cross-section hand, flat, half-round or round.
Cut single, double or rasp.
Grade rough, bastard, second cut or smooth.

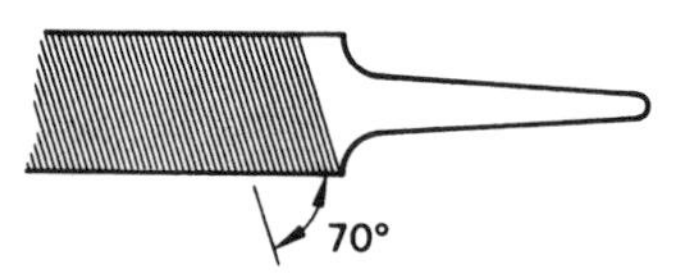

Figure 72 *File*

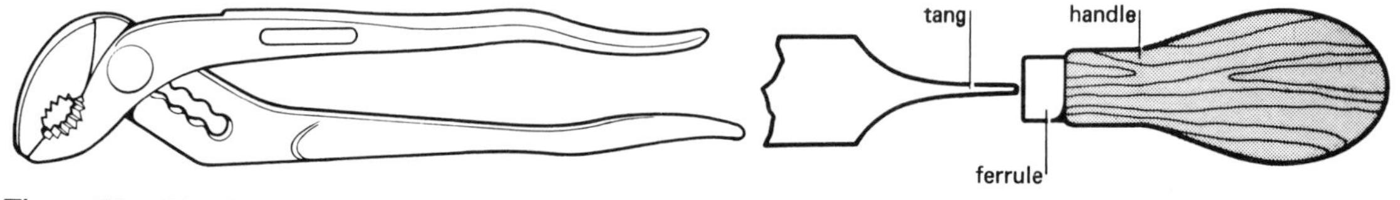

Figure 71 *Gland nut pliers*

Figure 73 *Handle for file*

The cut is standard for certain types of files. Flat files are double cut on the face and single cut on the edge. Hand files are similar, but have one edge uncut. Round files are usually single cut, and the half-round are double cut on the flat surface and single on the curved.

Spanners

Spanners are available in a variety of types. The most common are as follows:
Open ended
Ring
Box
Socket
Adjustable

Open ended spanners are usually double ended, with each end taking a different sized nut (see Figure 74). They are described by the size of the thread on which the nut screws, or by the distance across the flats of the nut.

Ring spanners fit completely round the nut to hold it very securely (see Figure 75). Ring spanners are safer to use than an open ended spanner as there is less risk of the spanner slipping off the nut. Also they are less likely to wear or open out. They are preferred for jobs where nuts must be tightened more securely.

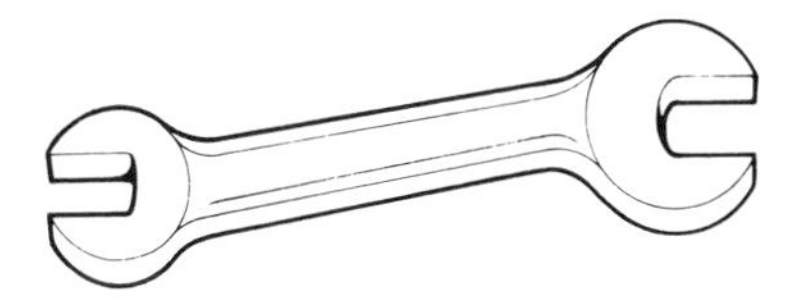

Figure 74 *Open ended spanner*

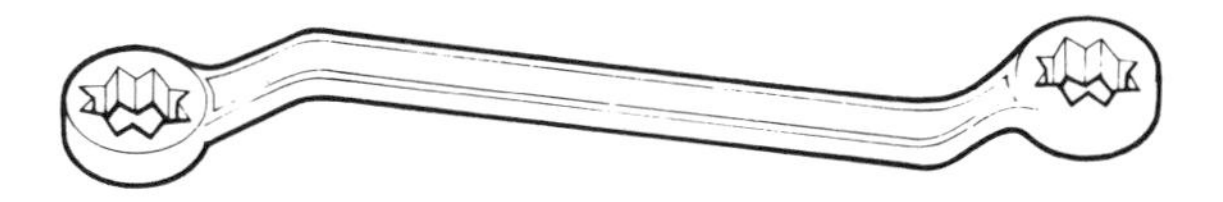

Figure 75 *Ring spanner*

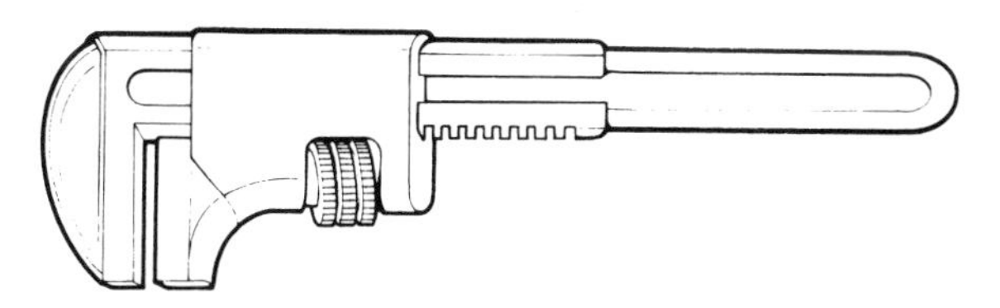

Figure 76 *Adjustable spanner*

Box spanners are most useful for releasing or tightening recessed nuts, or nuts in inaccessible positions such as those securing taps to wash basins, baths and sink units. Most box spanners are double ended and are turned by a steel rod called a tommy bar.

Socket spanners are a very robust type of tool. They may be used with a ring or open ended spanner or with a ratchet brace. Socket spanners are most useful for servicing work to boilers, water heaters etc.

Adjustable spanners are available in several sizes and different designs and most plumbers include at least two different lengths in their kit (see Figure 76). Thin jawed adjustable spanners are most suitable for assembly and disconnection work to pipework and components.

Pipe grips or wrenches

There are four main types of wrench in use:
Stillson pipe wrench
Footprint pipe wrench
Chain pipe wrench
Self grip wrench

The stillson wrench is a very robust tool and is most suitable for steel pipe work. These wrenches are available in a wide variety of lengths, ranging from 150 mm to 1.225 m. The most adaptable sizes for plumbers' work are 250 mm and 450 mm.

Footprint wrenches rely on hand grip pressure to secure the pipe or component. These are available in lengths ranging from 150 mm to 400 mm.

Chain pipe wrenches or chain tongs are usually associated with industrial work, but small models are available for domestic purposes. The length of lever handle may vary between 200 mm and 900 mm.

Self grip wrenches rely on hand grip pressure to secure the component although they also have a lock-on action to securely grip the wrench on to the component allowing the grip pressure to be

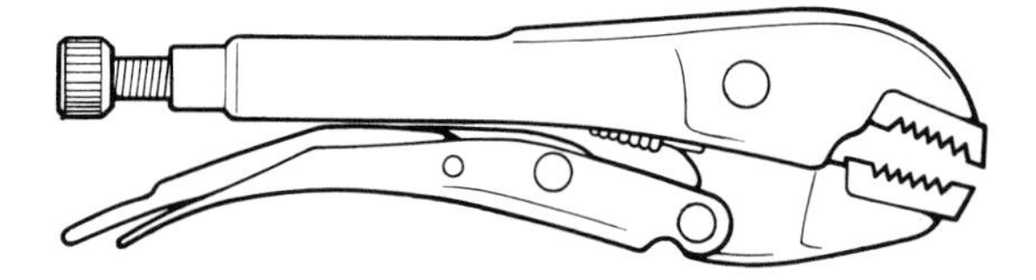

Figure 77 *Self grip wrench or mole wrench*

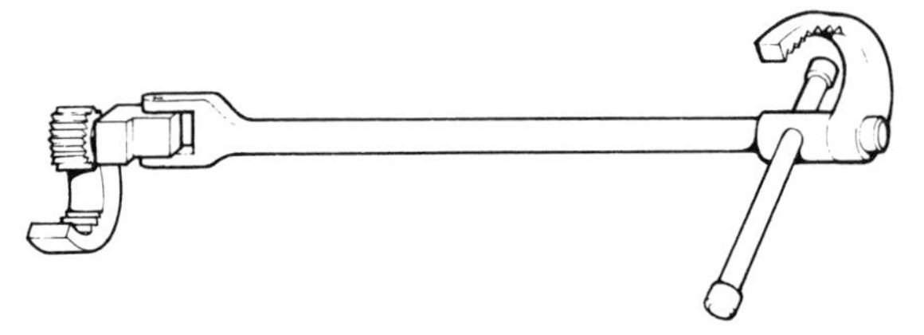

Figure 78 *Adjustable wrench*

released (see Figure 77). They are available with jaws of alloy steel and in lengths from 150 mm to 250 mm.

Wrenches for specialist tasks are produced by various manufacturers. The shetack basin wrench is specially designed for the difficult job of fitting back nuts and union nuts behind wash basins, baths and sink units. The tool may be used in the horizontal or vertical position enabling a nut to be tightened or loosened in the most inaccessible places. The spanner is approximately 250 mm long and fits standard size backnuts. A similar model is also available with a greater distance between the jaws for waste fittings and traps.

An adjustable wrench for use in similar locations as the shetack basin wrench overcomes the difficulty of non-standard size nuts and unions (see Figure 78). The serrated teeth give a ratchet action which is useful when space is limited.

Screwdrivers

These are available in many types and sizes. Blades are made from high carbon steel and handles of wood or plastic. Some screwdrivers include a ratchet for ease of operation. Some manufacturers produce a set of screwdrivers of varying length with interchangeable blades having various widths and blade pattern, all fitting into a common handle. Screwdrivers are identified by their type and the length of the blade. Sometimes the width of the blade is also stated.

A cabinet screwdriver is used for bigger screws and has a large handle to provide gripping power to turn the screw. The most useful sizes are 200 mm and 400 mm.

Stubby or dumpy screwdrivers are used for larger screws which are located in awkward places. The blades are very short, usually about 25 mm long and available in a variety of widths, the most popular being 6 mm.

Electricians' screwdrivers are made for smaller screws. These screwdrivers have a plastics handle and some are available with a plastics sheathed blade.

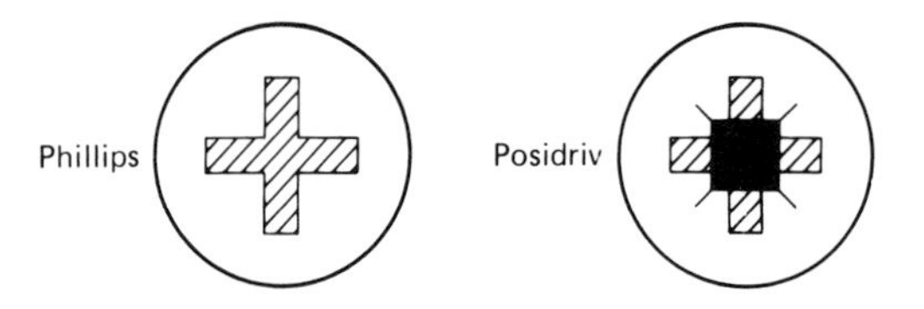

Figure 79 *Phillips and Posidriv screw heads*

Phillips and Posidriv screwdrivers both have cross-shaped blade points to fit cross-slotted head screws (see Figure 79). The Posidriv has superseded the Phillips type head, but the Phillips screwdriver is still retained because it will fit both screwheads, although the Posidriv cannot be used on Phillips screws. A Phillips screwdriver is more sharply pointed than the Posidriv, but the essential difference is in the square section between the slots of the cross which enables the Posidriv to fit closely between the blade and the screwhead. The corners of the square can be easily seen on the screwhead and are illustrated on the trade mark. Two sizes of blades are available. Cross headed screws are common on certain plumbing components such as water heaters and boilers.

Offset or cranked screwdrivers are very useful for getting at screws in awkward places. Blades may be flat or cross-slotted. These screwdrivers are usually all steel.

Gimlet and bradawl

These tools are used for forming holes in timber to enable a wood screw to start. The gimlet is

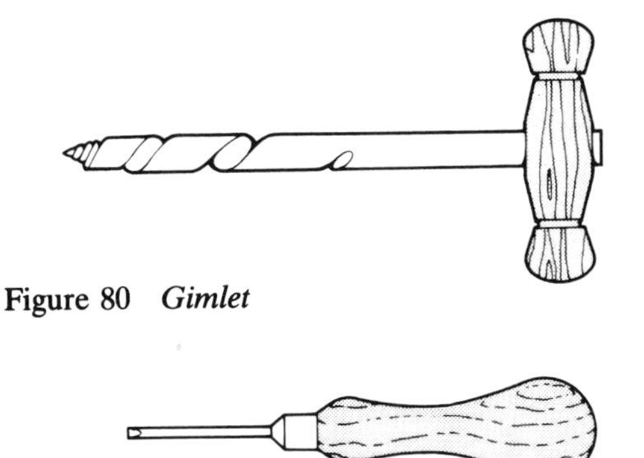

Figure 80 *Gimlet*

Figure 81 *Bradawl*

used on hardwoods (see Figure 80) and the bradawl on softwoods (see Figure 81).

Punches

Several types are available which vary in shape according to the job to be done.

The nail punch is for punching nails below the surface of the timber which is being secured (see Figure 82). This is essential practice on roof boarding which is to be covered with sheet

Figure 82 *Nail punch*

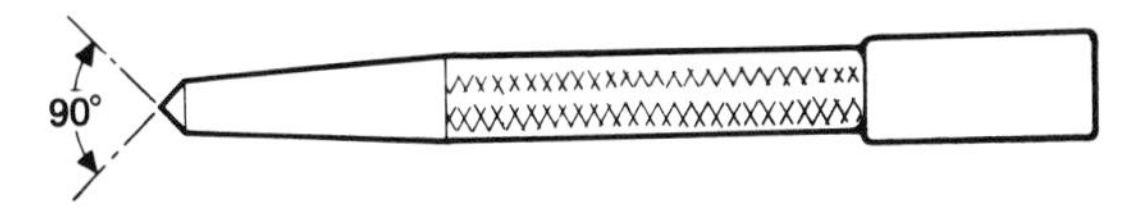

Figure 83 *Centre punch*

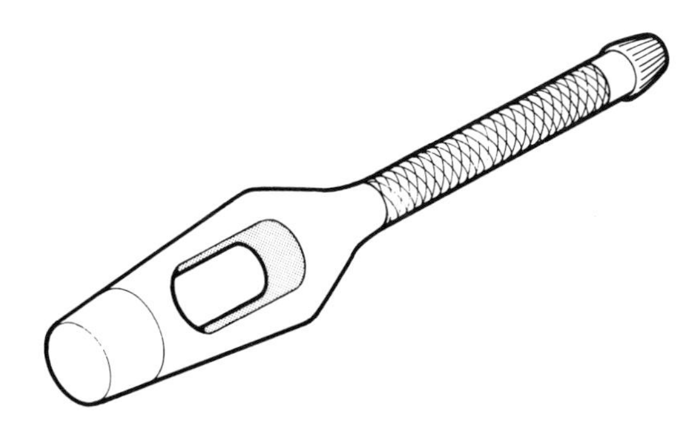

Figure 84 *Hole punch*

weatherings. Another function is to punch nails through floorboards so that the boards may be raised easily. The punch is usually about 100 mm long, with a 3 mm diameter head which is slightly recessed to prevent the punch from slipping off the nail head.

Centre punches are used to mark the centre of a hole to be drilled in metal (see Figure 83). The punch provides a small indentation which locates the drill in the correct position and prevents it from slipping. These punches are usually 100 mm to 200 mm long and the point is tapered at 90°.

Hole punches are used to cut a small circular hole in soft materials such as sheet lead, so enabling washers to be made (see Figure 84).

Taps

Taps are used for cutting internal or 'female' screw threads in metal. There are three taps to a set (see Figure 85).

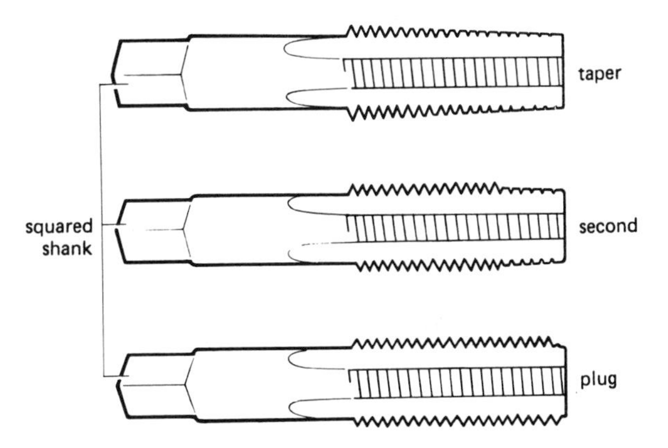

Figure 85 *Taps*

Taper tap

This is used to start the thread. It has no threads near its end so that it can enter the hole. This tap will cut a complete thread if it can pass right through the hole.

Second tap

This is used in a blind hole (a hole that does not pass right through the metal) to cut a thread near to the bottom of the hole, after preliminary cutting with the taper tap.

Figure 86 *Tap wrench*

Plug tap
This is used to complete the thread started by the taper and second taps, and cuts right through to the bottom of a blind hole.

Taps are turned by means of a tap wrench (see Figure 86).

Dies

Dies are used for cutting external or 'male' threads on circular sections of metal, plastic rod and pipe. Dies are available in a number of different forms and are held in 'stocks', so that they can be rotated around the rod or pipe. Figure 87 shows circular split pattern dies which are held in the stocks shown in Figure 88. The split allows for adjustment to be made to the size or depth of thread being cut.

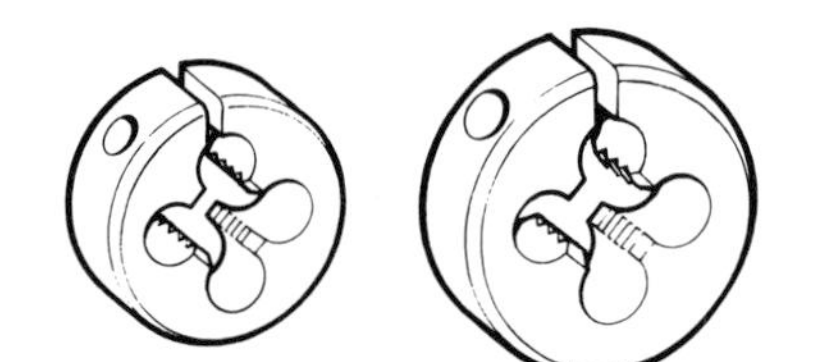

Figure 87 *Circular split pattern dies*

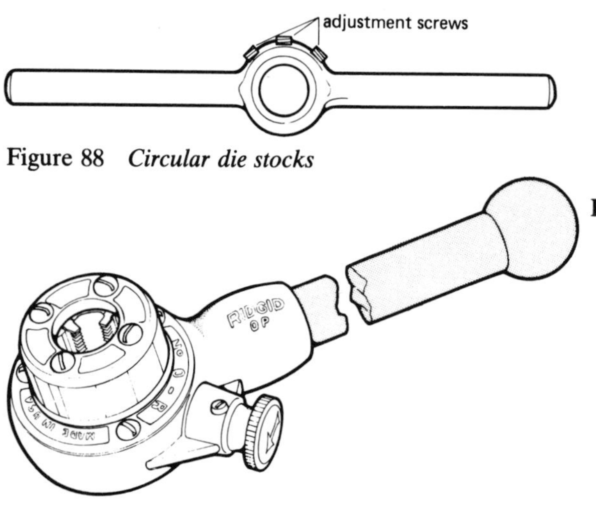

Figure 88 *Circular die stocks*

Figure 89 *Hand operated ratchet dies*

External threading of mild steel pipes is usually carried out using solid dies which have a ratchet included for ease of operation (see Figure 89).

Bending springs

Springs are used when bending copper tube or lead pipes by hand. The spring supports the tube and prevents it from kinking or losing its circular section. Springs for bending copper tube are made from square section spring steel and are available in two types: internal and external.

The internal spring is approximately 600 mm long and is most suitable for short lengths of pipe when a sharp tight bend is required (see Figure 90). The pipe is slightly over bent and then opened out to release the tension on the spring. The spring is then rotated in the direction required to reduce its diameter and withdrawn from the pipe.

External springs for copper tube are shorter in length than internal springs and have one or both ends opened slightly to assist tube insertion (see Figure 91). They are most useful for long lengths of pipe or when bending *in situ*. Generally external bending springs do not allow such sharp bends to be pulled as with the internal spring.

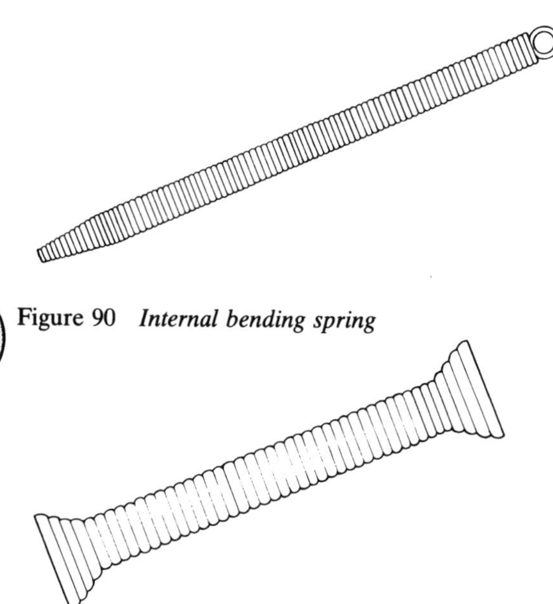

Figure 90 *Internal bending spring*

Figure 91 *External bending spring*

Pipe cutters

Pipe cutters operate by the rotation of a hardened steel cutting wheel around the outside of the pipe (see Figures 92 and 93). The wheel is gradually moved through the wall of the pipe as the adjustment is tightened until the pipe is cut. The cutting wheel is narrow and sharp, but the action of cutting produces a burr around the inside edge of the pipe which must be removed by a reamer (see Figure 94).

Cutters are generally used for cutting copper, stainless steel, mild steel and cast iron pipes. In the case of cast iron pipes, no internal burr is produced as the pipe shears at the cutting point before the cutting wheel reaches the inside edge of the pipe.

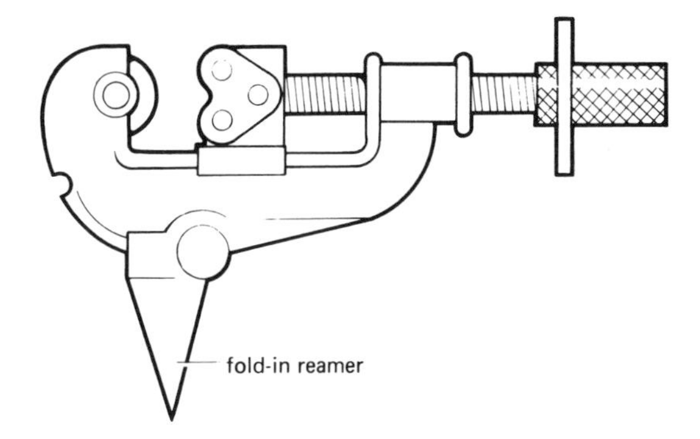

Figure 92 *Copper tube cutter*

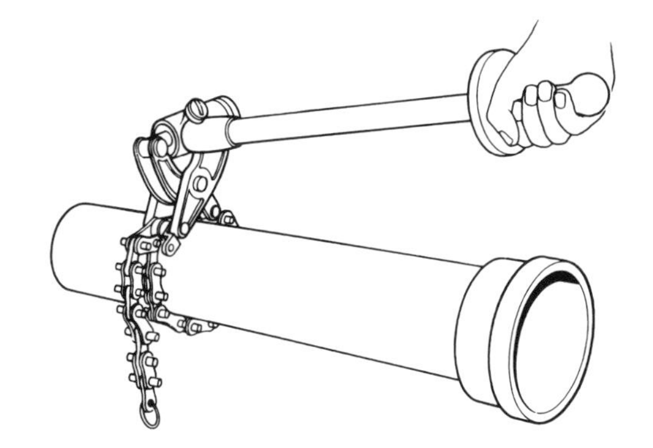

Figure 93 *Snap action cast iron pipe cutter*

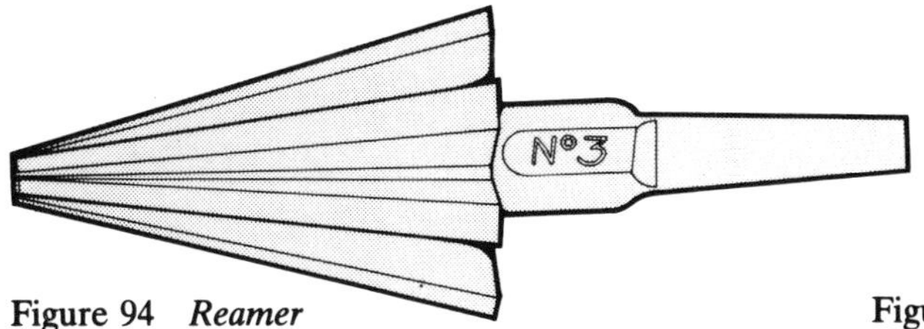

Figure 94 *Reamer*

Hole cutters

Hole cutters are used for cutting circular holes in cisterns, tanks and cylinders. The cutter shown in Figure 95 is used in a brace and the cutter is adjusted to the size of the hole required.

The cutter shown in Figure 96 is called a hole saw and resembles a circular hacksaw blade. These can be used in a brace or on an electric drill. They are available in a wide range of sizes and both the drill and blade are replaceable.

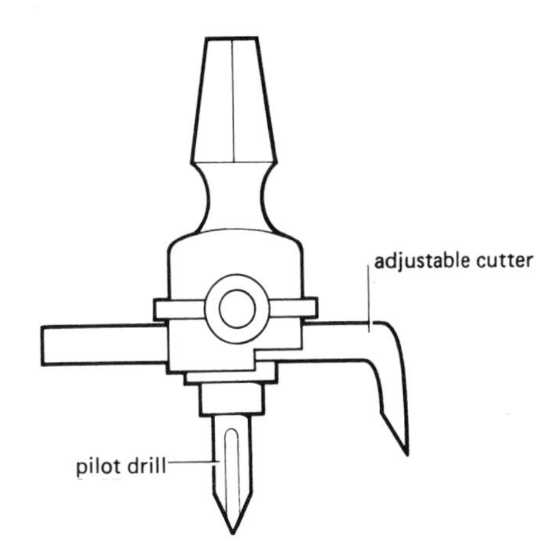

Figure 95

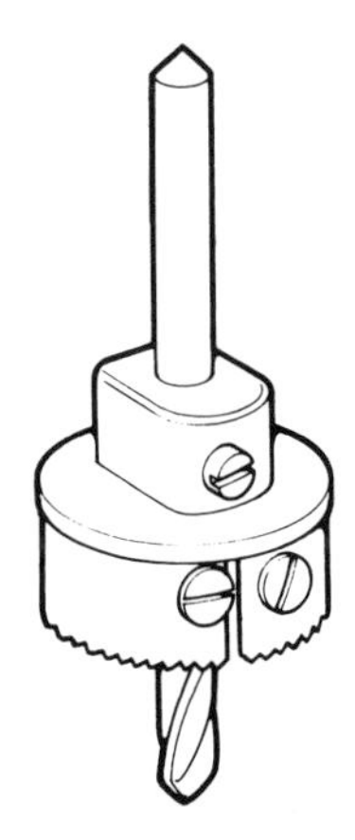

Figure 96

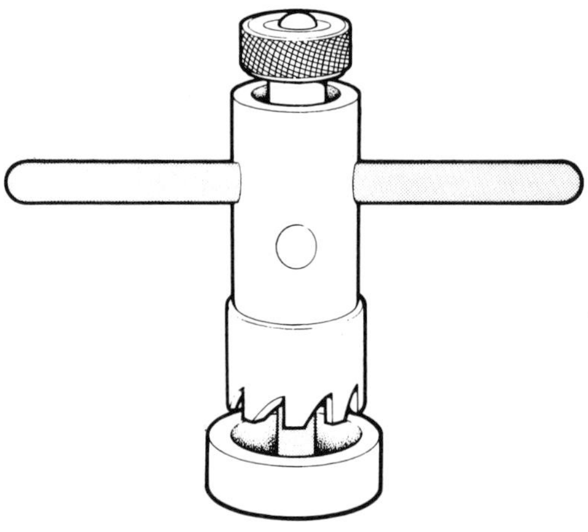

Figure 97

Figure 97 shows a hand operated cutter which is most commonly used for cutting galvanized mild steel cisterns and tanks; a ratchet handle is available for easier operation in confined spaces.

Tools for expanding the diameter of copper tubes

Mandrel

This is used to open the end of a piece of tube (BS 2871 Part 1 Table X) to a 15° taper, making the tube suitable for use on certain tube fittings. The tool may be suitable for one diameter of tube, or may be a combination tool as shown in Figure 98, incorporating two sizes in one tool.

Socket forming tool

This is driven into the end of the copper tube opening it out parallel and forming a socket to receive another piece of tube (see Figure 99). If the end of the tube is annealed the opening process will be simplified. The capillary joint formed is easily made by soldering or brazing. Suitable for use on tubes to BS 2871 Part 1 Table X.

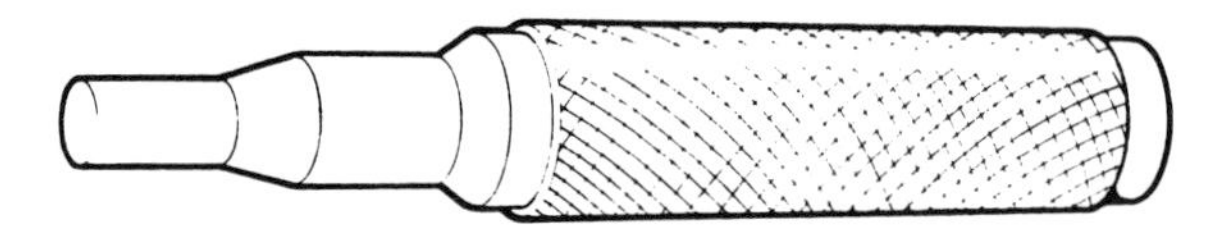

Figure 98

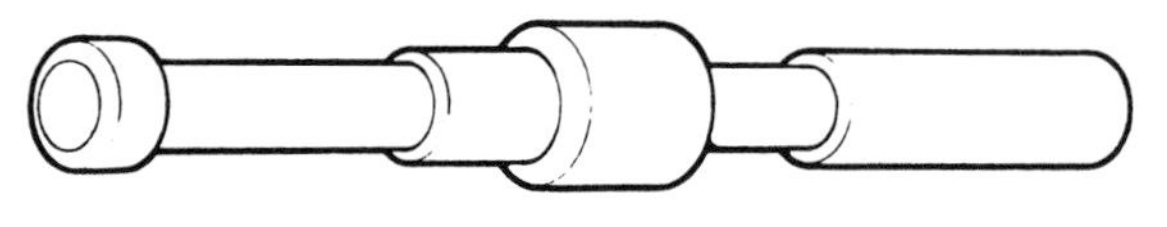

Figure 99 *Socket forming tool*

Wooden tools

Hardwood dresser

This tool is used for dressing copper, aluminium, zinc sheet and lead pipes and sheet (see Figure 100). Dressers are usually made from hardwoods or high density plastics.

Bossing mallet

This is a useful tool when working on sheet roofing materials. The mallet head is made from a hardwood and the handle from malacca cane (see Figure 101).

Tinmans mallet

This is used on sheet copper and aluminium weatherings to assist with joining together welts and seams, and may be used to strike another sheetwork tool, for example a setting-in stick, or dresser (see Figure 102).

Bossing stick

This is used mainly for bossing corners on sheet lead weatherings (see Figure 103).

Bending stick

This is an essential tool when bending lead waste pipes (see Figure 104). The stick is used to dress the lead around to the back of the bend. This tool may also be used on sheet lead weatherings.

Setting-in stick

This is, as its name implies, a tool used to reinforce or sharpen folds or angles which have been formed in sheet weatherings (see Figure 105).

Chase wedge

This is a tool used for setting-in the corner of a fold or crease and is available in widths from 50 mm to 100 mm (see Figure 106).

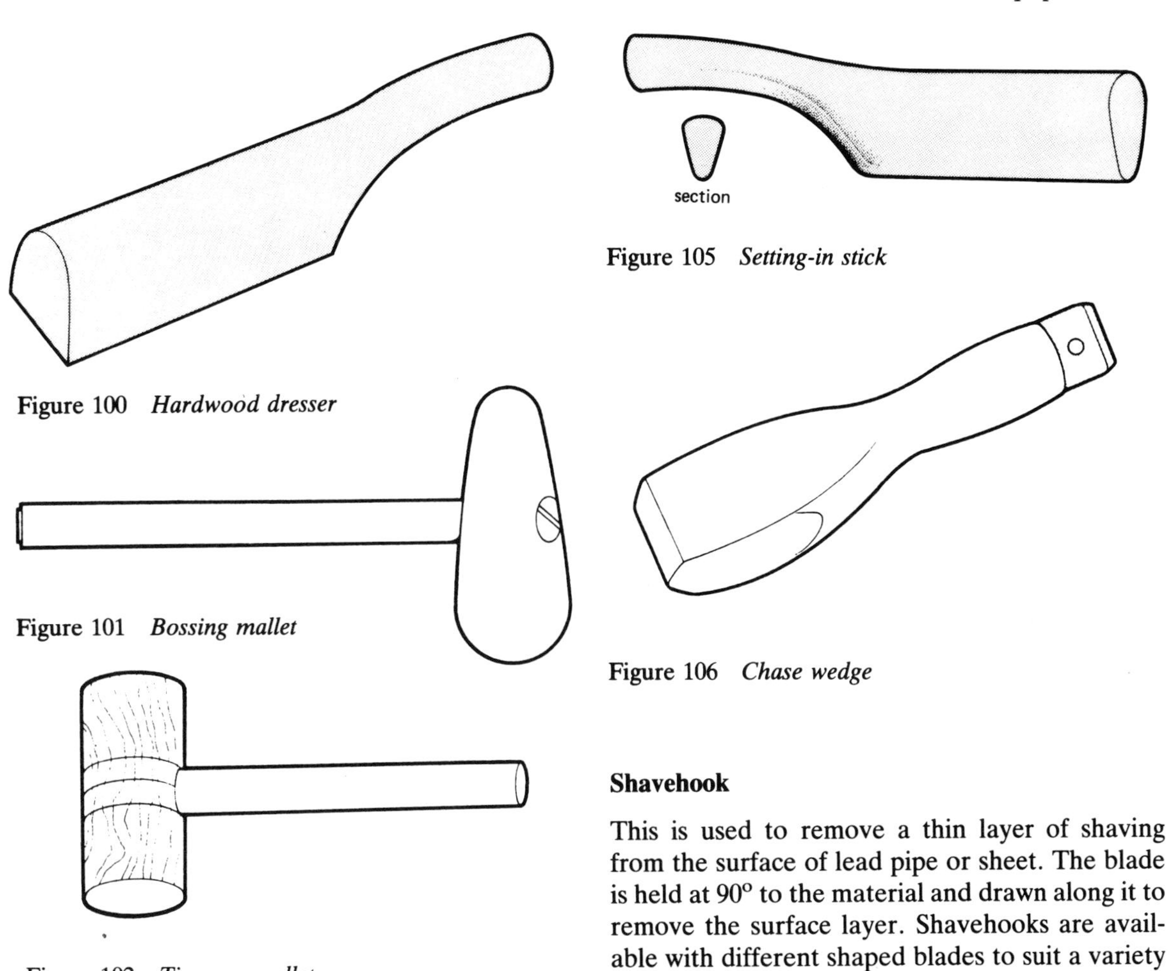

Figure 100 *Hardwood dresser*

Figure 101 *Bossing mallet*

Figure 102 *Tinmans mallet*

Figure 105 *Setting-in stick*

Figure 106 *Chase wedge*

Shavehook

This is used to remove a thin layer of shaving from the surface of lead pipe or sheet. The blade is held at 90° to the material and drawn along it to remove the surface layer. Shavehooks are available with different shaped blades to suit a variety of tasks (see Figure 107).

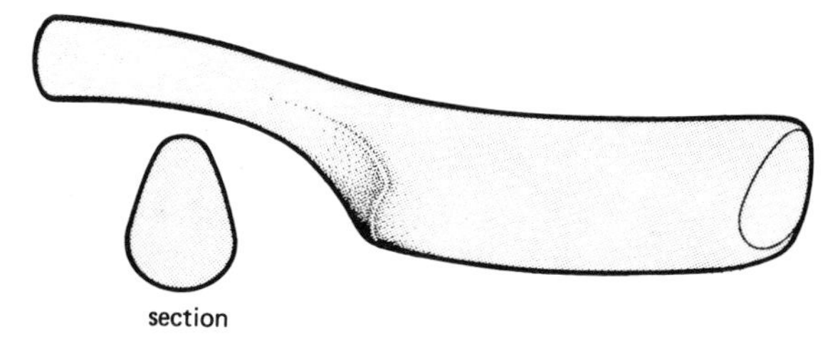

Figure 103 *Bossing stick*

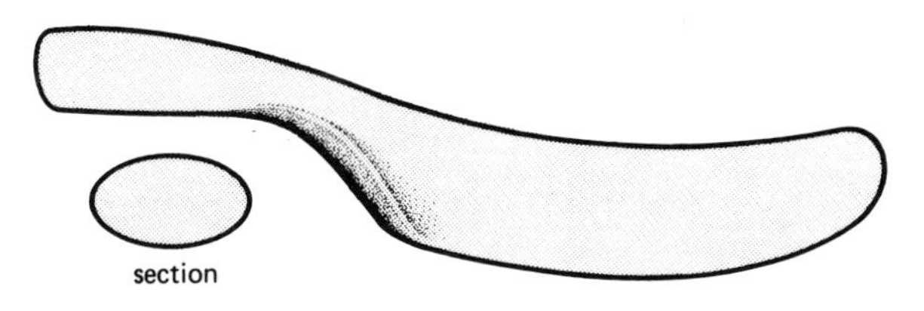

Figure 104 *Bending stick*

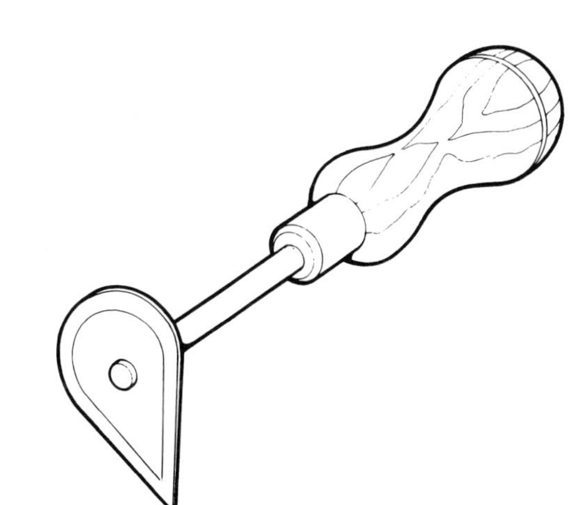

Figure 107 *Shavehook*

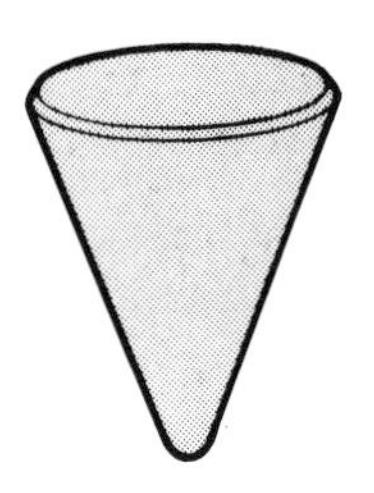

Figure 108 *Turnpin*

Turnpin or tan pin

This is a cone-shaped piece of boxwood used for opening out the end of a piece of lead pipe prior to jointing (see Figure 108). Steel turnpins are also available for use on copper tube.

Bent bolt

This is a cranked piece of steel rod approximately 230 mm long and 14 mm diameter, tapered to 5 mm at one end (see Figure 109). It is used for opening up a hole in lead or copper pipe to form a socket for a branch pipe. It is also used to provide leverage to enable small diameter lead pipes to be straightened.

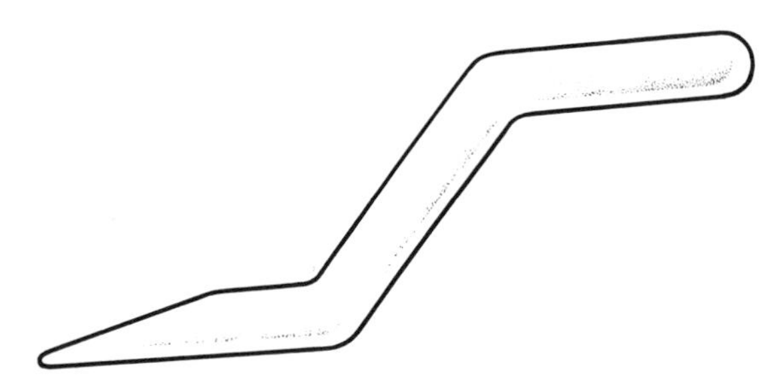

Figure 109 *Bent bolt*

Tin snips

These are used for cutting thin sheet metal and are available in several sizes, the most popular being 150 mm and 250 mm long (see Figure 110). Straight and curved pattern blades are available. Universal snips are heavy duty pattern to cut either curved or straight and have open-ended handles to prevent nipping.

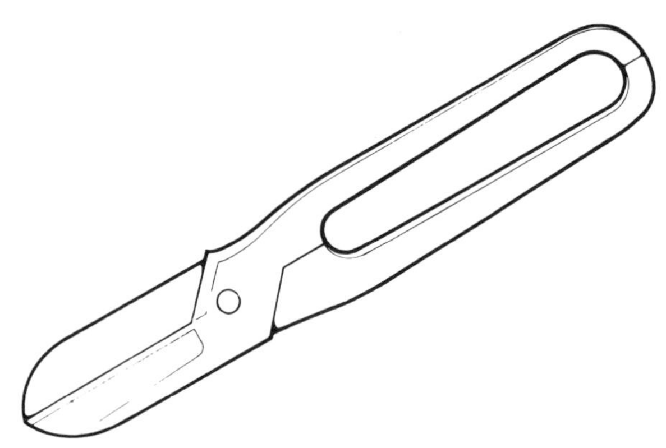

Figure 110 *Tinsnips*

Soldering irons

These are used for soldering and tinning purposes, for example copper and brass unions or zinc and copper sheet. Soldering irons have a forged copper 'bit', steel shanks and a whitewood handle and are usually sold by weight.

Small electric soldering irons are available and are most suitable for workshop use, for example making electric wiring connections.

Pipe vices

Pipe vices are needed to grip and secure mild steel pipes which are to be cut or threaded. They may be of the hinged type or chain type, and may be secured either to a work bench or to portable tripod stand.

Bending machines

These may be used for bending copper, stainless steel and mild steel pipes, and are necessary to bend pipes which are of large diameter or are too rigid to bend manually. They are also useful for prefabrication work, or when several bends have to be formed in a short length of pipe.

There are two types of machines: rotary (see Figure 111) and ram (see Figure 112).

Bending machines work by hand power or through a gear or ratchet action and employ special formers and back guides to ensure that the tube, when pulled to the required angle, maintains its true diameter and shape throughout the

Figure 111 *Portable rotary tube bender*

length of the bend. When using a machine it is advisable that the guides and formers should be lubricated and maintained in good condition.

Power tools

Although this title describes all kinds of tools driven by a variety of types of motor, so far as the plumber is concerned it is likely to include only a few tools, all of which are driven by electricity.

Electrical tools may be:

1 Mains voltage.
2 Low voltage, with a step-down transformer.
3 Double insulated.
4 All insulated.

Mains voltage

Portable tools, like drills, usually have single phase universal motors and operate on 240 V. These tools have a three core cable with the casing connected to the earth connection. If this earthing becomes faulty, particularly in wet situations, the tool can cause a lethal electric shock. When working in premises where the earthing cannot be guaranteed, mains voltage tools should not be used.

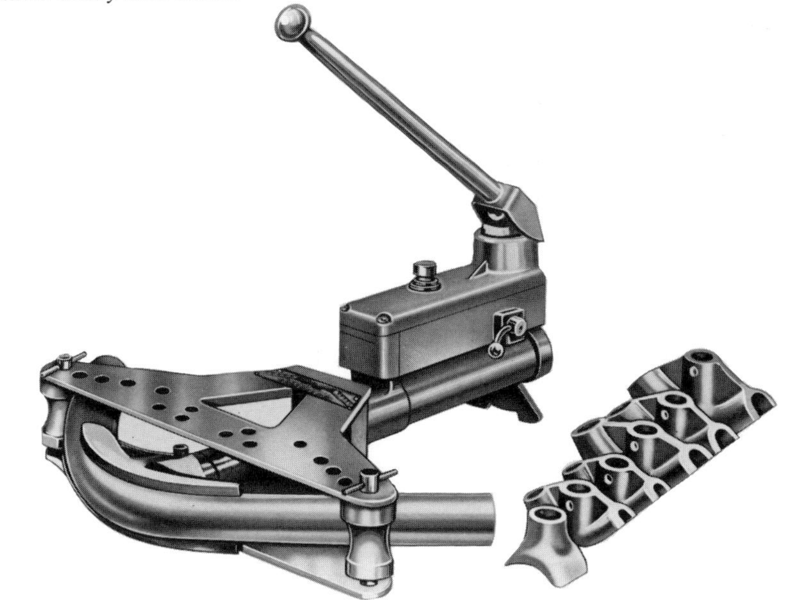

Figure 112 *Hydraulic ram bender for heavy duty pipe bending. This type of bender is operated by a hydraulic pump and is commonly used for mild steel pipes*

Low voltage
Transformers are used to step down the mains voltage from 240 V to 110 V for tools or 25 V for hand lead lamps. Usually both the live and neutral connections are fused on the transformer output. Low voltage tools can be fitted with special plugs so that they cannot be connected or used accidentally on full mains voltage.

Double insulated
These tools have additional insulation to eliminate risks from defective earthing. They are tested to 4000 V and may be used on 240 V supplies without an earth lead if they conform to BS 2769 and are identified by the appropriate BS symbols.

All insulated
This type of power tool is made entirely from shock-proof nylon and does not have a metal casing, therefore electricity cannot be conducted from any part, unless the casing becomes damaged. They are tested to 4000 V and may be used on 240 V without an earth like the double insulated tool.

The following notes are intended only as an introduction to power tools.

Drills
Portable electric drills generally have chucks to receive drills up to 10 mm in diameter. Two speed drills rotate at about 900 and 2400 revolutions per minute. Bench drills for larger work are usually fitted in workshops.

Percussion tools
These give the drill fast-hitting blows at the rate of about 50 per second. This hammer-action helps to penetrate hard materials such as concrete which are difficult to cut with an ordinary rotary drill. A special impact type of tungsten carbide tipped drill should be used with percussion tools to produce an accurate smooth hole.

Mechanical saws
These may be portable circular blade hand saws for cutting timber, or larger static machines for cutting mild steel pipes and rods. The item being cut must always be fixed or clamped securely and excessive pressure must not be used.

Screwing machines
There is a variety of types of machine for cutting similar to those on hand dies and many machines incorporate a pipe cutter and reamer. A suitable cutting lubricant must be used to keep the dies cool and assist with the cutting operation.

Cartridge tools
These tools act like a gun and shoot a hardened steel fixing stud into the material to which a fixing is required. Several types of fixing stud are available. Some are similar to wood nails for fixing timber, others have threaded ends to which a bracket, clip or nut can be screwed.

The foregoing tools and equipment comprise those used by the average plumber employed on work of a general nature. There are, however, several other tools and items of equipment which might be used for specific or specialist tasks. These will be illustrated in their respective chapters.

Self-assessment questions

1 Which saw would be best for cutting a floorboard?
(a) junior hacksaw
(b) hacksaw
(c) tenon saw
(d) shetack saw

2 A claw hammer is used for:
(a) welting sheet copper
(b) removing nails from timber
(c) lifting floorboards
(d) chasing brickwork

3 The pointed end of a rasp or file which fits inside the handle of the tool is called a:
(a) snipe
(b) pein
(c) shoulder
(d) tang

4 A tommy bar is used in conjunction with a:
(a) box spanner
(b) ring spanner
(c) pipe wrench
(d) chain wrench

5 In order to form a hole to give screws a start the best tool to use would be a:
(a) screwdriver
(b) bradawl
(c) centre punch
(d) plugging chisel

6 Before drilling a hole in metal a tool is used to mark the position of the hole. This tool is called:
(a) centre punch
(b) nail punch
(c) hole punch
(d) round file

7 Cutting tubes with a wheel cutter produces a burr around the inside edge of the pipe. This burr should be removed with a:
(a) die
(b) reamer
(c) tapered tap
(d) plug tap

8 The tool used to remove a thin shaving or layer from the surface of lead sheet or pipe is called a:
(a) screwdriver
(b) gimlet
(c) lead knife
(d) shavehook

9 Which tool is used for cutting thin sheet metal?
(a) cold chisel
(b) junior hacksaw
(c) tinsnips
(d) wheel cutter

10 The type of chisel used for cutting out joints between bricks is called a:
(a) wood chisel
(b) plugging chisel
(c) cold chisel
(d) bolster chisel

4 Communication: sketching, drawing and geometry

After reading this chapter you should be able to:

1. Recognize basic drawing instruments and state their use.
2. Sketch details as a method of communication.
3. Understand and apply drawing practice and symbols to BS 1192.
4. Set out and use scales and templates.
5. Understand and apply basic principles of geometry and proportional sketching.
6. Use methods of development.

Introduction

Communication is the imparting of information and facts. There are several ways in which we achieve this: one is the spoken word, another the written word, yet another one drawings or sketches. Most plumbers would agree that the ability to transfer information by means of drawings is of paramount importance. It is therefore very necessary for the student to be familiar with the instruments and have the knowledge required to perform the task.

The ability to understand BS 1192 Building Drawing Practice, read drawings, interpret symbols and use scales, all form part of a skilled plumber's attributes.

The student will find that much of his or her technology work will consist of drawings, geometry and sketching. The skill of expression through one of these methods of communication is often more accurate and effective than the written or verbal description.

Draughtsmanship

The basic techniques of draughtsmanship are described in this chapter. Although valuable information can be obtained from reading the following pages, it must be remembered that the skill of drawing can only be obtained with practice. If the basic skills are set out at this stage, then with guidance and practice, any willing person will be successful in this area.

Drawing materials and equipment

The following notes outline the basic materials and equipment that a student would be expected to use at craft level. When purchasing drawing equipment, you would be advised to purchase equipment which is manufactured to a British Standard, as good tools help produce a better standard of work.

The following list outlines some of the basic

drawing equipment you would be expected to provide for a technical drawing lesson:

Drawing board
T-square
Metric scale rule
45° and 60–30° set squares
A selection of pencils
An eraser (rubber)
A compass
Drawing board clips
A protractor
A pencil sharpener or knife
A duster
A set of templates for drawing circles and arcs.

Drawing boards

Many educational establishments throughout the country have the facility of adequate drawing offices, which will include drawing boards (see Figure 113). The majority of drawing done at craft level will be done on either A1 or A2 size drawing boards, which are 950 mm × 650 mm and 650 mm × 465 mm respectively, with the appropriate A1 and A2 drawing paper (see Figure 114 and Table 11 for paper sizes). The board should have a completely flat surface, and drawing pins should *not* be used to hold drawing paper in position, because these will damage the board. When not in use, the boards should be stored away and not used as an every day flat surface.

Some drawing boards have the facility of being able to alter the angle of the flat surfaces. Choose the angle to which you are most suited for comfort and ease of working. The board may also include an adjustable drawing ruler, which should be used as an alternative to the conventional T-square.

Table 11

Identification code	*Width mm*	*Length mm*	*Area m^2*
A0	841	1189	1
A1	594	841	1/2 (0.5)
A2	420	594	1/4 (0.25)
A3	297	420	1/8 (0.125)
A4	210	297	1/16 (0.625)
A5	148	210	1/32 (0.03125)

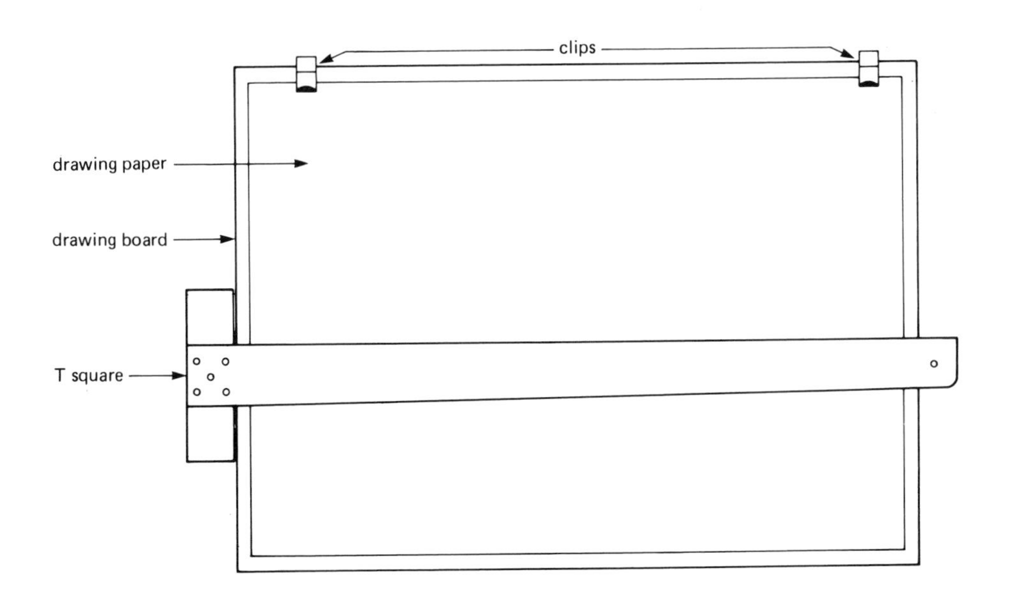

Figure 113 *Ordinary wood drawing board*

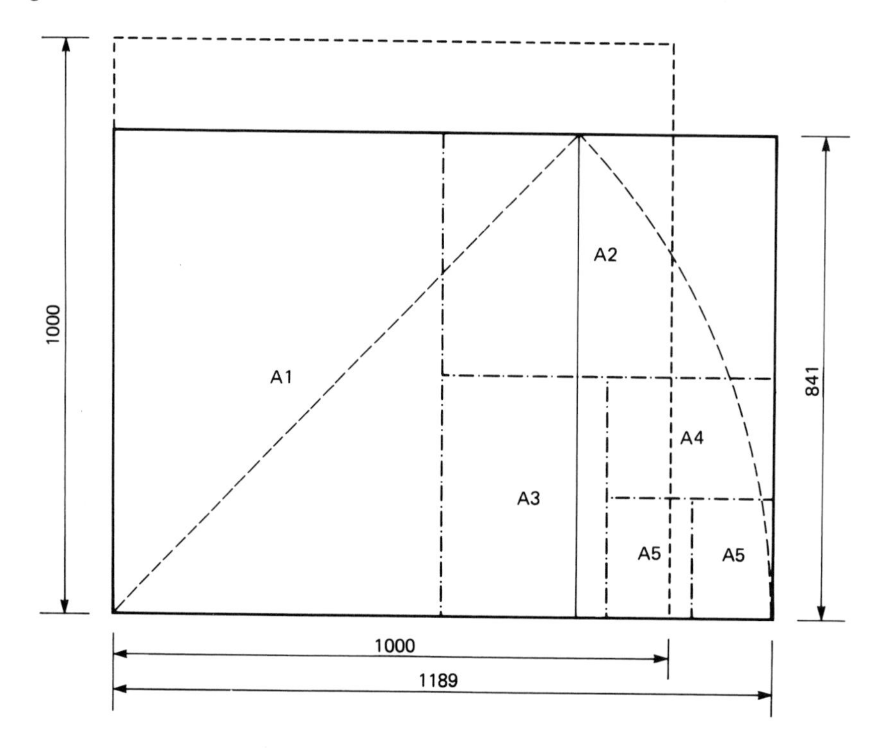

Figure 114 *Sheet paper: sizes and identification codes*

Pencils

The majority of drawing work done by craft students can be carried out quite adequately by using an ordinary hexagonal cedarwood pencil, with a lead composed of a mixture of compressed clay and graphite. You will find that the 175 mm hexagonal pencil is much easier to control than a round pencil.

The leads of pencils range from very soft to very hard and are graded as in Table 12.

As with all drawing equipment, you should purchase only the best pencils, as leads which tend to be gritty and break only make good draughtsmanship difficult.

Table 12

	Harder *Softer*
Hard	9H, 8H, 7H, 6H, 5H, 4H, 3H, 2H, H
Standard hard black	HB
Fine	F
Black soft	6B, 5B, 4B, 3B, 2B, B

Selecting the right pencil is very important, and will affect the quality of the finished product. Never use a pencil which is harder than an H if you are drawing on cartridge paper. A hard pencil tends to bite into the soft paper and makes drawing more difficult. Use the following as a guide to your selection:

General drawing work – use HB or F pencil
Sketching – use 6B–B pencil
Setting out and fine work – use H pencil

For the more advanced draughtsman a clutch pencil may be preferred to the ordinary pencil. These are available in grades from 6H to 4B. The advantage of the clutch pencil is the constant balance and feel of the pencil which enables one to produce a better standard of work.

Set squares

Set squares are manufactured from clear plastic. They are triangular in shape and available in three basic types:

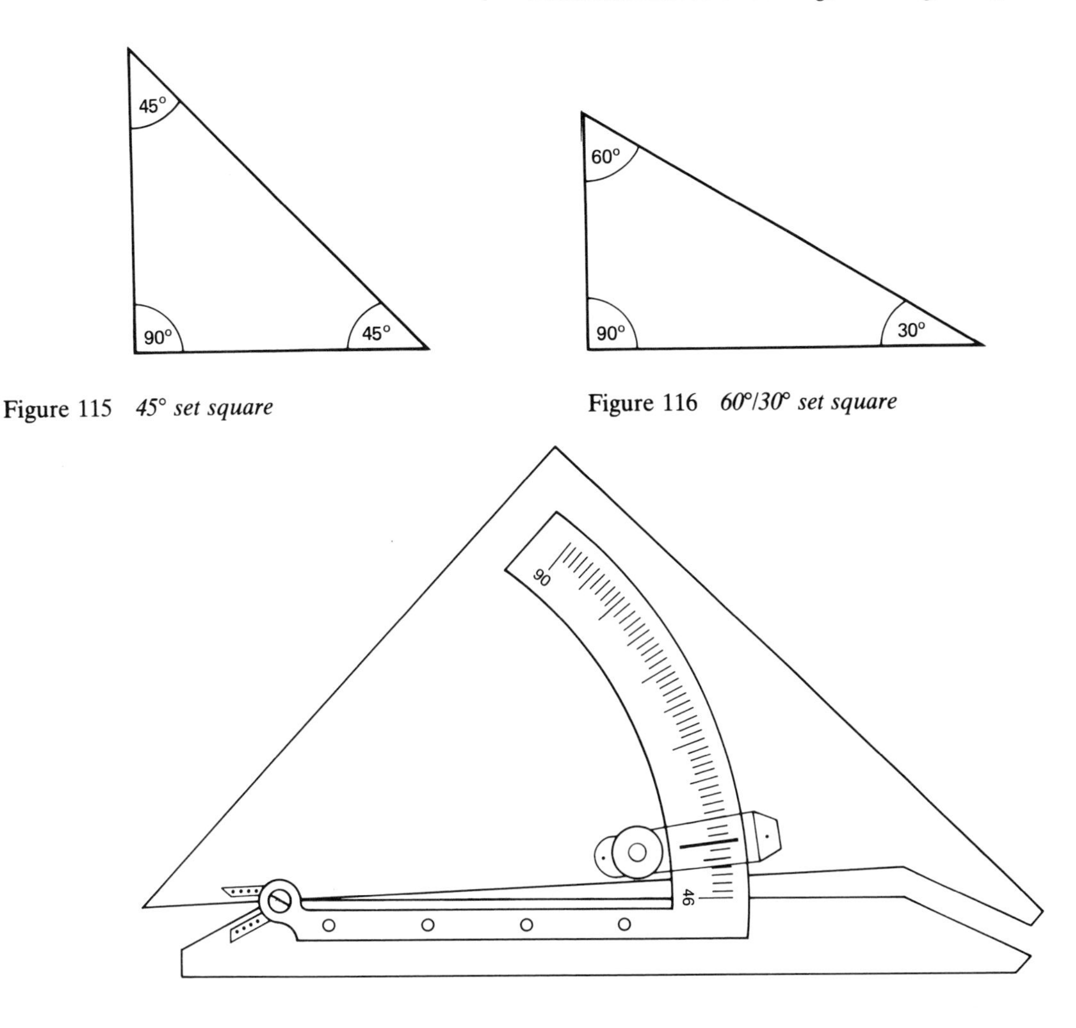

Figure 115 *45° set square*

Figure 116 *60°/30° set square*

Figure 117 *Adjustable set square*

1 45 degrees (see Figure 115)
2 60/30 degrees (see Figure 116)
3 adjustable (see Figure 117)

Set squares are used for drawing vertical and inclined lines when used in conjunction with a T-square. The advantage of the adjustable set square is that it enables you to draw lines at any angle you may require.

Compasses
Used for drawing circles and arcs, these are available in several sizes. A compass with a maximum radius of 125 mm will generally be adequate for students at craft level.

There are several types of compass available, but the spring bow compass enables you to produce work which is both fine and accurate (see Figure 118). The compass is set by adjusting a small finger screw which may be located on either side of the compass or between the legs of the

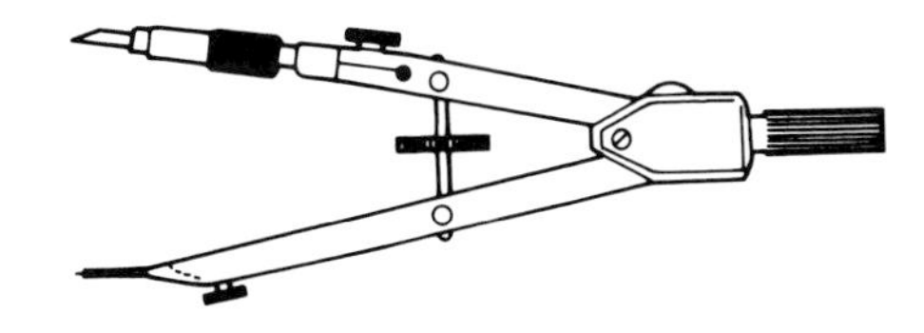

Figure 118 *Springbow compass*

compass. The pencil leg and the pointed leg are also adjustable.

The type of compass which requires you to insert a cedarwood pencil is not suitable for accurate drawing. The compass should be selected with care, for cheap compasses could lead to poor draughtsmanship. It is advisable to consider the purchase of a set of drawing instruments. In the early stages of study a pair of compasses may be all that is required but as the student progresses through his or her studies he or she may need or wish to have additional instruments such as attachments for converting them into dividers or pencil or ink compasses.

Layout of drawing sheets (BS 1192)

All drawings should include the following:
A margin
A title panel
An information panel

Margins
All the drawings should have a margin on all four sides of the sheet. The margin on the left hand should be 20 mm for filing purposes, while a 10 mm margin is adequate on the remaining sides.

Title panel
The title panel should be placed in the bottom right-hand corner of the sheet and should include the following:
Job title
Subject of drawing
Scale
Date of drawing
Job number
Drawing number
Revision suffix
Name of the architect, engineer or surveyor
Address and telephone number of the architect
Any other relevant information

Information panel
The information panel is drawn as part of the title panel and may include a record of renumbering of drawings and general revision notes.

Figure 119 shows one recommended method of layout of a drawing sheet for use by craft students.

Lettering

All drawings require a title, and in most cases additional descriptive words are needed so that the information on the drawing can be clearly understood. The objective of lettering is to convey this information, so it is vital that one should employ neat and uniform letters and figures, that will ensure good legible reproductions from ink or pencil originals. Lettering should meet the following requirements:

1 It should be legible
2 It should be of a suitable size
3 It should be correctly spaced
4 It should be correctly positioned

Legibility
You are strongly recommended to practice both vertical and sloping letters and figures for use on drawings, and with practice and experience you will adopt the style which is most suited to you.

Sizing
All letters and figures should be of a uniform character and easy to read. Stencils may be used to obtain uniformity, especially when dealing with larger lettering for titles and headings. The recommended sizes for letters and figures are as follows (see Figure 120):

Titles and headings	5–8 mm
General notes	1.5–4 mm

Spacing of words
Letters should be evenly spaced within the words, with not less than twice the line thickness between each letter. Words should be so placed to allow a sentence to be read without difficulty.

Positioning of words
Words should appear on the drawing in a neat, logical manner. They should either be written horizontally, or, if written vertically, they should be the right way up when viewed from the

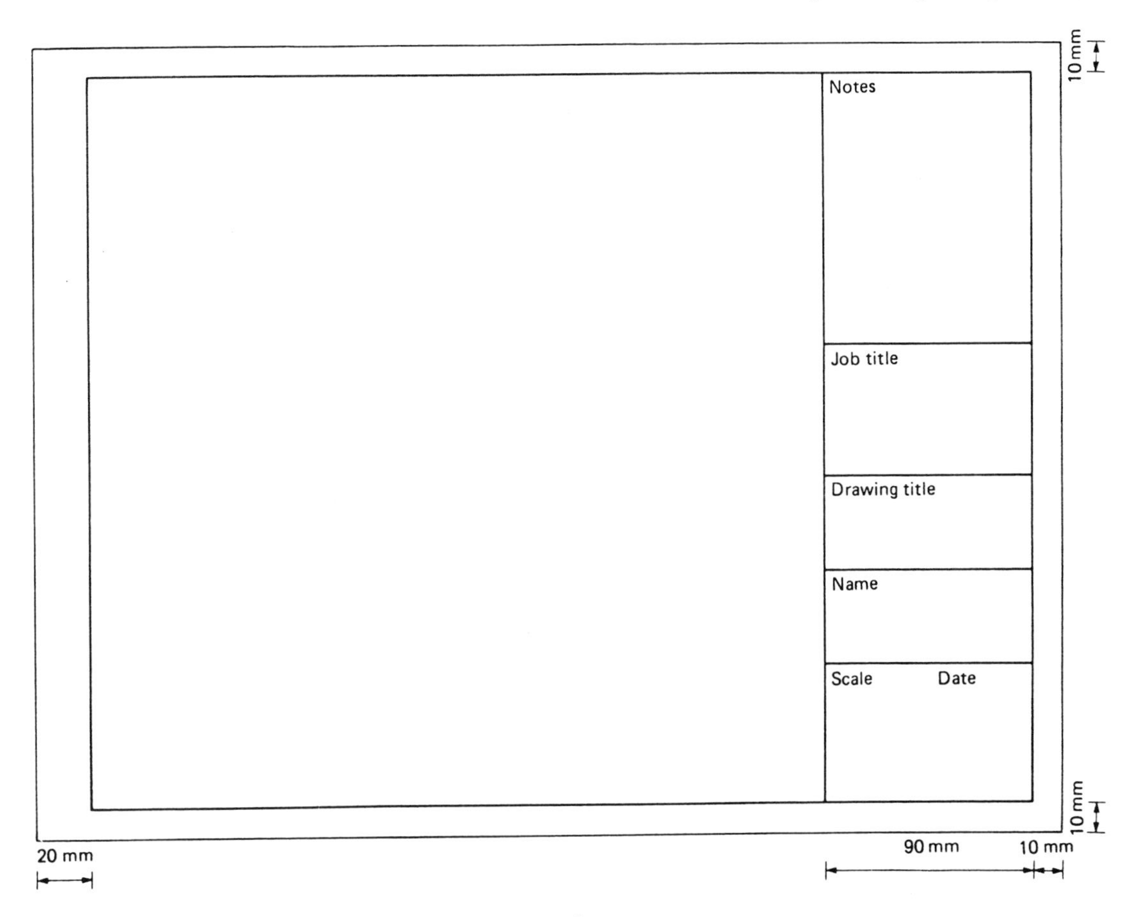

Figure 119 *Suggested method of laying out a drawing sheet*

Figure 120 *Sizes of lettering*

right-hand edge of the drawing. Words should never be written diagonally across the page, and certainly not upside-down.

Drawing technique

Having considered and obtained the drawing equipment, we should now look at the technique of producing a good drawing. The first and most important point is that we keep the drawing as clean as possible; even if the draughtsmanship is of a good standard the finished drawing will look messy if it is covered with finger marks, erasure marks, and general smudging of the pencil lead. If you consider the following points they will help you keep your drawing clean.

1 Make sure your hands are clean.
2 Clean all your drawing equipment with a clean duster both before you start and during the process of drawing.
3 When you have to use your rubber, make sure you remove all the rubber crumbs by blowing them off the paper.
4 Sharpen all your pencils away from the drawing board.
5 Avoid rubbing the surface of the paper with your T-square or set squares, as this will tend to smudge the lead. Lift them off the paper before you move them.
6 Use a light pencil, as this will rub out more easily than a hard black pencil.

Drawing lines
All lines should be firm and clean and of uniform thickness. They should not be coarse and uneven or double.

The following points will help you provide uniform lines. Make sure your pencil is sharp. If you are right handed then draw on the paper from left to right; if left handed, from right to left. Hold the T-square in position and draw a line close to the rule edge, keeping an even pressure on the pencil. Turn the pencil round in your fingers as you draw it along. This tends to keep the pencil sharp and so avoids continuous resharpening. With a little practice you will produce a clean, uniform line (see Figure 121).

As you continue to draw, remember to sharpen your pencil (away from the drawing). Adopt a comfortable position throughout the exercise and, above all, practise regularly.

Figure 121

Dimensioning
Drawings should be fully and clearly dimensioned. Dimension lines should be lightly drawn and terminate against short thin lines. Small arrowheads should indicate the points between which the dimension runs, and should terminate at the line (see Figure 122).

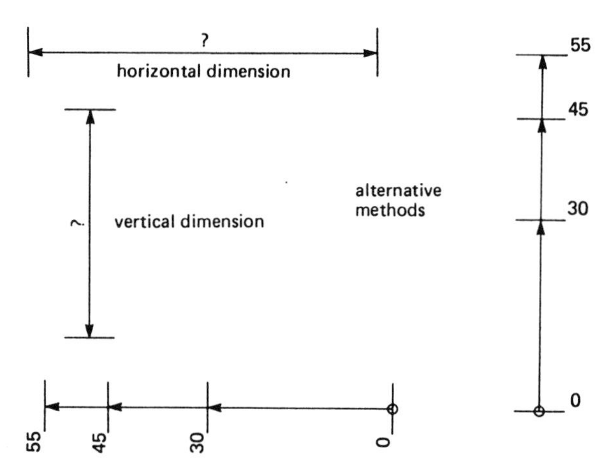

Figure 122 *Positioning of labels*

Scales

A suitable scale will often be necessary to either show the detail of a small object, or to accurately represent a large object on a smaller sheet of paper. The factors which govern the choice of scale (from BS 1192) are as follows:
1 The need to communicate both adequately and accurately the information necessary to enable the work to be carried out.
2 The need to achieve economy of effort and time in preparation of drawings.
3 The character and size of the subject.
4 The desirability of keeping the sheets for one project to one size as far as possible.

Scale rules
Scale rules are manufactured from boxwood or plastic, and have accurate divisions along each edge (see Figure 123). The size of each division will depend upon the size of scale.

Examples
The following examples explain the relationship between the object's actual size and its scaled size.

Example A: scale 1:100
If an object is 10 metres in length and is measured to a scale of 1:100 and drawn on a sheet, it will be represented by a line of 0.1 m or 100 mm.

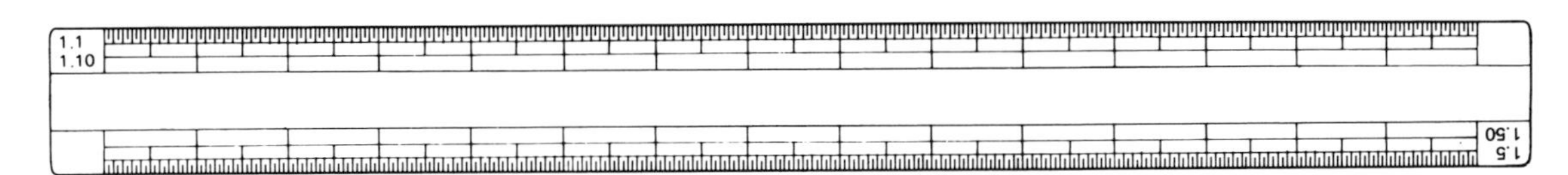

Figure 123 *Scale rule*

The formula used to calculate this is shown below:

$$\frac{\text{length of object}}{\text{scale}} = \frac{10}{1:100}$$

$$= \frac{1}{100} \times 10$$

$$= 0.1 \text{ m} = 100 \text{ mm}$$

Example B: scale 1:5
If an object is 850 metres in length, and is measured to a scale of 1:5 and drawn on a sheet, it will be represented by 0.17 m or 170 mm.

$$\frac{\text{length of object}}{\text{scale}} = \frac{850}{1:5}$$

$$= \frac{1}{5} \times 850$$

$$= 5\overline{)850}$$

$$= 0{\cdot}17 \text{ m}$$

$$= 170 \text{ mm}$$

Most of the scaled work done in the building industry will be to reduce objects to a smaller, more suitable size for drawing sheets. However, it will become necessary from time to time to increase the size of an object, so that detail may be shown more clearly, e.g. a section of a 15 mm compression fitting for copper tube could be drawn to a scale of, say, 3:1 which would increase the size of the joint by 3 times:

The list below shows the preferred scales used for building drawings.

1 Location drawings 1:2500
2 Block plan 1:2500
3 Site plan 1:500, 1:200
4 General location 1:200, 1:100, 1:50
5 Component drawings 1:100, 1:50, 1:20
6 Detailed drawings 1:10, 1:5, 1:1 (full size)
7 Assembly drawings 1:20, 1:10, 1:5

One of the more common scale rules used in the construction industry is the Royal Institute of British Architects (RIBA) approved scale which incorporates the following scales: 1:1, 1:5, 1:10, 1:20, 1:50, 1:100, 1:200, 1:250, 1:2500.

Metric measurement

Table 13 shows the main Système International units (SI units) and symbols used in the building industry.

Table 13

Quantity	*SI unit*	*Symbol*
Length	Metre	m
	Millimetre	mm
Area	Square metre	m^2
	Square millimetre	mm^2
Volume	Cubic metre	m^3
	Cubic millimetre	mm^3
Capacity	Litre	l
Force	Newton	N
Power	Watts	W
Mass	Kilograms	kg
Volume flow rate	Cubic metres per second	m^3/s
Heat flow rate	Watts	W
Temperature	Degree Celcius	°C
Time	Second	s
U value	Watts per square metre per degree Celcius	W/m^2 °C
Pressure	Newtons per square metre	N/m^2
	Kilonewtons per square metre	kN/m^2
Calorific value	Kilojoules per cubic metre	kJ/m^3
Energy	Joules	J

Graphical symbols and abbreviations

Graphical symbols and the appropriate abbreviations are used on building drawings in order that components installed within the building may be clearly identified. Figure 124 outlines the main symbols and abbreviations used to show the plumbing and heating components. (For a complete list refer to BS 308.)

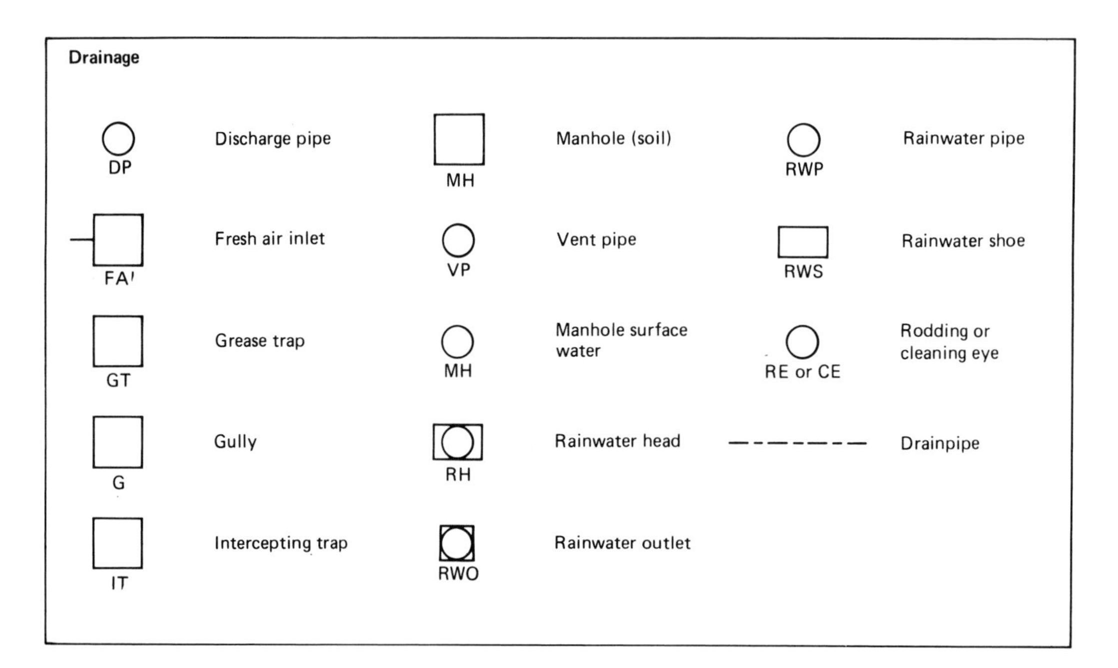

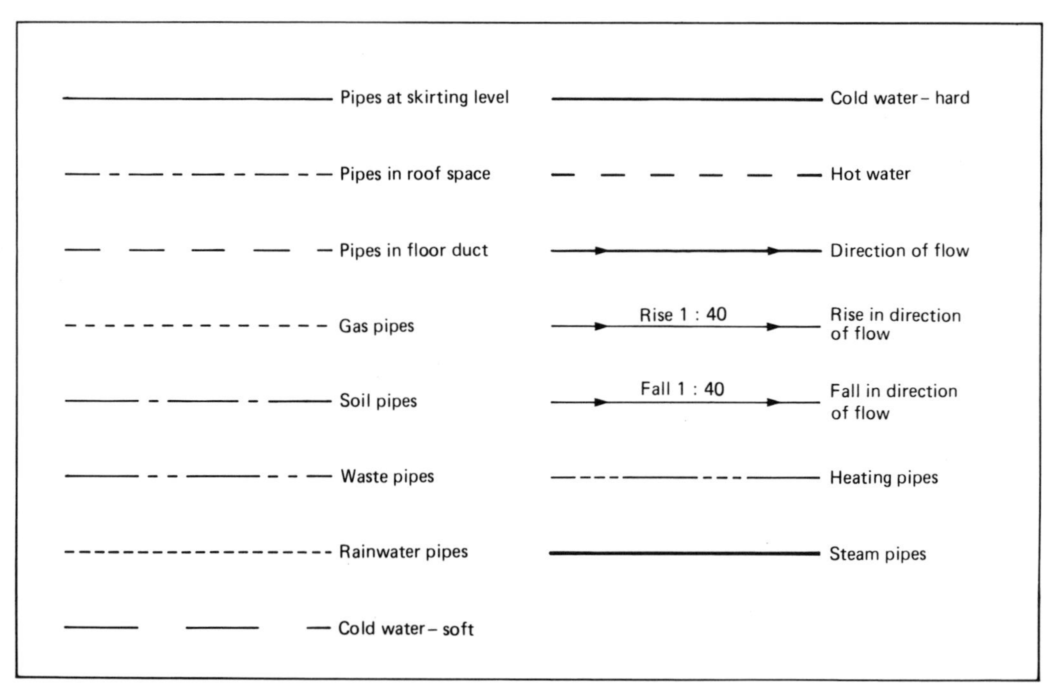

Figure 124 *Graphical symbols*

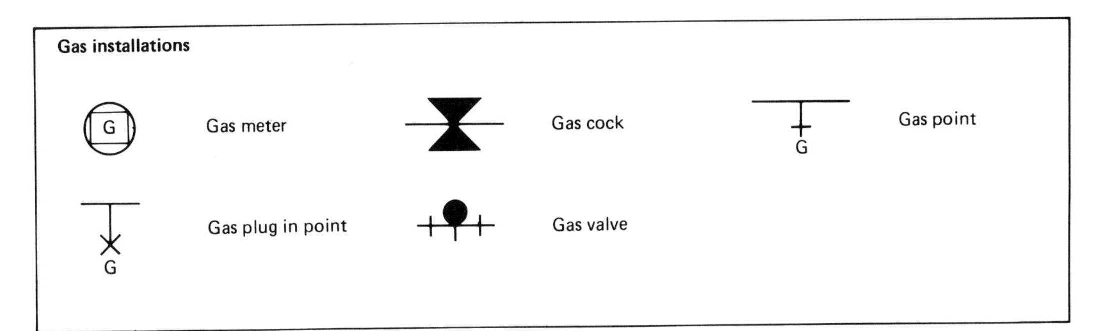
Gas installations
G
Gas meter
Gas cock
G
Gas point
G
Gas plug in point
Gas valve

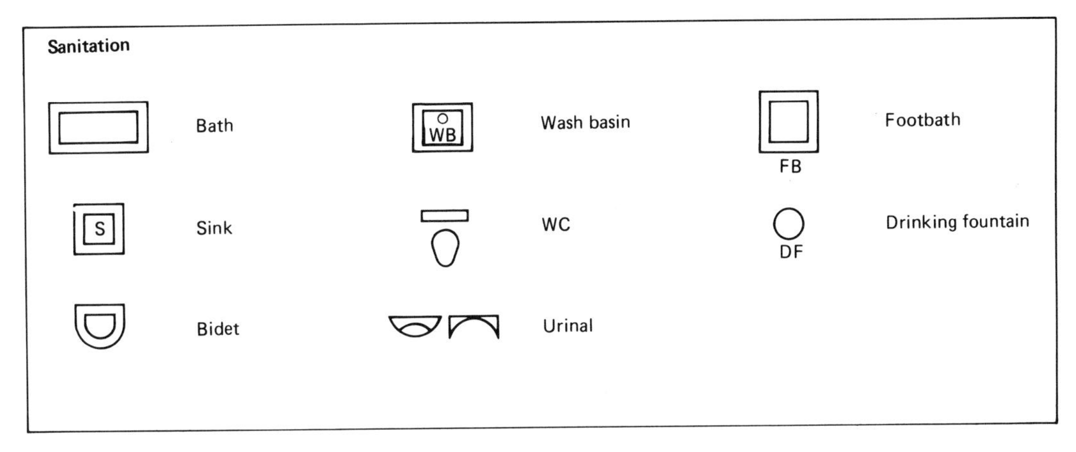
Sanitation
Bath
WB
Wash basin
FB
Footbath
S
Sink
WC
DF
Drinking fountain
Bidet
Urinal

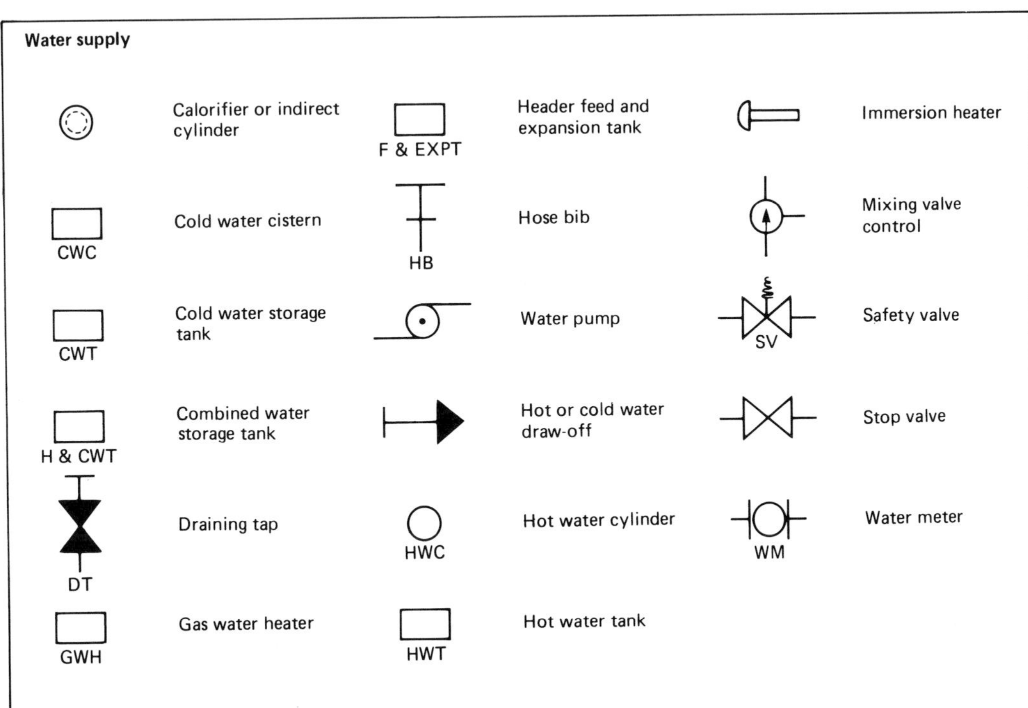
Water supply
Calorifier or indirect cylinder
F & EXPT
Header feed and expansion tank
Immersion heater
CWC
Cold water cistern
HB
Hose bib
Mixing valve control
CWT
Cold water storage tank
Water pump
SV
Safety valve
H & CWT
Combined water storage tank
Hot or cold water draw-off
Stop valve
DT
Draining tap
HWC
Hot water cylinder
WM
Water meter
GWH
Gas water heater
HWT
Hot water tank

Geometry

Parts of the circle

Figure 125 illustrates the following parts of the circle:

Centre A point from which the circle is drawn.

Radius A straight line drawn from the centre of the circle to any point on the circumference.

Diameter A straight line which passes through the centre of the circle and terminates on the circumference.

Circumference The outside boundary of a circle.

Quadrant A quarter of a circle.

Semi-circle Half of a complete circle.

Sector Part of the circle bounded by two radii and the circumference.

Segment Part of the circle between the circumference and the chord.

Arc Part of the circumference.

Normal A straight line drawn from the circumference of the circle in an outwards direction.

Tangent A straight line touching the circumference at one point and at a 90° angle to a normal at that point.

Chord A straight line drawn touching the circumference in two places but not through the centre of the circle.

Concentric circles Circles drawn from the same centre.

Annulus The portion of a circle contained between concentric circles (i.e. pipe section).

Figure 126 shows the number of degrees in (a) a full circle (360°); (b) a semi-circle (180°); and (c) a quarter-circle (90°).

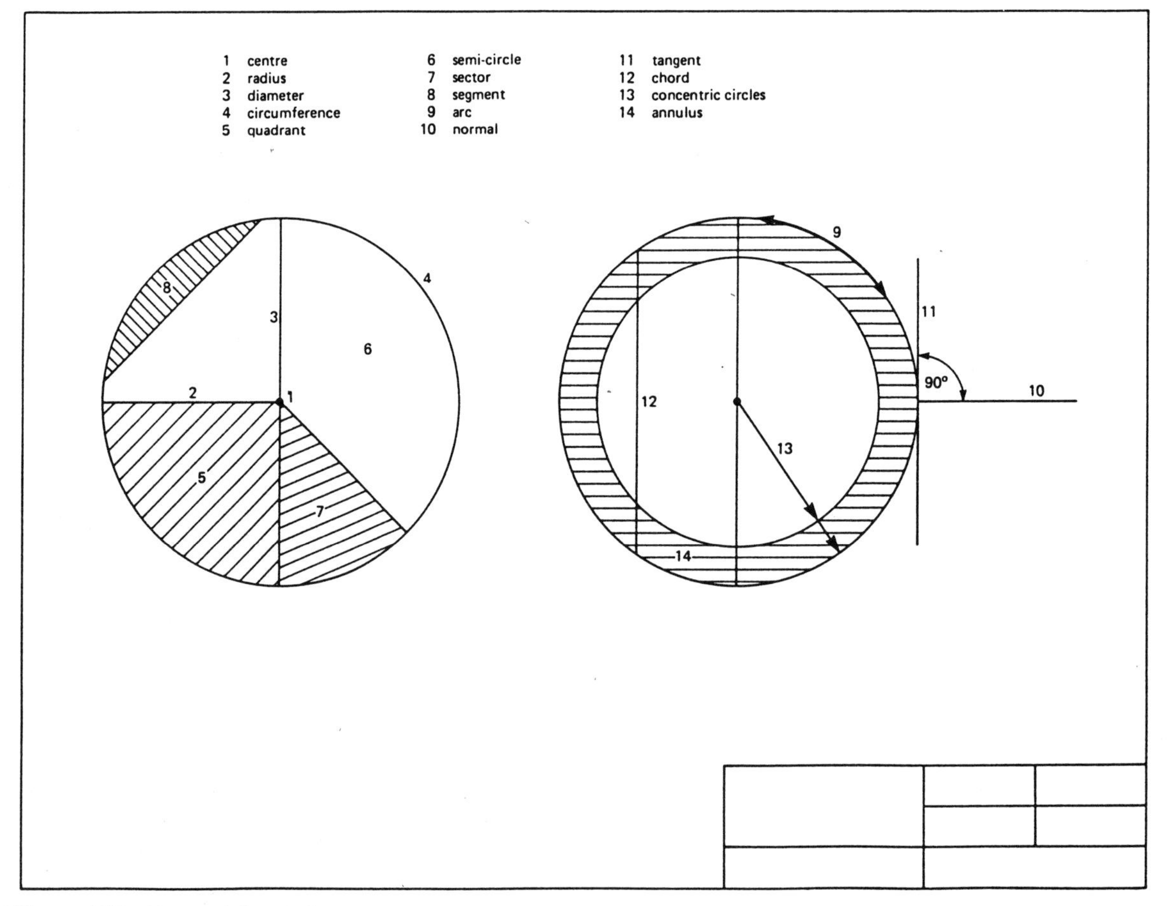

Figure 125 *Parts of the circle*

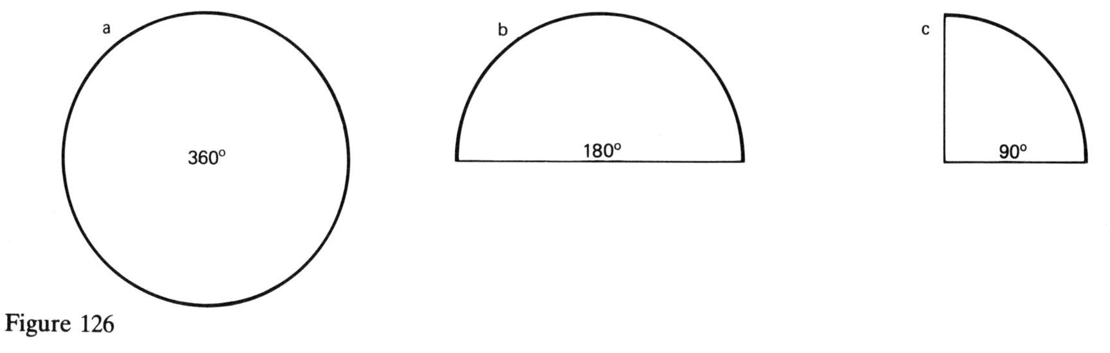

Figure 126

Triangle
This is a figure bounded by three sides. The sum of its three angles must equal 180°.

Right angled triangle
In this triangle one angle must be a right angle, i.e. 90° (see Figure 127).
Sum of angles 90 + X + Y = 180°.

Equilateral triangle
In this triangle all the angles are 60° and all sides are of equal length (see Figure 128).
Sum of angles 60 + 60 + 60 = 180°.

Acute triangle
In this triangle all angles are less than 90° and the lengths of the sides are unequal (see Figure 129).
Sum of angles X + Y + Z = 180°.

Obtuse triangle
In this triangle one angle (in Figure 130, angle Y) is greater than 90°.
Sum of angles X + Y + Z = 180°.

Figure 127

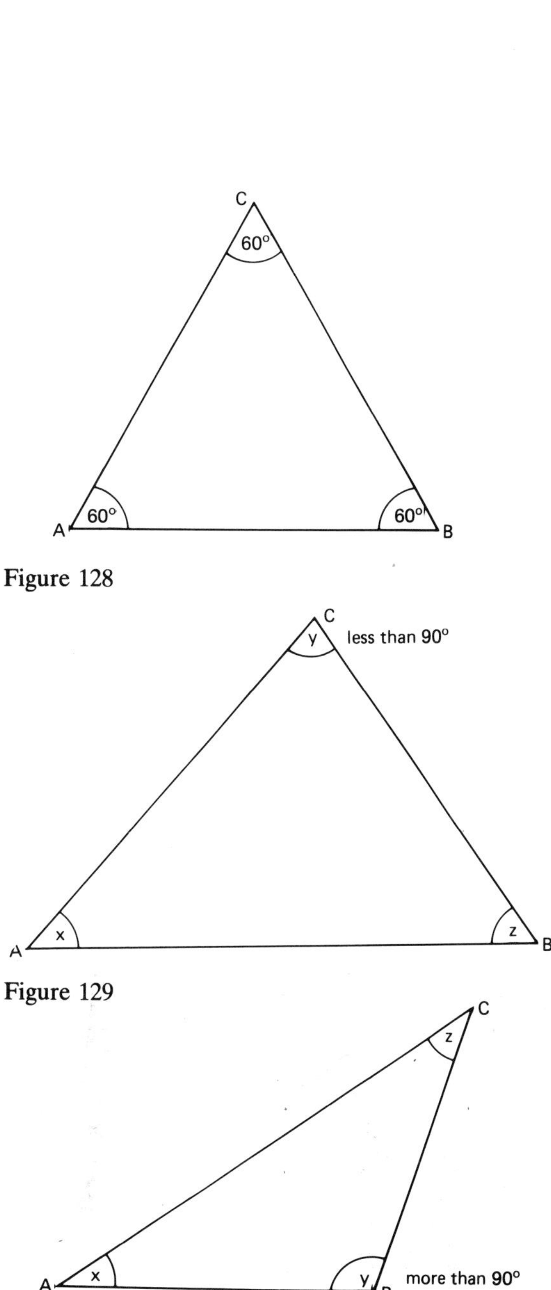

Figure 128

Figure 129

Figure 130

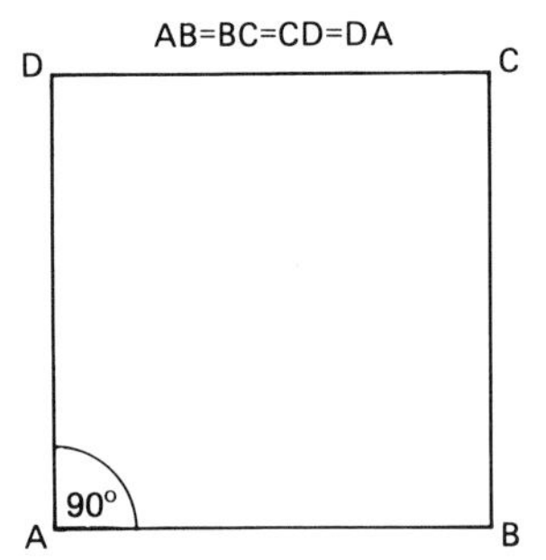

Figure 131 *Square*

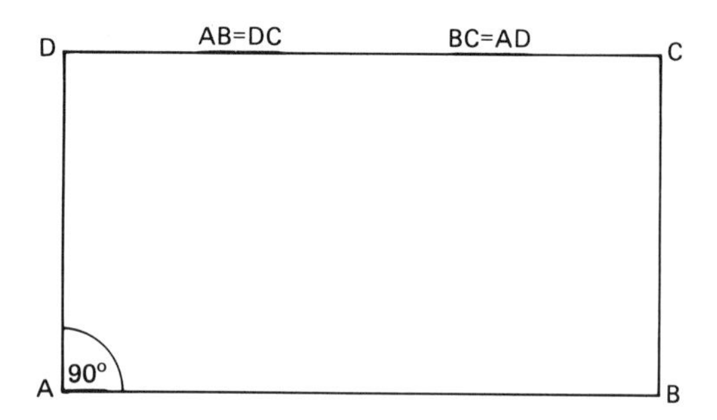

Figure 132 *Rectangle*

Square

This is a figure bounded by four sides where all the sides are of equal length (see Figure 131). All the angles must be 90°.

Rectangle

This is a figure bounded by four sides but in this case opposite sides are equal in length (see Figure 132). All angles must be 90°.

Parallelogram

Similar to a rectangle, but in this case the angles are not 90° (see Figure 133).

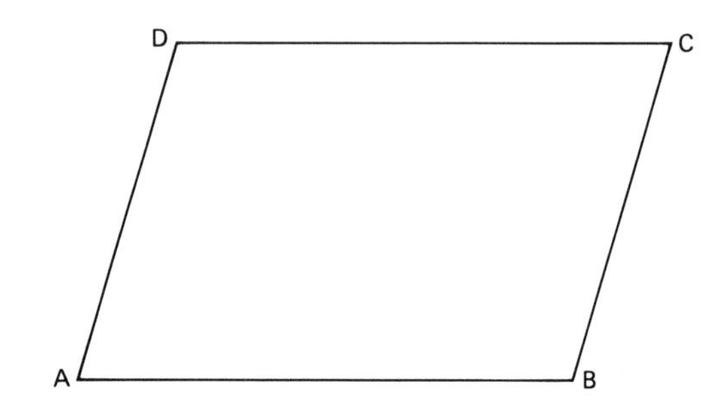

Figure 133 *Parallelogram*

Polygon

Any figure with more than four sides is known as a polygon. The most commonly used are:

(a) Pentagon (five sides)
(b) Hexagon (six sides)
(c) Heptagon (seven sides)
(d) Octagon (eight sides)
(e) Nonagon (nine sides)
(f) Decagon (ten sides)

There are several methods of constructing these figures. Figure 134 shows one method of constructing a pentagon which can also be applied to any figure with more than three equal sides.

Draw line AB to given length.
Bisect AB to give centre line and point C.
Using C as centre, draw arc BD to cut centre line at D.
Using A as centre, draw arc BE to cut centre line at E.
Bisect space between DE to obtain F.
Using F as centre, AF as radius, draw circle and step off AB into circumference 5 times.

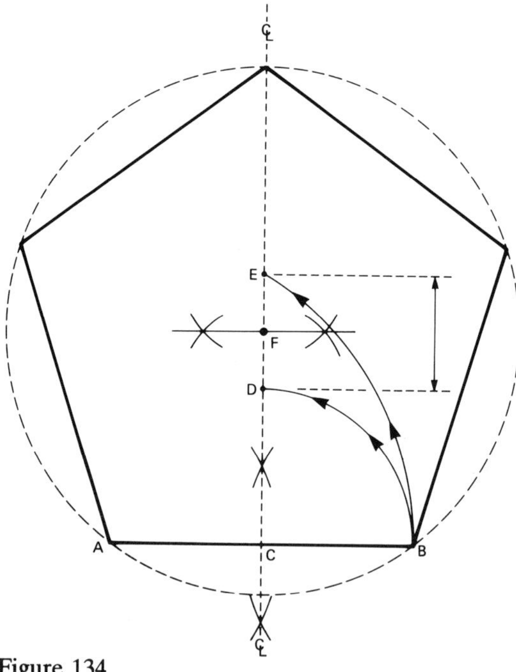

Figure 134

Note: It will be readily observed that it is a simple matter to step off on the centre line distance EF to give the centre for any required polygon.

Setting out angles
One of the easiest ways of setting out angles is to use a watch or clock face as an aid (see Figure 135).

There are 360° in any circle.

There are 60 minutes on the clock face.

Therefore each minute $= \frac{360}{60}$

each minute = 6°

To construct an angle of 30°:

$6° = 1$ minute

$30° = \frac{30}{60}$ minutes

$30° = 5$ minutes

To construct an angle of 45°

$6° = 1$ minute

$45° = \frac{45}{6}$ minutes

$45° = 7.5$ minutes

To construct an angle of 60°

$6° = 1$ minute

$60° = \frac{60}{6}$ minutes

$60° = 10$ minutes

An angle of 90° = 15 minutes, and so on.

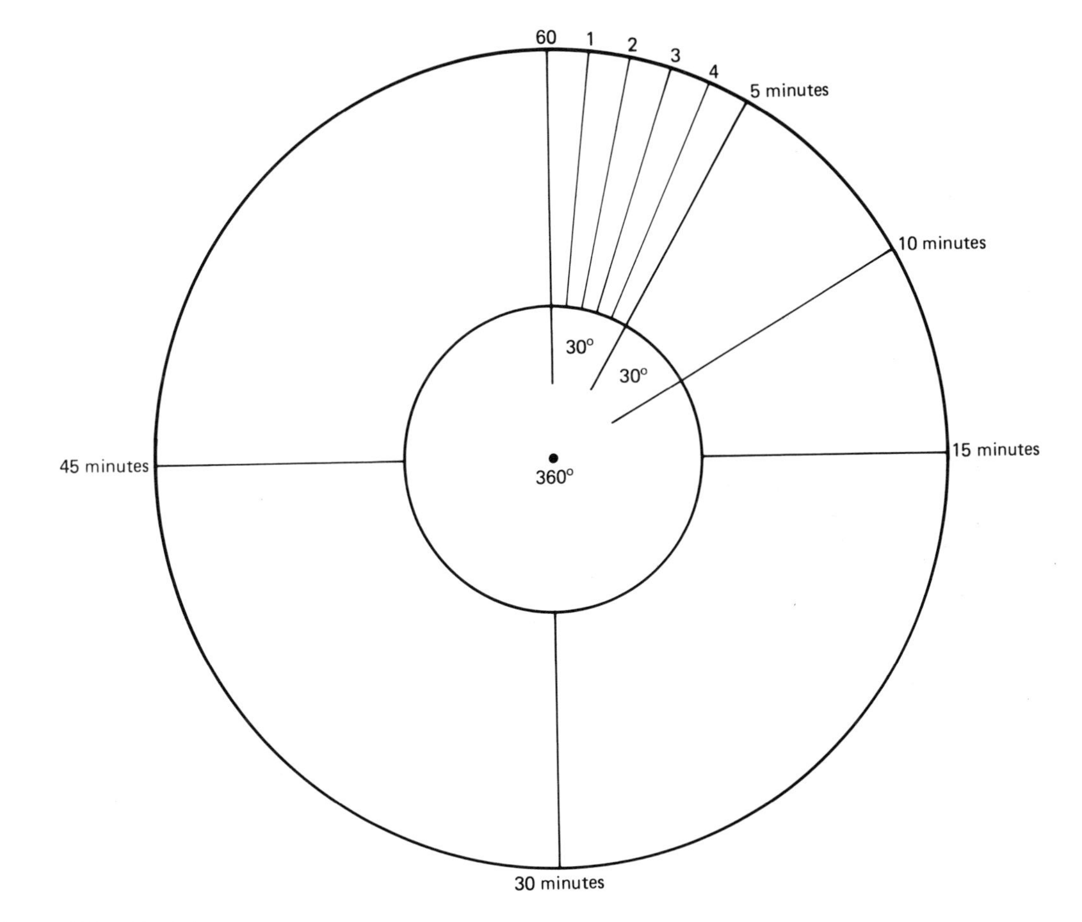

Figure 135 *Clock face*

Most of the angles required in plumbing work can be obtained by the use of the 45° and 60°–30° set squares by simply adding or subtracting these amounts, e.g. an angle of 135° would be obtained by: 90° + 45° = 135°

An angle of 120° would be obtained by 90° + 30° = 120°.

Method of erecting a perpendicular (Figure 136)

Draw base line any convenient length.

From point A mark two equal distances on base line BC.

Using B and C as centres and compasses at any size greater than AB draw arcs to obtain point D.

Join AD to obtain the required perpendicular.

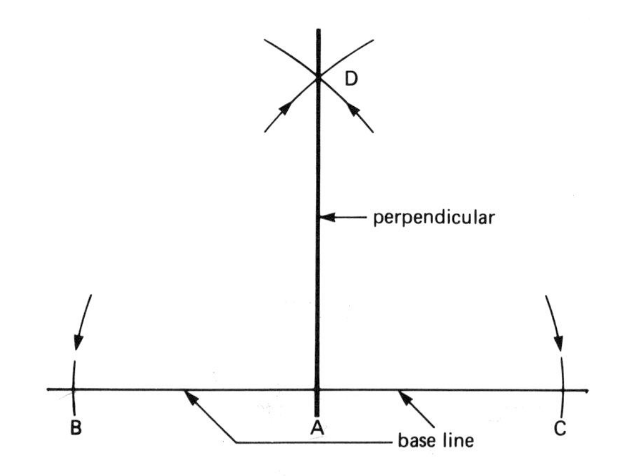

Figure 136 *Method of erecting a perpendicular*

Alternative method of using 45° set squares (Figure 137)

Draw base line, mark A.

Mark points B and C at equal distances from A.

Using 45° set squares project lines to bisect at D.

Join AD to obtain required perpendicular.

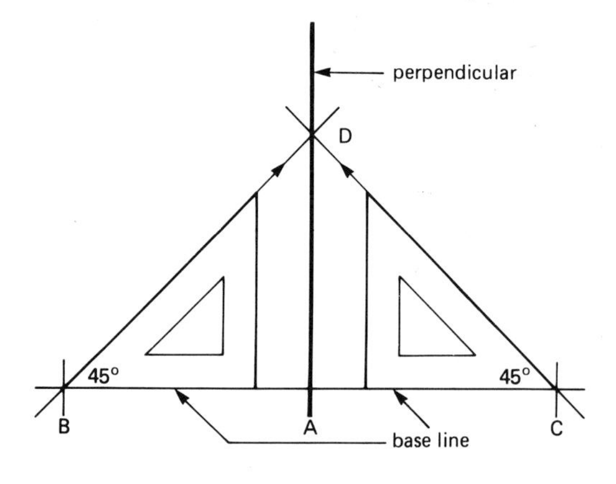

Figure 137 *Alternative method using 45° set squares*

Method of erecting a perpendicular from a given point A (Figure 138)

Draw horizontal base line.

Set compasses at any convenient distance with A as centre and draw arc to obtain points B and C.

Using B and C as centres draw arcs to obtain point D below base line.

Join AD to obtain required perpendicular.

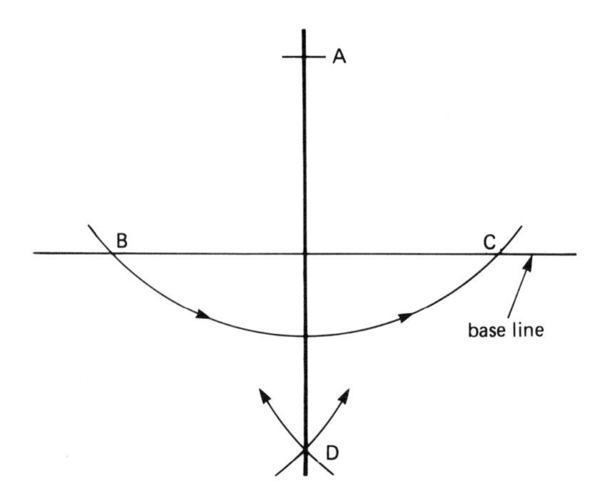

Figure 138 *Method of erecting a perpendicular from a given point*

Method of drawing parallel lines (Figure 139)

Draw base line.

Mark A and B on the base line.

Using A and B as centres describe arcs to obtain C.

Using B and C as centres describe arcs to obtain D.

Join points C and D to obtain the required line parallel to the base line.

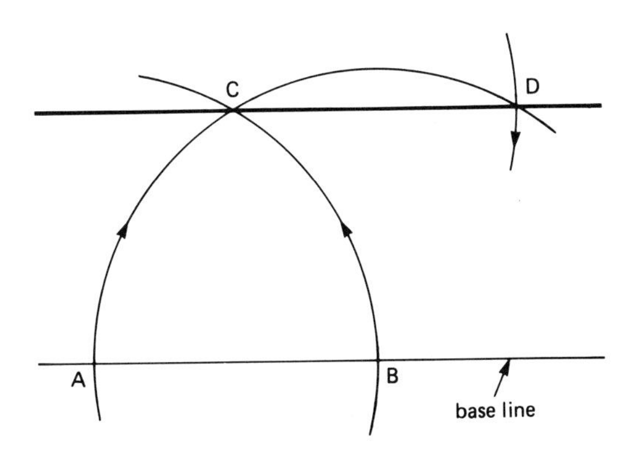

Figure 139 *Method of drawing parallel lines*

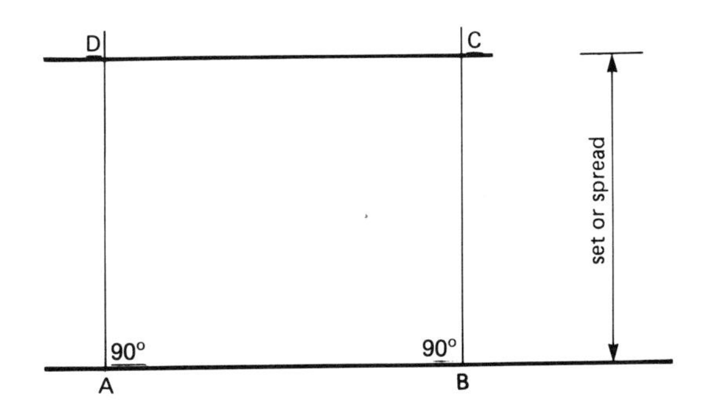

Figure 140 *Method of drawing parallel lines a specified distance apart*

Method of drawing parallel lines a specified distance apart (Figure 140)

Draw base line.
Mark points A and B any convenient distance apart on the base line.
Erect two perpendiculars from A and B by any of the methods shown.
Mark on perpendiculars the required distance to give C and D.
Join C and D to give the required parallel line.

This method is used to make an offset, where the set or spread is the required distance between the parallel lines.

Method of bisecting any angle (Figure 141)

With A as focal point draw arc BC (any size).
Using B and C as centres draw two arcs which bisect to obtain D.
Draw line AD. The angle has now been equally bisected.

Method of dividing 90° angle into three equal 30° parts (Figure 142)

With A as focal point draw arc BC (any size).
With B as centre, draw arc AD.
With C as centre draw arc AE.
Draw lines AD and AE to give three 30° angles.

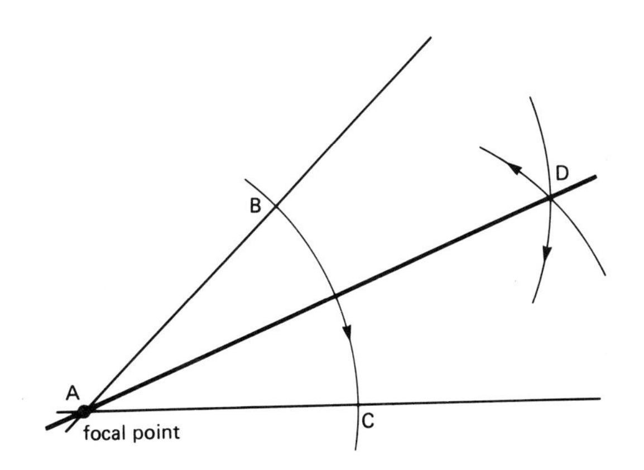

Figure 141 *Method of bisecting any angle*

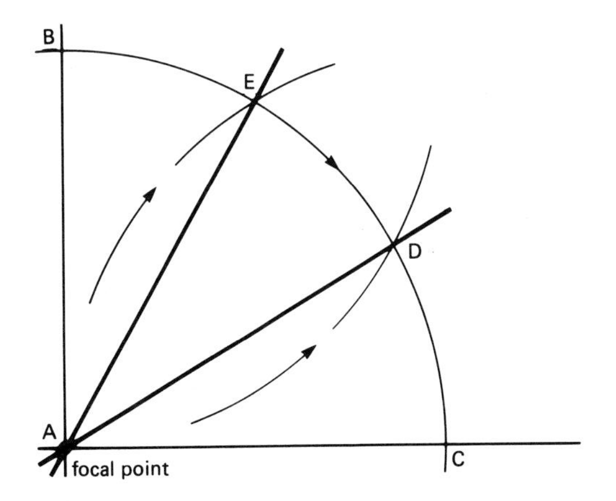

Figure 142 *Method of dividing 90° angle into three equal parts*

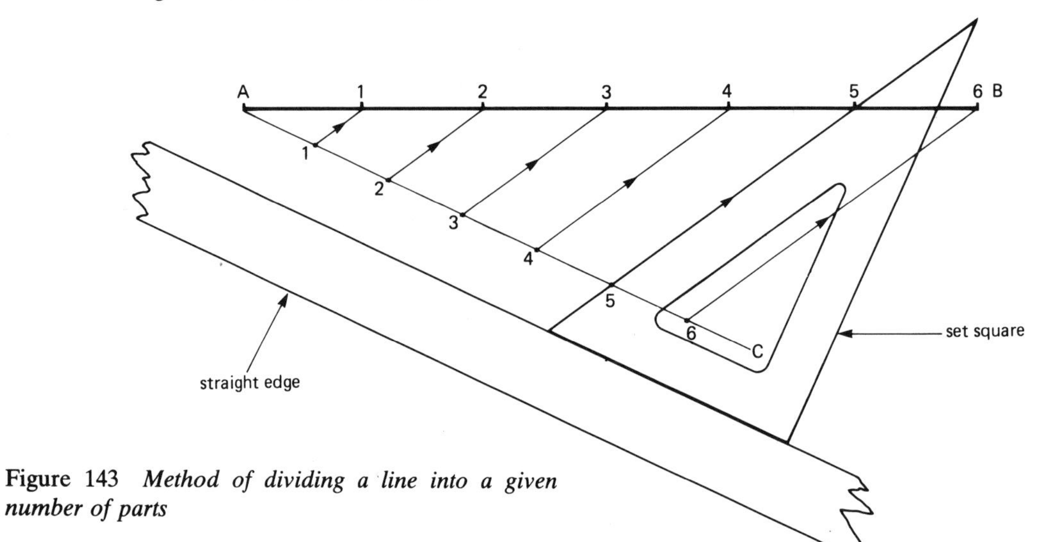

Figure 143 *Method of dividing a line into a given number of parts*

Method of dividing a line into a given number of parts

Figure 143 shows a line AB divided into six equal parts. The method is easily adapted for any number of parts.

Draw line AC at any angle to AB and any length.

Set compasses at any length and mark on line AC the number of equal parts into which AB is to be divided (in this example, six).

With the aid of a set square and a straight edge line up point 6 with point B.

Hold straight edge still while moving set square progressively along marking each point on line AB as shown.

Setting out and use of templates

You will be required from time to time to set out details of pipework for plumbing installations. This is usually done using a template. A template is an aid to making or forming a piece of work. It can be in the form of a piece of wire, bent to the required angle, or a piece of metal cut and formed to the required shape. The following factors should be considered when bending pipes:

1 The diameter of the pipe.
2 The material being used.
3 The radius required.
4 The method of bending the pipe.
5 The length of the pipe.
6 The type of bends required.

The diameter of the pipe will often determine the method to be used to bend the pipe and also the radius of the bends. Use the following formula as an approximate guide:

radius = diameter × 4

The following examples show the method of setting out various bends in pipework. These examples show all the necessary detail required to produce these templates. You are advised to practise these setting out details in the workshop and also in your technical drawing classes.

Method of making a square

Nail together two straight pieces of wood to form an L shape.

On one leg (side) mark a point 4 units in length.

On the other leg mark a point 3 units in length.

Cut another piece 5 units in length and line up with the other points.

Provided the marking and setting up is done with precision a 90° angle is obtained. The size of the square will depend on the size of the unit adopted.

Method of setting out template for square bend

Draw lines AB and CD at right angles (see Figure 144).

Bisect angle.

Mark on centre line (℄) the length of radius.

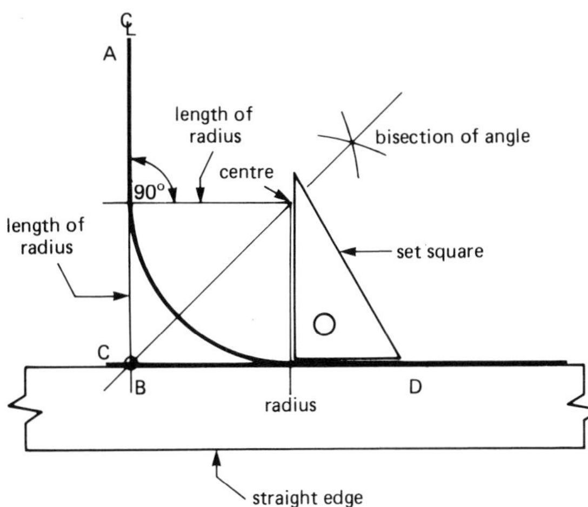

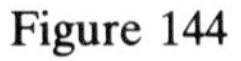
Figure 144

Place straight edge on centre line. Slide set square along until in line with the radius. Project line to cut bisection line.
This operation can be repeated on the other leg of the bend.
This intersection is the centre for inscribing bend.

Method of setting out template for obtuse bend
Draw ℄ to angle required (see Figure 145).
Draw parallel lines at a distance equal to radius required.
Slide set square along straight edge until it lines up with intersection of parallel lines.
Draw in radius line. This marks the beginning and end of the bend.

Method of setting out template for return (U) bend
Extend arms of bend to focal point (see Figure 146).
Bisect angle.
Using set square mark 90° right angles.
Intersection gives centre for return bend.

Figures 147, 148 and 149 illustrate the three methods described above.

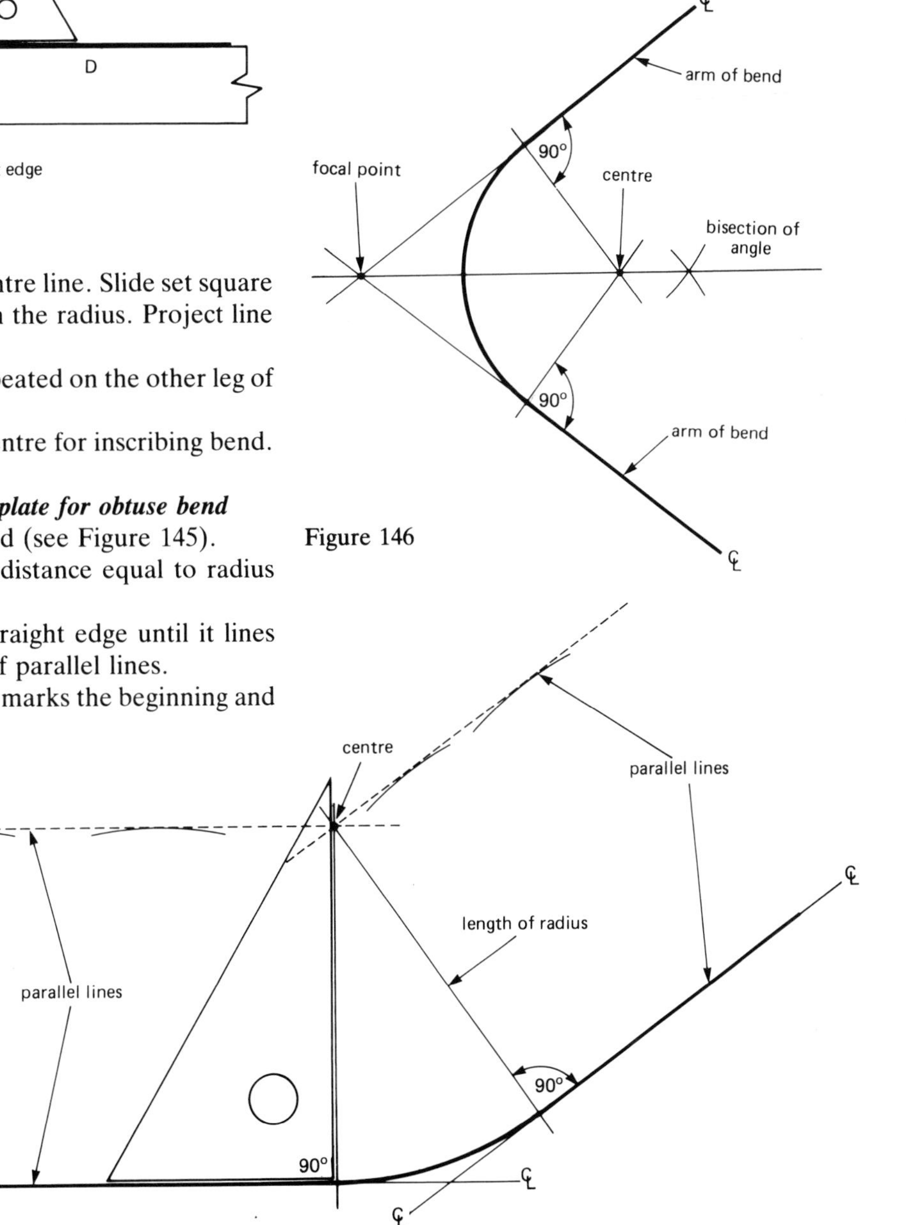

Figure 146

Figure 145

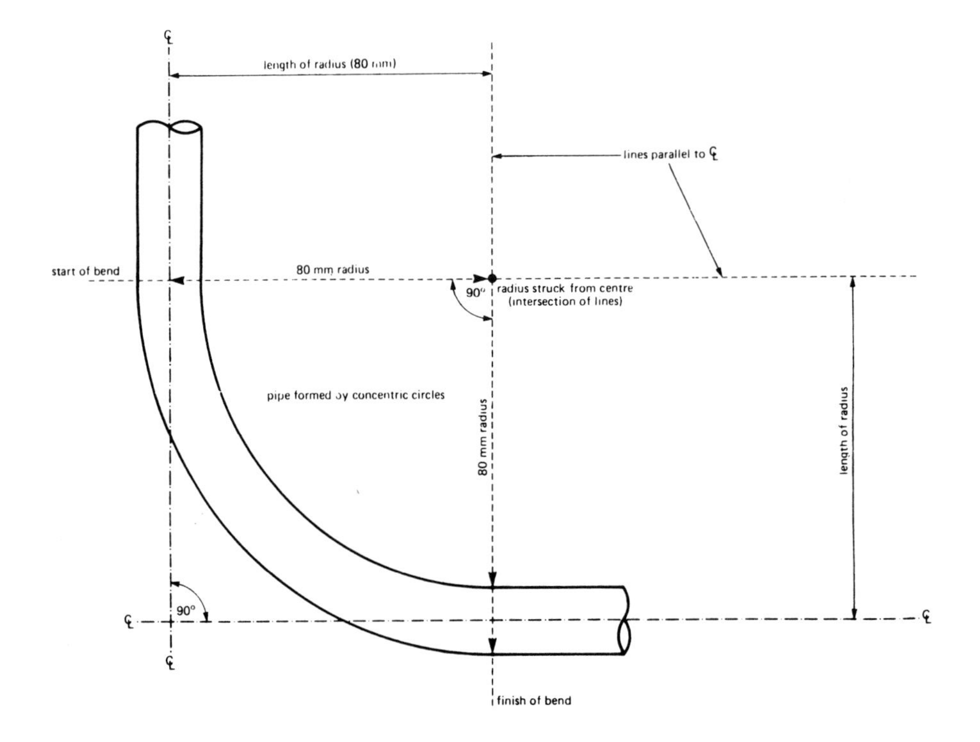

Figure 147 *Method of setting out 90° bend with 80 mm radius*

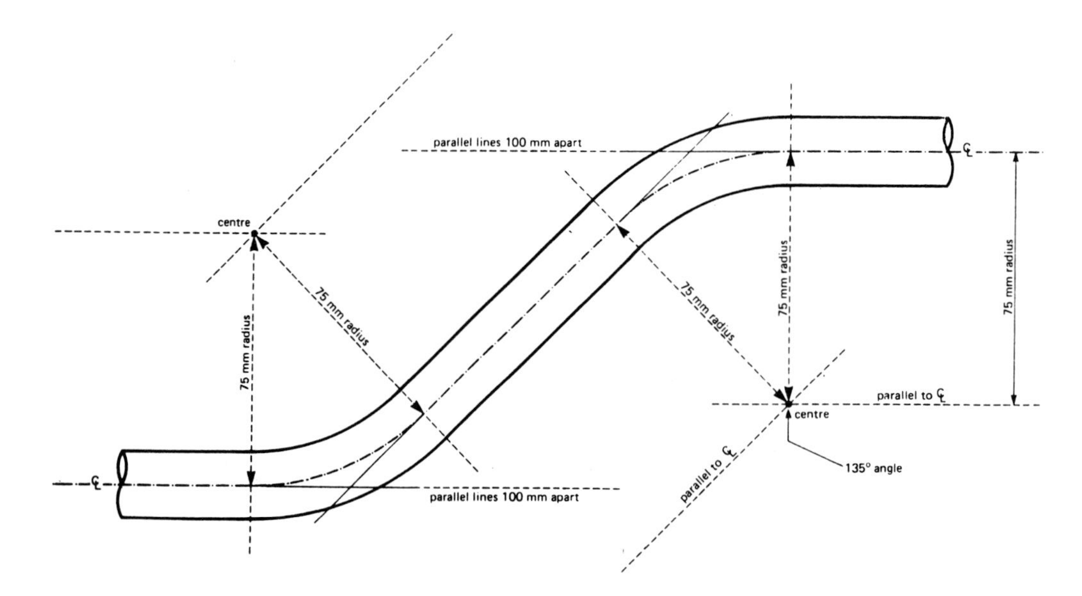

Figure 148 *Method of setting out 100 mm offset with 75 mm radius*

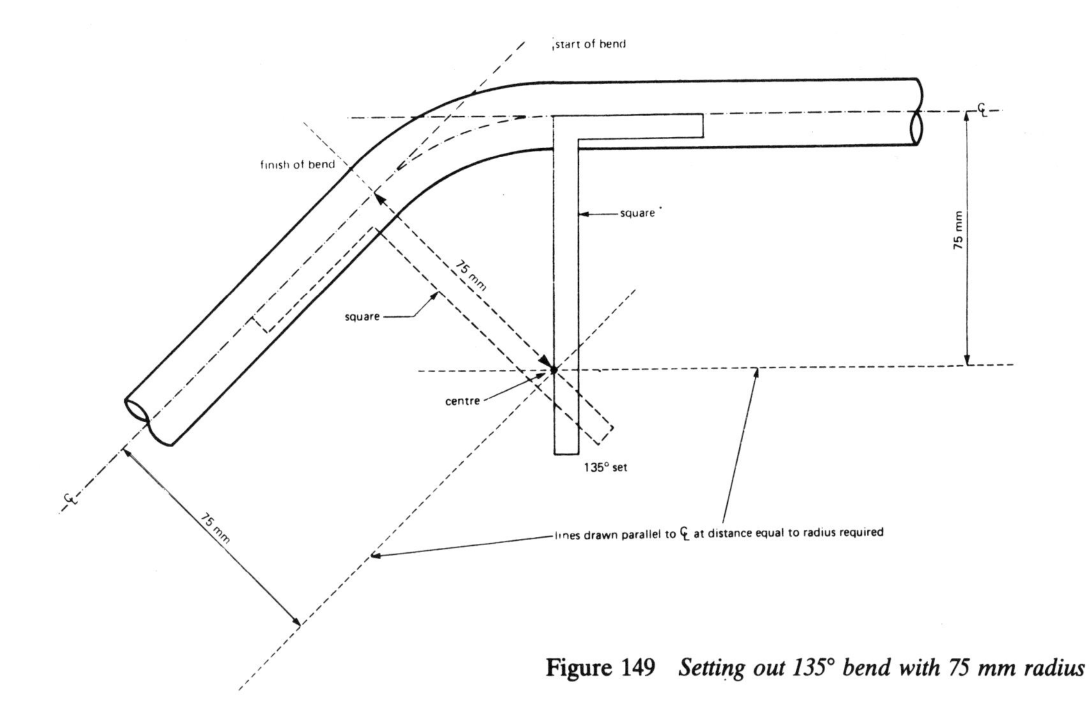

Figure 149 ***Setting out 135° bend with 75 mm radius***

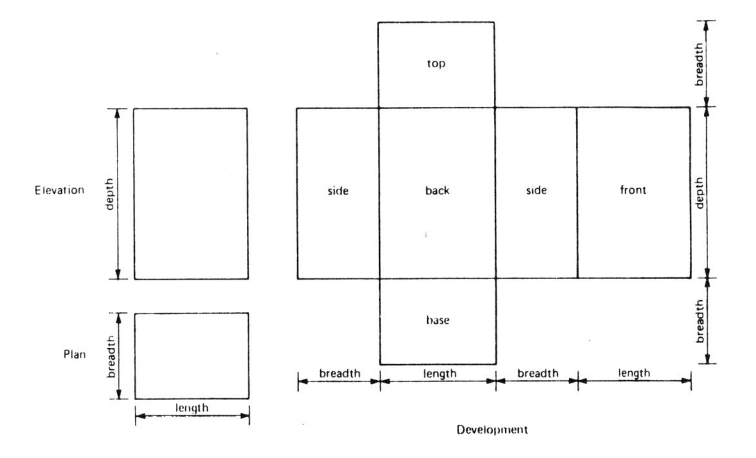

Figure 150 *Development of rectangular tank*

Development

Development is a geometrical method by which the whole of the surface area of a solid is shown on one plane, as on a sheet of metal or drawing paper. Figure 150 illustrates the development of a rectangular tank.

The development of surface areas plays a big part in plumbing. It aids understanding of the shape so that calculations can be performed with accuracy. It is also necessary for setting out, cutting and forming to make items such as tanks or linings, cylinders, slate pieces and pipe intersections.

Development of cylindrical tank (Figure 151)

Draw elevation (rectangle ABCD).

Project plan (circle).

Using 60°–30° set square divide plan circumference into twelve equal parts.

Project AB to form line EF equal in length to circumference of circle. This is achieved by stepping off one of the equal distances on the plan twelve times along this line.

Draw perpendiculars at 0 and 12 to terminate at line HG.

Rectangle EFGH is the required development.

Development of a pyramid (Figure 152)

Draw elevation the given length and height.

Project plan.

Using C as centre (see plan) and AC as radius draw arc until it intersects horizontal line.

Project this point up to base line at F

Join point FC on elevation. This is then the true length of one the hips.

Draw line GH equal in length to four times length of base side, and mark it into four equal lengths.

Using true length of hip as radius draw arcs as shown to obtain the required development.

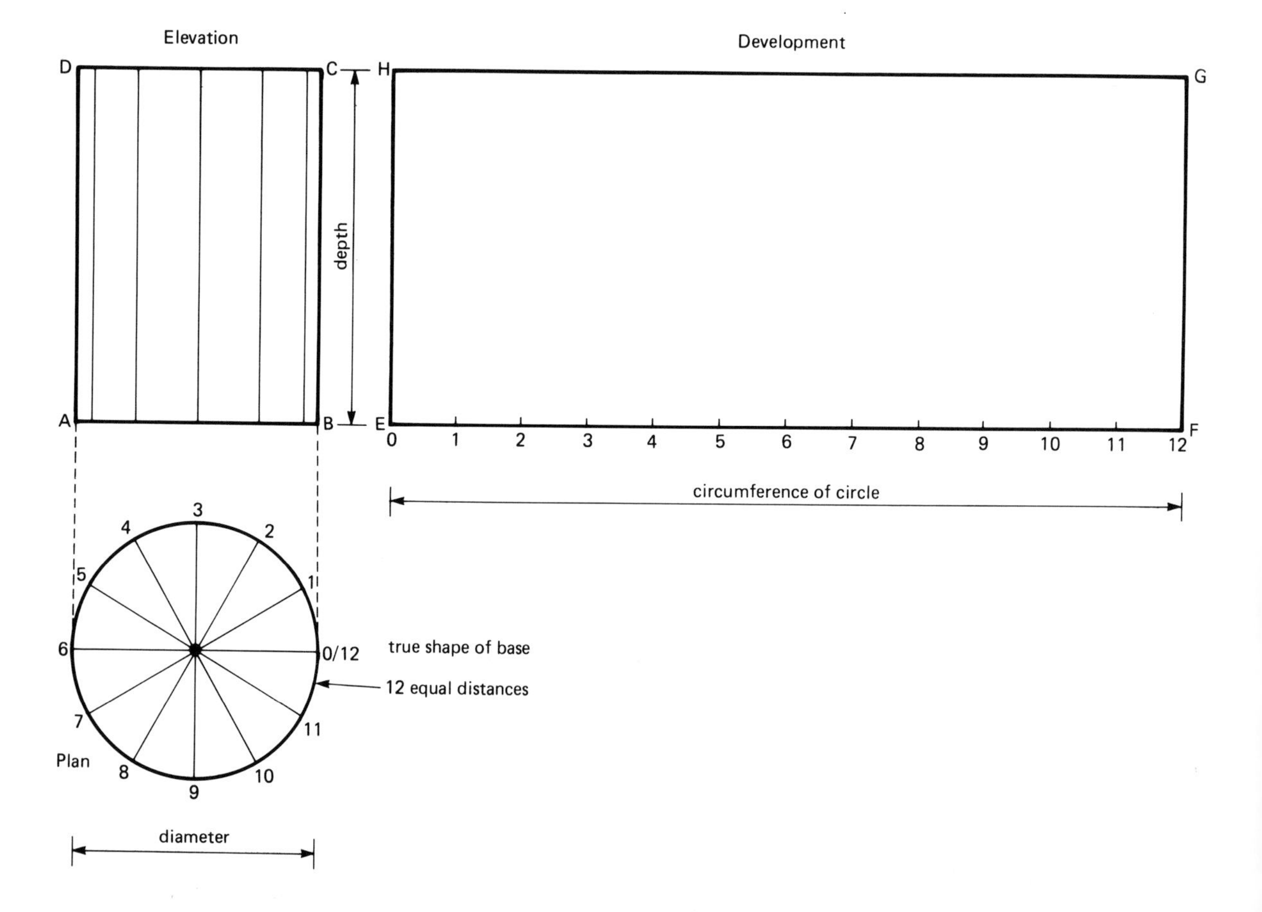

Figure 151 *Development of cylindrical tank*

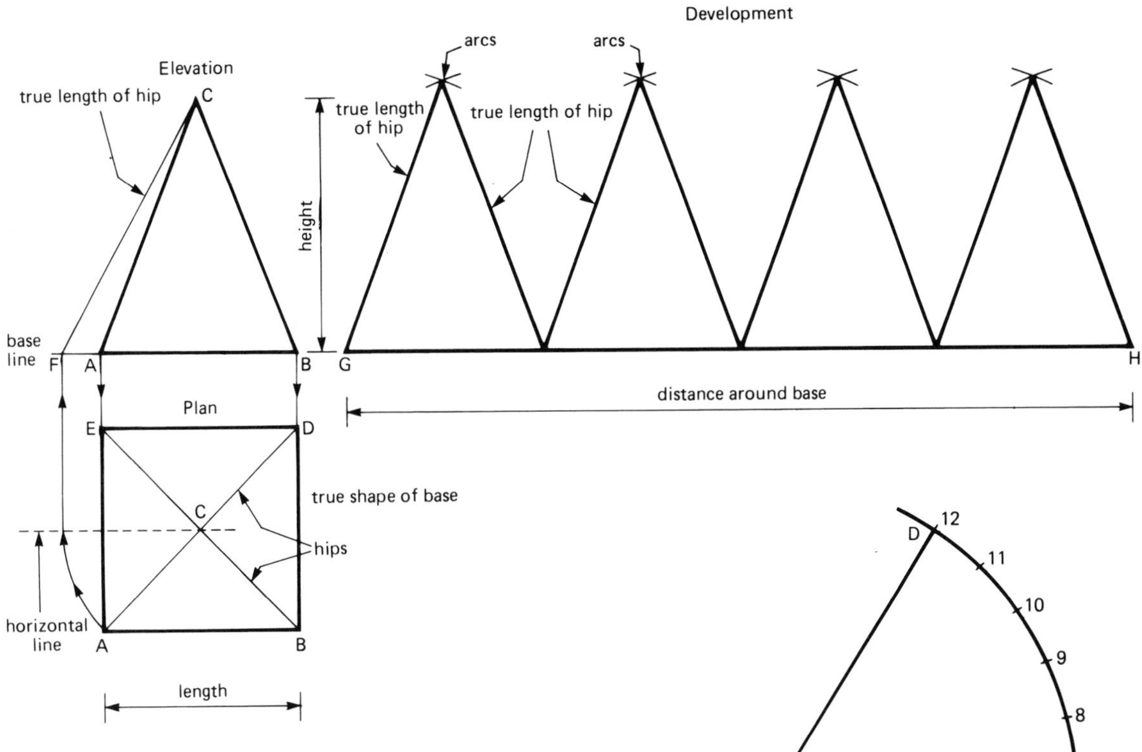

Figure 152 *Development of a pyramid*

Development of cone (Figure 153)

Draw elevation the given base and height.
Project and draw plan.
Divide base into twelve equal parts.
Using C as centre and BC as radius draw arc BD.
Mark off on arc BD twelve equal distances from plan.
Draw line DC to give required development BCD.

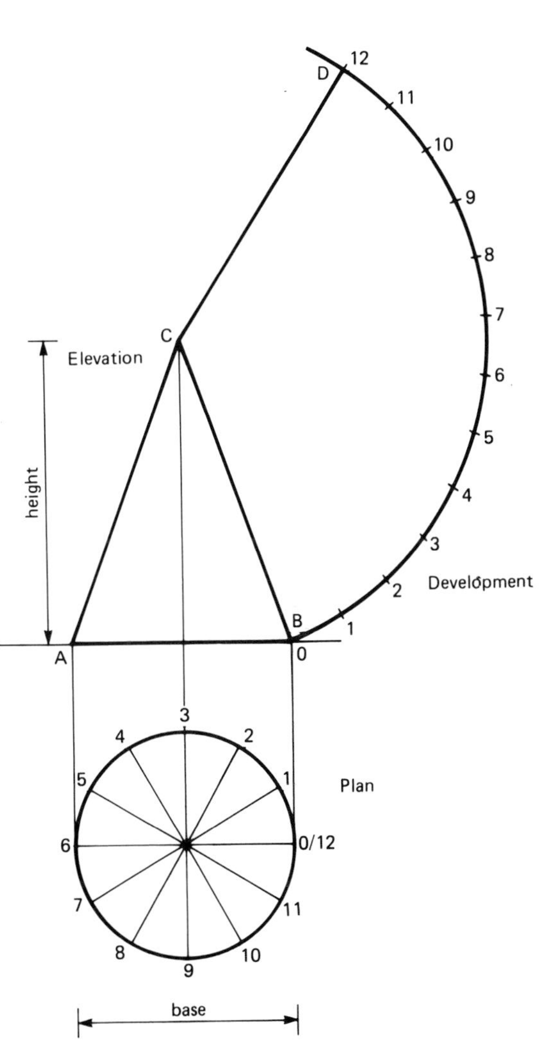

Figure 153 *Development of a cone*

Development of half of obtuse elbow (Figure 154)

Draw elevation of elbow.
Draw semi-circle and divide circumference into twelve equal parts (using 60°–30° set square).
Project points on plan up on to elevation to give intersection at mitre.
Draw line AB and mark off twelve divisions equal to divisions on circumference of circle. Drop vertical lines from these points.
Project lines horizontally from the intersections on the mitre to cut the vertical lines.

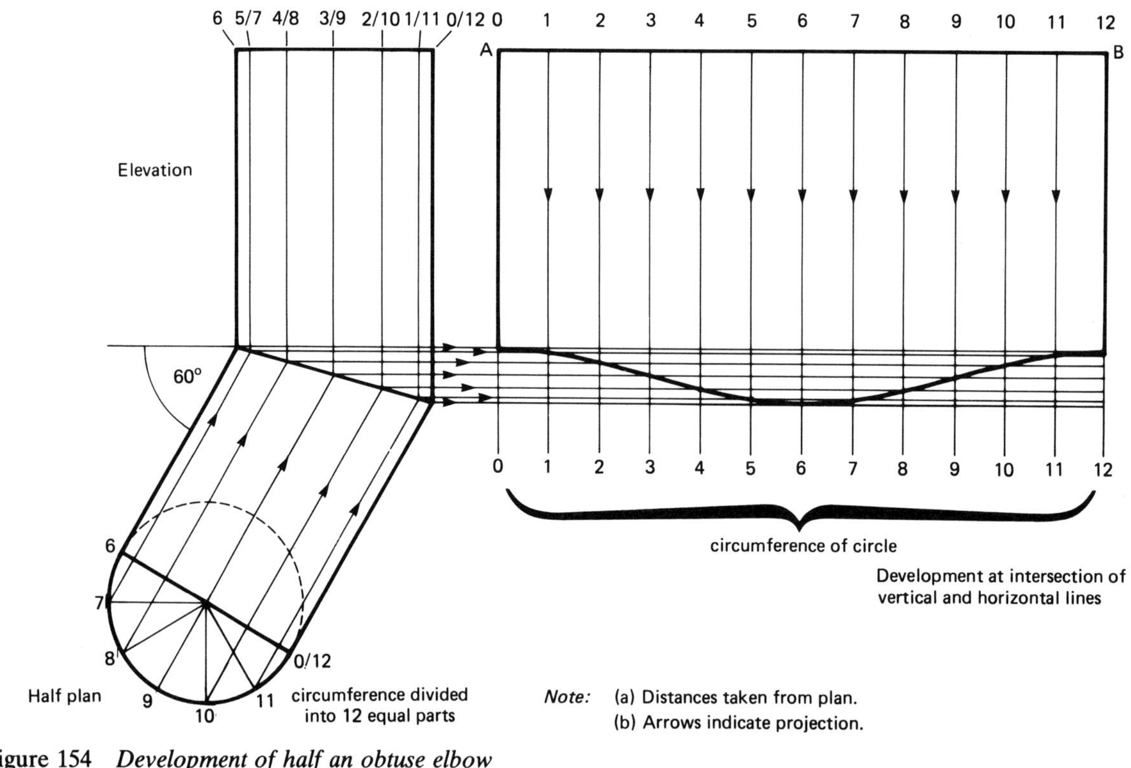

Figure 154 *Development of half an obtuse elbow*

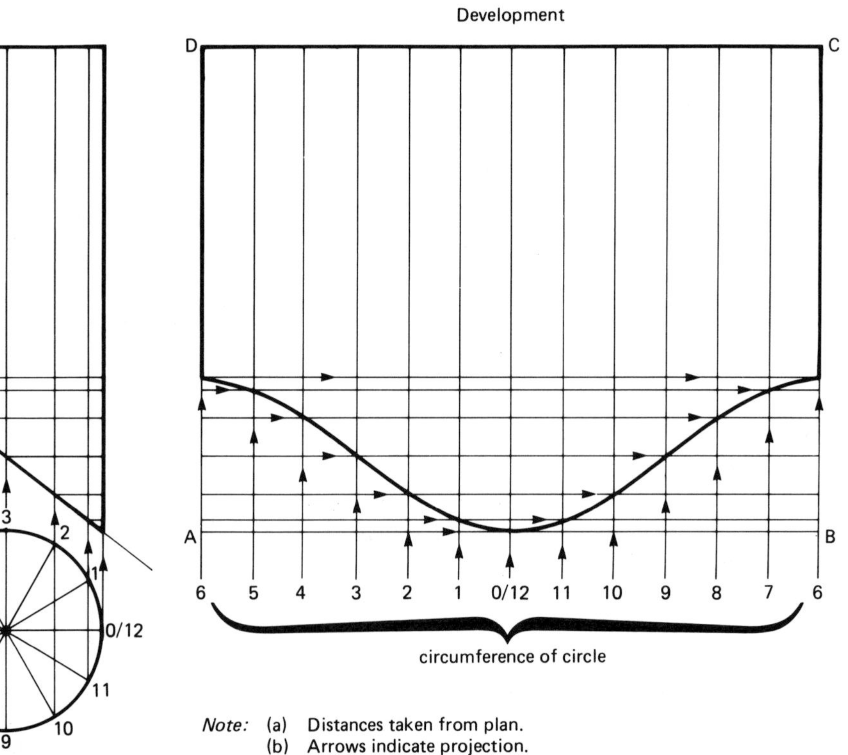

Figure 155 *Development of slate piece*

Carefully plot the intersections to corresponding vertical lines, i.e. 0/12 to 0/12, 1/11 to 1/11, 2/10 to 2/10 and so on.

The true shape of the required development will be obtained by joining the intersection points by a freehand line as shown.

Note: For the complete development two pieces of metal would be required.

Development of slate piece (Figure 155)

Draw the elevation of slate piece.

Draw plan and divide circumference into twelve equal parts (using 60°–30° set square).

Project points from plan up on to elevation to give intersection at roof line.

Draw rectangle ABCD, so that AB is equal in length to twelve divisions as shown on plan, and AD is at least equal to height of slate piece.

Project horizontal lines from intersections at roof line and vertical lines from divisions on AB to obtain intersections marked by crosses (follow arrows).

The true shape of the development is obtained by joining the intersecting points by means of a freehand line as shown.

Development of true shape of hole through roof (Figure 156)

Draw elevation of slate piece.

Draw plan. Divide circumference into twelve equal parts as shown.

Project points from plan to elevation.

Project points at right angles to roof angle to line AB.

To obtain true shape of hole, step off distances X.Y.1, X.Y.2, X.Y.3 etc. on to A.B.1, A.B.2, A.B.3 and so on, as shown by crosses.

To obtain true shape join intersections by freehand line as shown.

Methods of projection in drawing

Methods of projection have to be adapted when drawing three-dimensional objects, so that their appearance from different viewing angles can be accurately conveyed.

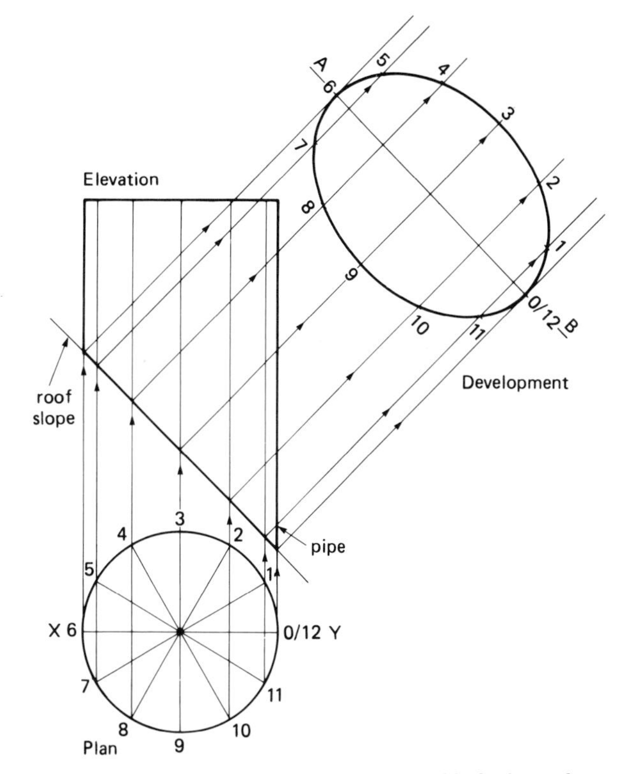

Figure 156 *Development of true shape of hole through roof*

There are several methods of projection of which the three most common are illustrated here. It is not intended to deal with the subject in great depth but only to give the practical application to building problems.

The three most common methods are:

1 Orthographic projection
2 Isometric projection
3 Oblique projection

Orthographic projection

This is the method adopted for working drawings and comprises: plan, elevations and sections. They are laid out on the drawing sheet in a combined first angle and third angle projection.

The front elevation is drawn first just as it is viewed. The plan is drawn directly below the front elevation. The side or end elevations are drawn to the sides of the front elevation so that what is seen on the left hand side of the object is drawn on the right hand side of the front

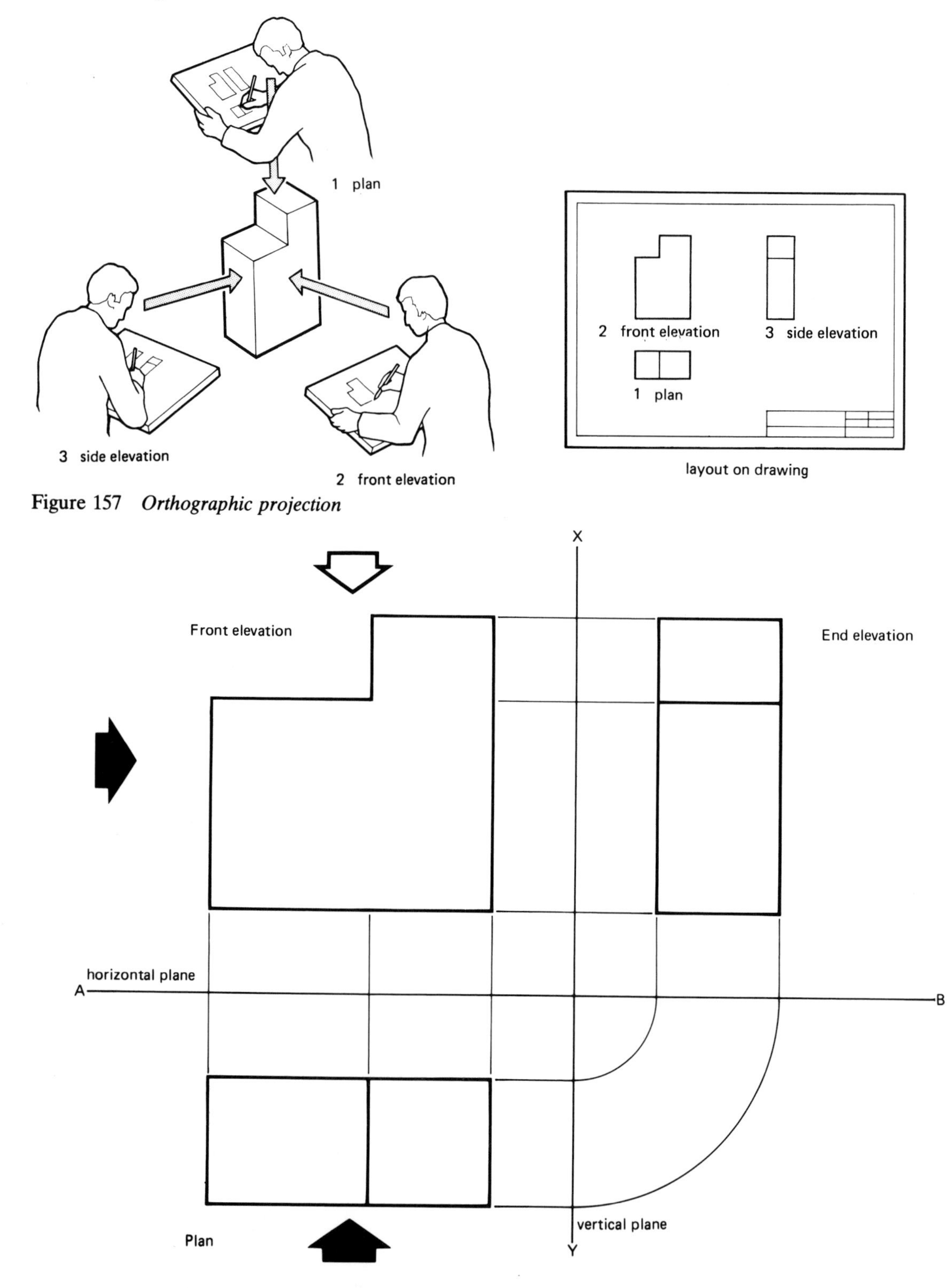

Figure 157 *Orthographic projection*

Figure 158 *Orthographic projection: first and third angle projection*

elevation, and what is seen of the right hand side of the object is drawn on the left hand side of the front elevation.

Plan
This is the view when seen directly from above and should show the true shape of the object.

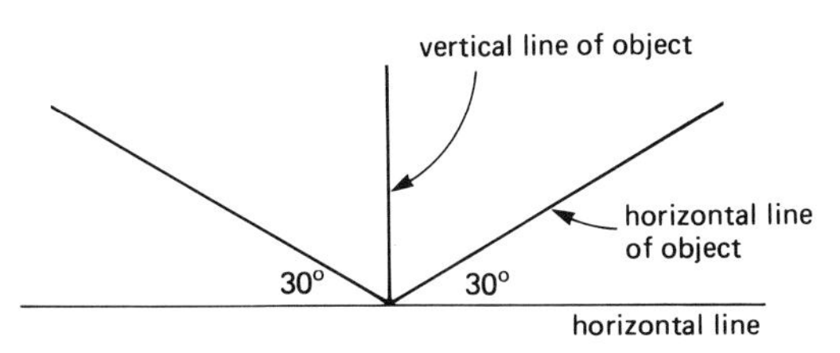

Figure 159

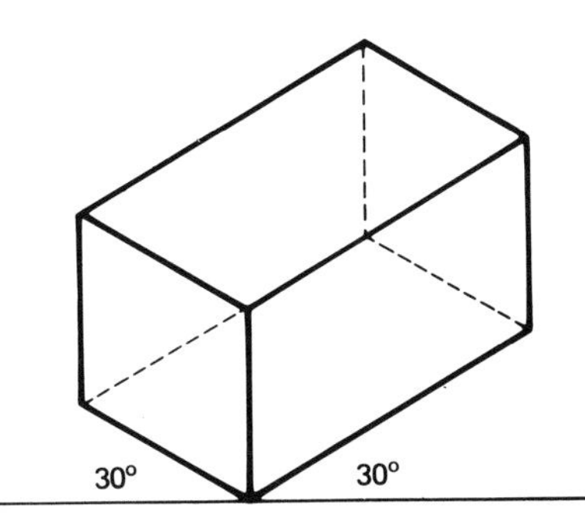

Figure 160 *Isometric view*

Elevation
This is the view as seen when looking in a horizontal direction at the front or back of the object. Horizontal and vertical lines are true.

End elevation
This is the view as seen when looking at the end of the object in a horizontal direction. Horizontal and vertical lines are true.

Figures 157 and 158 show examples of orthographic projection.

Isometric projection
This is a pictorial view embracing the three drawings comprised in the orthographic projection and is perhaps the most popular method of presentation. In isometric projection all horizontal lines are drawn 30° to the horizontal while vertical lines are drawn vertical (see Figures 159 and 160).

Much of the plumber's work involves circular work such as cylinders, pipe and slate pieces. It is therefore essential for the student to learn how to draw circular objects in isometric projection.

Figure 161 shows the method of drawing a circle in isometric projection. Figures 162 and 163 show two examples of cylindrical objects portrayed in this way.

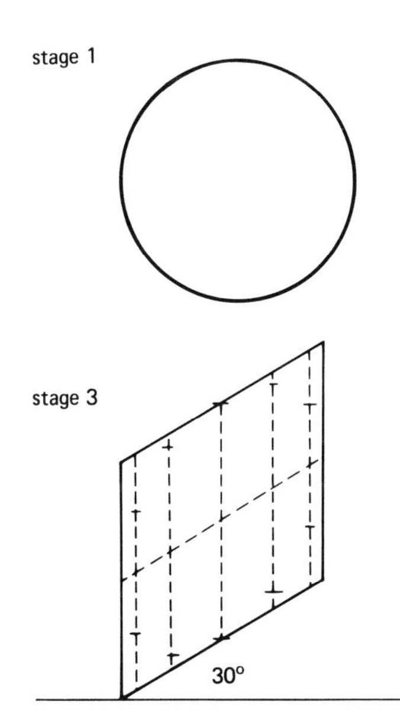

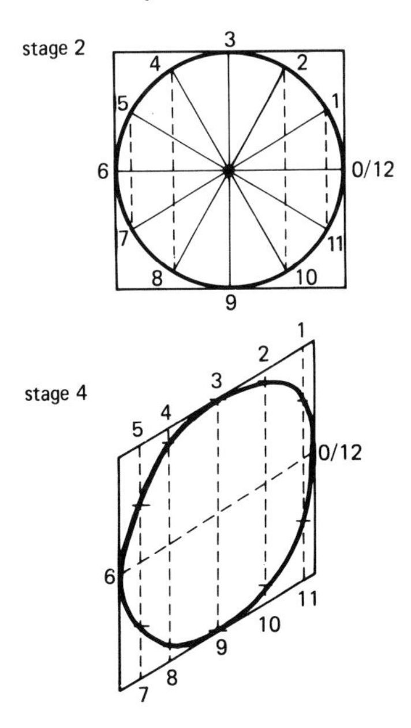

Figure 161 *Method of isometric projection*

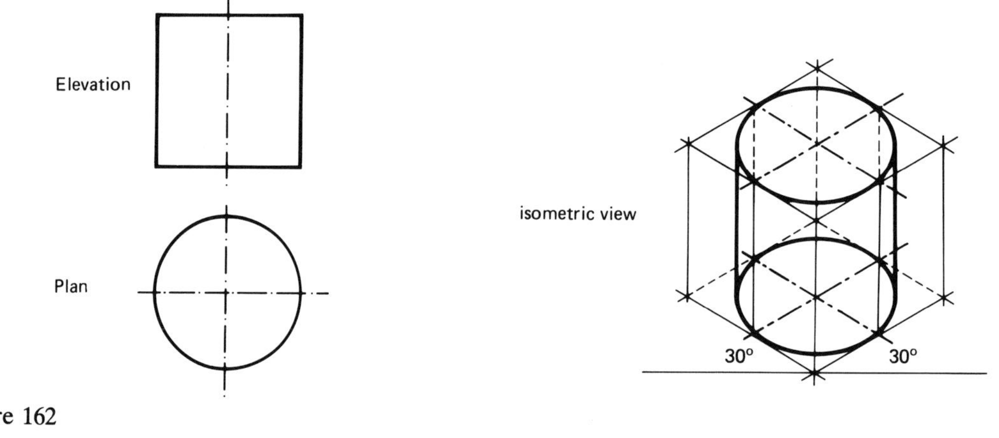

Figure 162

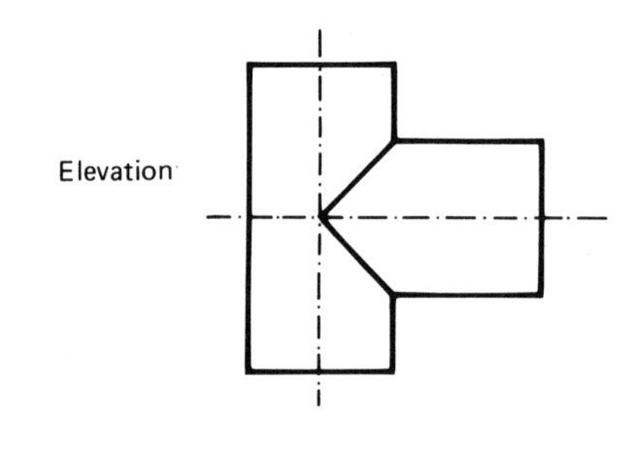

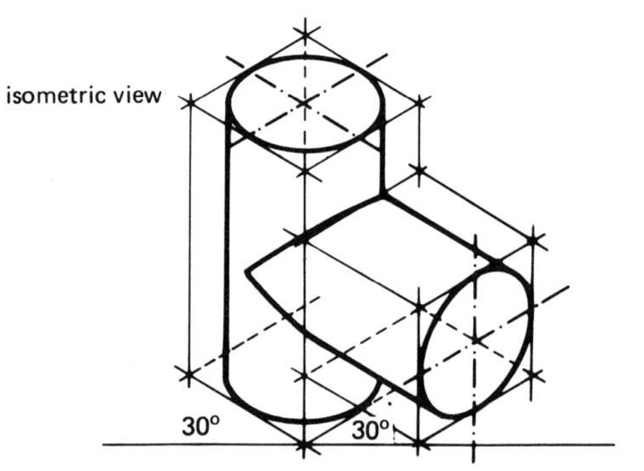

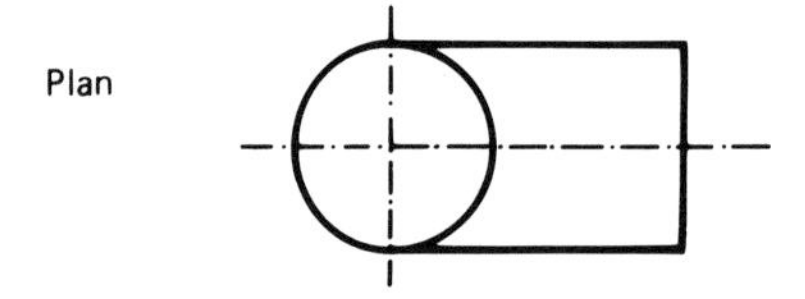

Figure 163

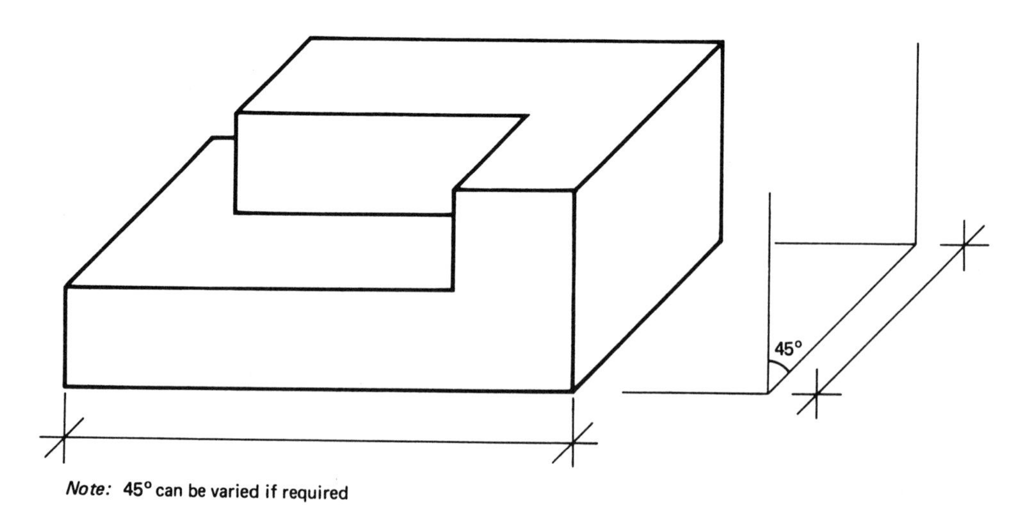

Note: 45° can be varied if required

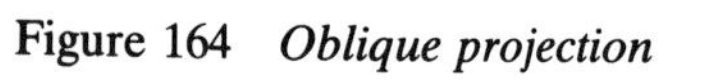

Figure 164 *Oblique projection*

Oblique projection

Oblique projection is illustrated in Figure 164.

Sketches of tools and components

It is often necessary for plumbers to sketch the tools and components used in the industry, so that they are able to explain clearly to other people the points they are trying to make.

The sketches in Figures 165 and 166 are examples of the tools and components in general use in the industry. This technique can be used to draw any other tools used. There are three basic steps to take when using this technique:

1 Draw a faint outline of an isometric box large enough to contain the object.
2 Indicate in a faint line the main dimensions of the object.
3 Sketch the object using the box and dimensions as a guide to the main outline of the object.

Isometric projection of a flat dresser (Figure 165)

1 Lightly draw an isometric box which will just contain the object to be sketched.
2 Mark important points for guides.
3 Draw in isometric projection the tool required.

Note: The special shape of the dresser is to allow the user to set in the sheet near the wall.

Isometric projection drawing of a basin wrench (Figure 166)

1 Lightly draw isometric box which will contain the basin wrench.
2 Mark important guide points.
3 Complete drawing of tool in isometric projection.

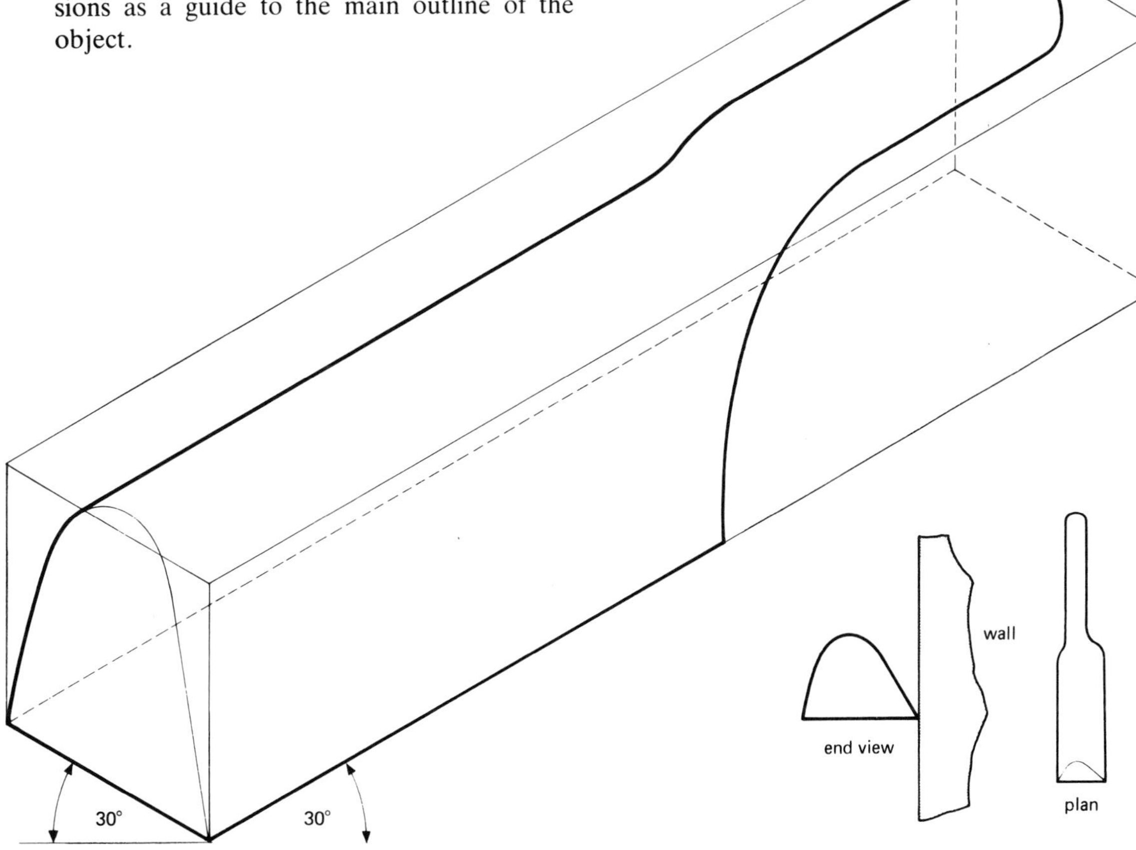

Figure 165 *Isometric projection drawing of a flat dresser*

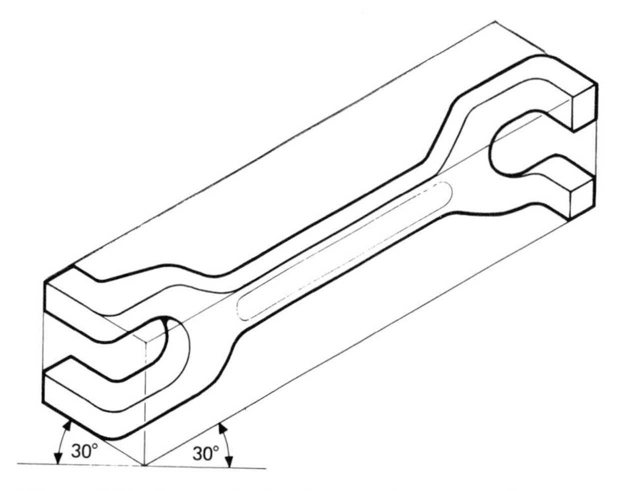

Figure 166 *Isometric drawing of a basin wrench*

Rule assisted sketches

Rule assisted sketching is the ability to reproduce an object which is not to scale but at the same time is drawn in proportion to a fair degree of accuracy.

Traps (Figure 167)
When drawing traps the most important detail is the water seal. Figure 167 shows a method of drawing traps in five stages.

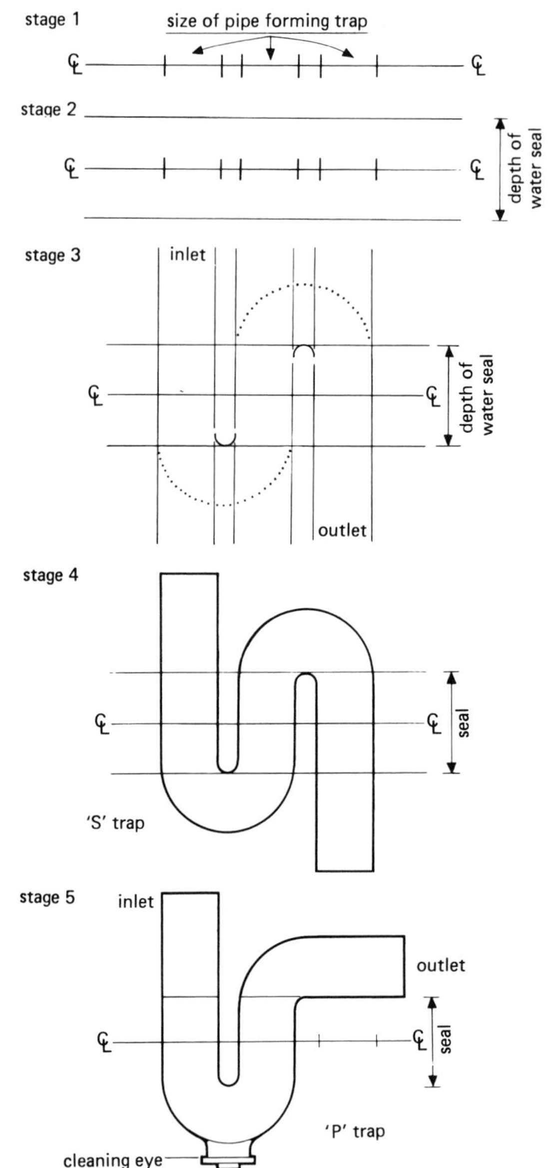

Figure 167 *Method of sketching traps*

Water closets (Figure 168)
Water closet pans can be drawn quite easily with the aid of a few simple guide lines. The most important detail is the trap and once this is correctly drawn the rest of the appliance will fall nicely into place.

Stage 1
Draw a rectangle in the proportion of 3:2.

Stage 2
Draw vertical or near vertical line to form back of pan.
Draw bottom of pan (semi-circle).
Extend this line to form outlet.
Complete outlet.
Draw 50 mm rim.

Stage 3
Draw freehand the front of pan joining rim to semi-circle at base.
Draw inlet
Draw in the base.
Draw water line.

Stage 4
Draw parallel lines for the thickness of the material.
Complete details.

Note: The same principle applies for both 'S' and 'P' trap pans.

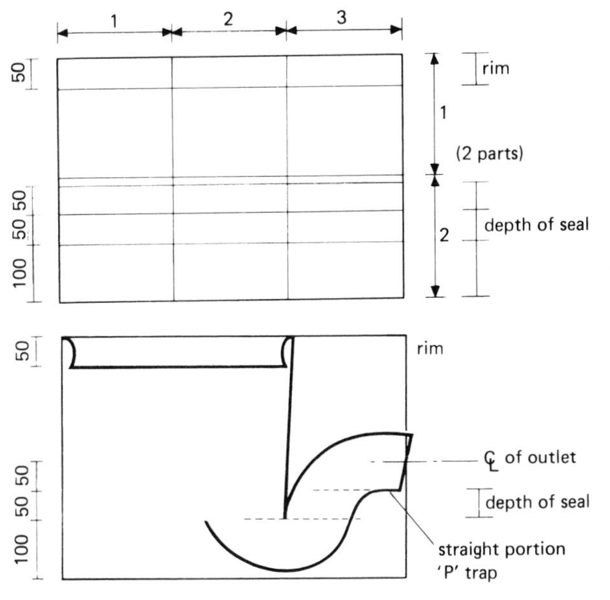

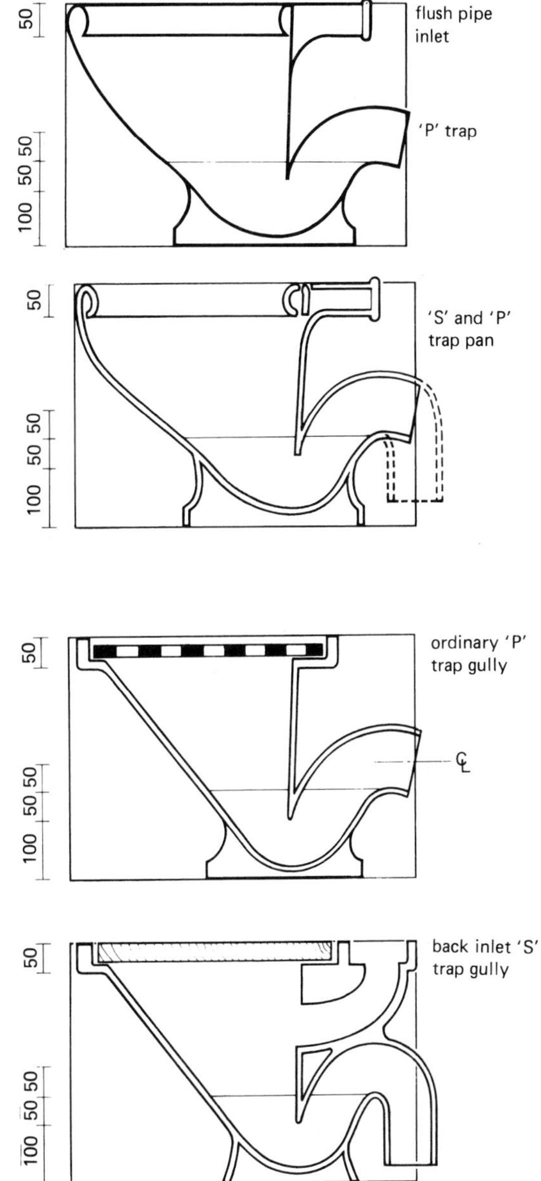

Figure 168 *Method of sketching water closet pans*

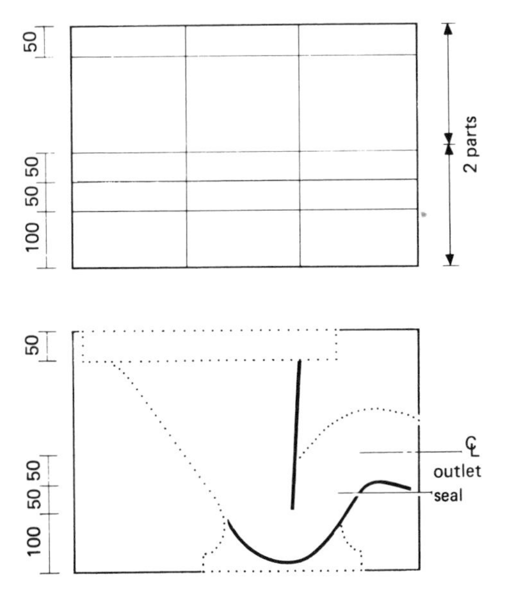

Figure 169 *Method of sketching gully traps*

Gully traps (Figure 169)

The basic designs of water closets and gully traps are very similar whether they be ordinary type or the back-inlet variety.

The method of setting out and sketching is, therefore, almost identical and provided the basic guide lines are adhered to no problems should be encountered.

Stage 1

Draw rectangle in the proportion of 3:2.

Stage 2

Draw vertical or near vertical line to form back of gully.

Follow instructions as for WC pan.

Belfast sink/wash basin

Figure 170 shows a section through a Belfast sink. Superimposed is a section of a wash basin showing how with a slight alteration the section through the sink can be changed into that of the basin.

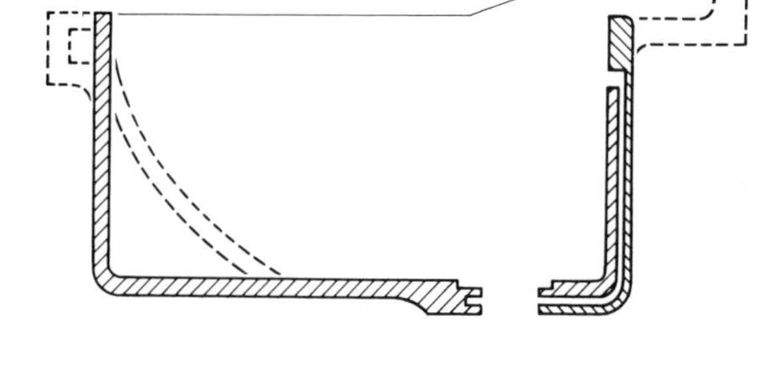

Figure 170 *Rule assisted sketch of a Belfast sink*

Self-assessment questions

1 A line drawn across a circle from the circumference at two points and cutting the centre is the:
(a) radius
(b) chord
(c) diameter
(d) arc

2 The portion contained between two concentric circles is the:
(a) annulus
(b) segment
(c) sector
(d) quadrant

3 The number of degrees in a full circle is:
(a) 45°
(b) 90°
(c) 180°
(d) 360°

4 How many degrees must an obtuse bend have?
(a) less than 90°
(b) less than 45°
(c) more than 90°
(d) any number

5 An equilateral triangle has:
(a) all angles equal
(b) all angles unequal
(c) one angle at 90°
(d) one angle at 45°

6 A polygon is a figure bounded by:
(a) three sides
(b) four sides
(c) a circle
(d) more than four sides

7 The true shape of a hole in a sloping roof to allow a circular pipe to pass through will be:
(a) circular
(b) elliptical
(c) square
(d) semi-circular

8 Which of the following statements is true of isometric projection?
(a) vertical lines are vertical, horizontal lines at 30
(b) vertical lines are vertical, horizontal lines at 45
(c) vertical lines are inclined at 60, horizontal lines at 30
(d) all vertical lines are inclined at 45

9 The symbol used to indicate a stop valve is:
(a)
(b)
(c)
(d)

10 The part of a circle shown as is known as
(a) segment
(b) quadrant
(c) arc
(d) sector

5 Cold water supply

After reading this chapter you should be able to:

1 Understand the requirements of the relevant model and local bye-laws.

2 Define and classify water and its sources of supply.

3 Describe the method of mains connection for each of the different materials used.

4 Recognize and name each type of cold water system and draw its component parts.

5 State the functional requirements and working principles of each system.

6 Describe the properties of materials used in the manufacture of cisterns.

7 List the factors relevant to siting a cistern.

8 Describe methods of jointing all materials used for conveying cold water.

9 Describe the features of taps, cocks and valves.

Bye-laws

All the materials, i.e. pipes, fittings, appliances, as well as the manner in which they are installed, are covered by *model water bye-laws* and/or *local authority bye-laws*. Before any work is carried out the plumber must be conversant with all aspects of the work and the relevant regulations.

Model water bye-laws are framed by various government bodies and give the guidance about materials and the method of installation. These are generally accepted throughout the whole country.

Local bye-laws are framed by the local authority. Usually they are copies of the model water bye-laws but local bye-laws can embody changes to take into account circumstances and irregularities which are only met in that district. Once local bye-laws have been approved they become the legally enforcable laws for that area and must be worked to irrespective of the model water bye-laws.

Copies of bye-laws can be obtained from:
Government offices;
Retail distributors;
Public or college libraries;
Local council offices.

Reference should also be made to all relevant British Standard Codes of Practice, i.e. CP 310 and CP 99.

Definition of water

Water is a chemical compound of the two gases hydrogen and oxygen. It is formed when the gas hydrogen or any substance containing hydrogen is burned.

One of the most important properties of water is its solvent power. It dissolves numerous gases and solids to form solutions. The purest natural water is rainwater collected in the open country. This contains small amounts of dissolved solids, mainly sodium chloride (common salt) dissolved from the air, and also dissolved gases: nitrogen, oxygen, carbon dioxide, nitric acid, and ammonia.

Rainwater collected in towns contains higher percentages of dissolved substances, and also soot, etc. In particular it may contain acids – sulphuric and carbonic – which may cause damage to certain stones and metals, particularly limestone and zinc.

Spring water usually contains more dissolved solids, the amount and kind depending on the type of soil and rocks through which it has passed.

River water is water which has passed through and over the ground and also contains some rainwater. Sea water contains the most dissolved matter.

Classification

Water can be classified as soft or hard.

Soft water

A water is said to be soft when it lathers readily. It is not very palatable as a drinking water, and has a detrimental effect on most metals. It can cause rapid corrosion particularly of those metals which contain organic solutions.

Hard water

A water is said to be hard if it is difficult to obtain a lather. A hard water can be temporarily hard, permanently hard or both, depending upon the type or earth strata through which the water has passed.

Measurement of hardness

Hardness of water is expressed in parts per million (weight/volume) or in milligrams per litre, which is the same figure. At one time, degrees of hardness were classified according to Clarks scale in which 1° represents one grain of calcium per 4.5 litres.

Table 14 gives the accepted classifications of water according to their total hardness (both permanent and temporary).

Table 14

Designation of water	*Parts per million or mg/litre*
Soft	0 to 50
Moderately soft	50 to 100
Slightly hard	100 to 150
Moderately hard	150 to 200
Hard	200 to 300
Very hard	over 300

Sources of water supply

Rainfall

Rainfall is the source of all natural fresh water. When rain falls on the surface of the earth part immediately runs off to ditches or natural water courses (streams and rivers). The remainder soaks into the ground, some to remain underground for the whole of its journey until it mingles with salt water in the seas, some to break forth as springs, and some to be artificially extracted from wells (see Figure 171).

Whatever source of water is used – lakes, rivers, springs or wells – the available supply depends on the nature and size of the catchment area and the amount of rain that falls on it. The greatest supply is usually obtained where moisture-laden winds from the ocean pass over the mountain range which deflects them into higher and cooler regions where the moisture is condensed.

The average rainfall in England is about 0.875 m per annum, in Wales and Scotland 1.25 m, in Ireland 1.17 m and in the British Isles as a whole 1.15 m.

For purposes of water supply, rain which falls in autumn, winter and spring is the most important. A large proportion of the summer rainfall is lost by evaporation. Summer supplies from wells are not reliable because summer storms are so short and intense that more runs off the surface, instead of collecting in the way that long steady winter rainfall does.

Sources of water supply may be grouped under two headings:

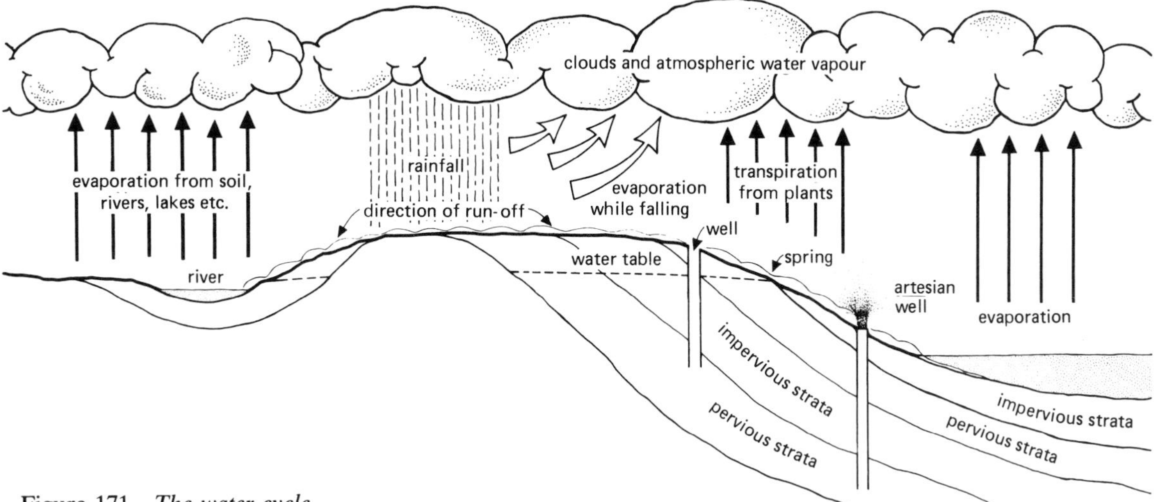

Figure 171 *The water cycle*

Surface sources

These include:

1 Run-off from natural catchments into natural or artificial lakes, reservoirs formed by damming valleys, and the run-off of small catchments into ponds (dew ponds).
2 Intake of water from streams and rivers.
3 Collection of rainwater direct from roofs or areas specially paved for the purpose.

Underground sources

These include:

1 Springs
2 Shallow wells
3 Artesian and deep wells

Surface sources

Upland catchments Water from upland catchments is usually excellent quality, being soft and free from sewage or animal contamination or suspended matter. It is liable to be acid in character if it contains water from peat areas, and corrosive to lead pipes (a condition known as plumbo-solvency). As lead is poisonous such water must not be conveyed in lead pipes in an untreated state.

Lakes Supplies from natural mountain lakes are similar in nearly all respects to those from upland catchments with artificial storage. Pond water cannot usually be regarded as a fit source of domestic water supply.

Rivers and streams The quality of river water varies. A moorland stream which is not unduly peaty will generally be wholesome. Downstream a river will usually collect drainage from manured fields, farmyards and roads and become progressively more contaminated. Most lowland rivers receive sewage from towns and industrial effluent from factories and have become heavily polluted.

Outside manufacturing districts most large rivers furnish water for the supply of one or more towns past which they flow, where the amount of impurity is not too great.

River water varies in hardness according to the proportion of water reaching it from springs and surface run-off. Water in the upland reaches of rivers is naturally soft, that in the lower reaches is generally hard.

Collection tanks

Rainwater can be collected from roofs of buildings into tanks, in order to be stored over dry periods.

Slate roofs are best for collecting rainwater: tile or galvanized iron roofs will also serve. Roofs covered with lead, copper or tarred material are unsuitable.

Rainwater is the softest natural water, but after collection on a roof it may contain leaves, insects and bird droppings. If required for domestic purposes it should always be filtered, and before drinking it should be subjected to a suitable purification process – boiling is sufficient where small quantities are involved.

Underground sources

Public water supplies from underground sources are usually drawn from water-bearing formations deep in the earth and covered by impervious strata. The water is often derived from gathering grounds several miles away, and the long journey underground provides a very thorough filtration.

Springs

Spring water is water that has travelled through the ground and come to the surface as a result of geological conditions. Its qualities are similar to those of a well in the same circumstances.

Spring water may vary considerably in quality. When it has travelled long distances through a stratum of rock it may be free from contamination, but hard. When a spring is fed by local rainfall organic pollution is possible, but the water may be soft.

Wells

Wells are classified as shallow or deep according to the water-bearing strata from which they derive their water. A well that is sunk into the first water-bearing strata is classified as shallow. A deep well is one sunk into the second water-bearing strata. Since the earth's strata vary in thickness, it is possible to have a shallow well deeper than a deep well.

Table 15 *River pollution commissioners' classification table*

Wholesome	1	Spring water	Very palatable
	2	Deep well water	Very palatable
	3	Uplands surface water	Moderately palatable
Suspicious	4	Stored rainwater	Moderately palatable
	5	Surface water from cultivated land	Moderately palatable
Dangerous	6	River water	Palatable
	7	Shallow well	Palatable

Artesian wells

An artesian well or borehole is one which pierces an impervious stratum and enters a lower porous zone from which water rises as a 'gusher' above ground level. A similar borehole where water rises part of the way but does not reach the surface is called 'sub-artesian',

Table 15 gives a summary of the quality of water from various sources.

Connection to company main

Water authorities are able to run a water main from their storage reservoirs to supply wholesome water to most premises except those in very isolated areas. Water mains are constructed of asbestos cement, steel, PVC or cast iron and are laid to as convenient a place as possible. The method of making the connection to the main will differ according to the type of material and its thickness and strength (see Figures 172 and 173).

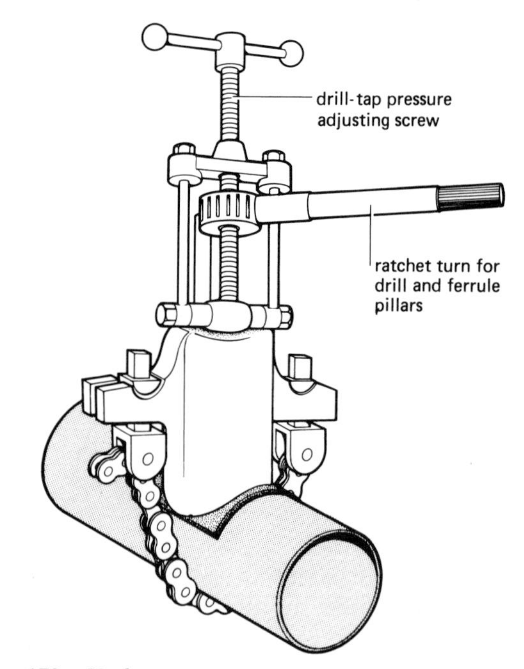

Figure 172 *Under-pressure tapping machine in place on main*

The connection is generally a screw-down gunmetal stop tap ferrule to which the communication pipe serving the house is connected (see Figure 174). Figure 175 shows an exploded view of a screw-down valved ferrule.

The LA (local authority) mains are buried to great depths depending upon local conditions but they must have a minimum cover of 900 mm. The service pipe must have a minimum cover of 900 mm beneath roads and 750 mm elsewhere in order to prevent damage owing to settlement and/or frost.

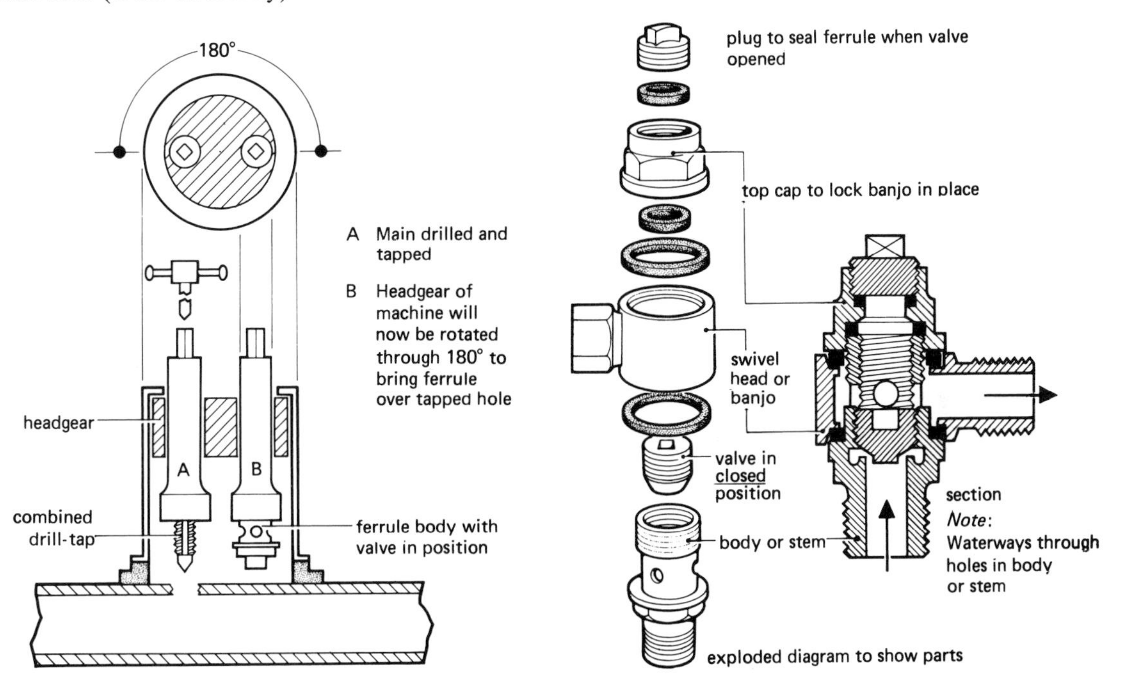

Figure 173 *Diagram of tapping machine showing principle of design and operation*

Figure 175 *Exploded view of screw-down valved ferrule*

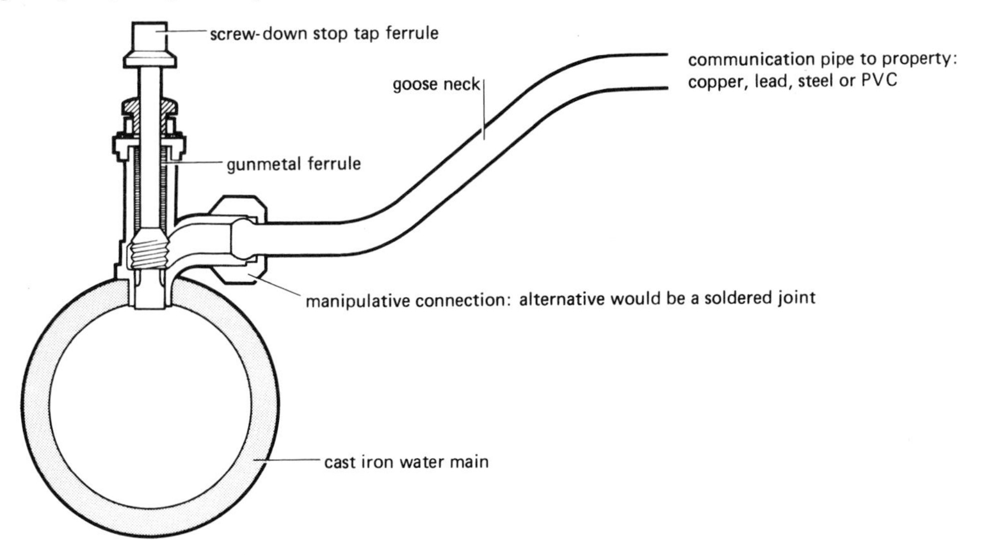

Figure 174 *Connection to communication pipe*

Asbestos cement mains
These have been used with safety for conveying water which could corrode iron or steel pipes. It can be manufactured in various thicknesses and can withstand the same internal pressures as cast iron pipes. Owing to the much lower mechanical strength of asbestos, special cladding devices are necessary when making service connections.

Steel mains
These are not usually used as distribution mains so the problem of tapping them is seldom encountered. Where tapping is required it can be satisfactorily performed by bolting on the clamp as for asbestos or by welding on a thick pad of metal.

PVC
Polyvinyl chloride is now becoming more widely used and in some instances has sufficient thickness and strength to make drilling and tapping a possibility.

Cast iron
Cast iron mains are by far the most commonly encountered and time has proved them to be the most satisfactory. The only real problem is their susceptibility to deterioration in certain corrosive soils, but this can be overcome by taking protective measures.

Cold water supplies
At the boundary of the premises a screw-down stopcock is provided. This control should be housed in a properly constructed stop tap box that is accessible from ground level. The box is covered with a hinged cover and frame.

A 150 mm 'seconds' glazed stoneware drain-pipe makes an excellent stop tap box and provides adequate freedom of movement for the loose key

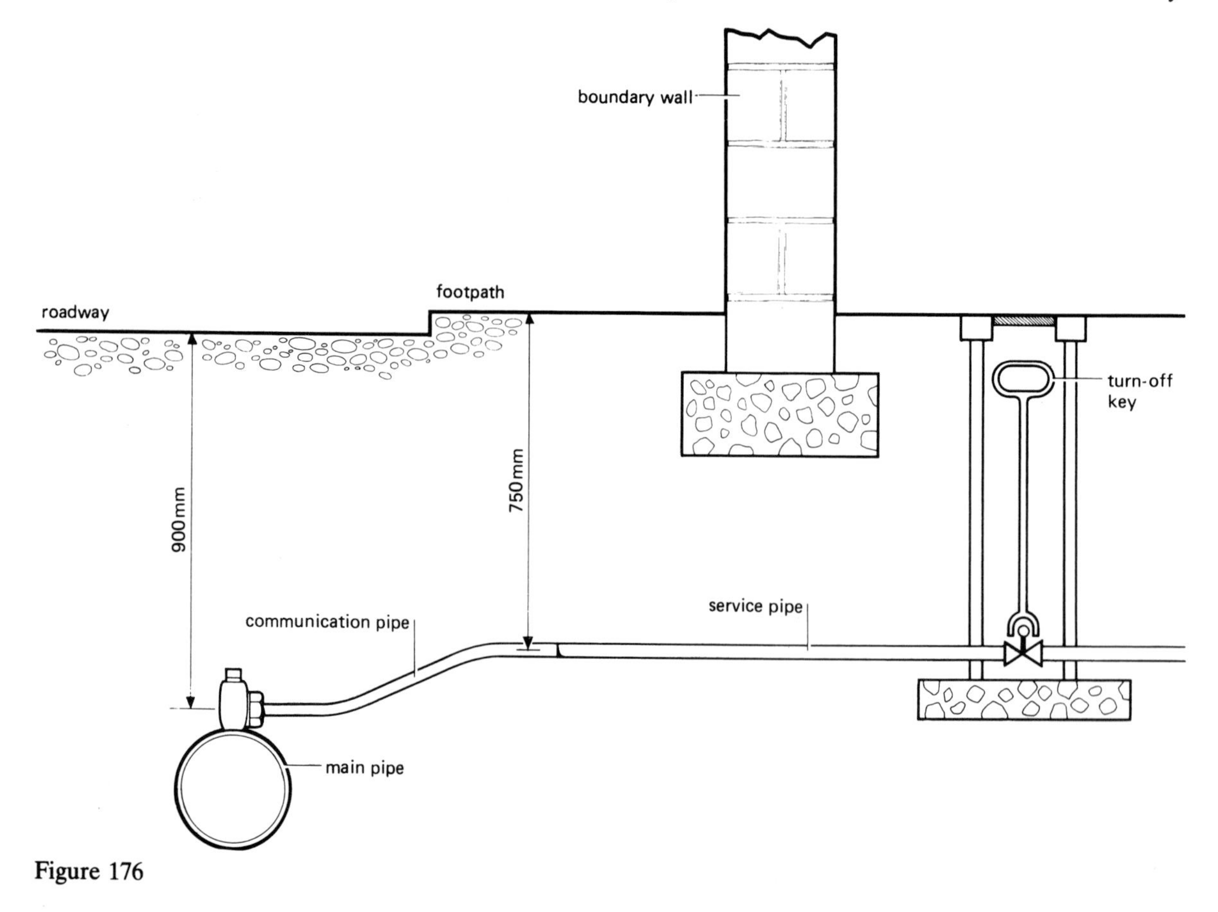

Figure 176

which, when fitted to the crutch of the stopcock, permits ready and easy control of the supply (see Figures 176 and 177).

As the service pipe continues and enters the premises, a combined stopcock and draincock is fitted just above the finished floor level (see Figure 178). This permits the supply to be controlled easily from within the premises. It also provides for a complete draining down of the installation for alteration or repair work or when frost damage is likely during periods of absence etc.

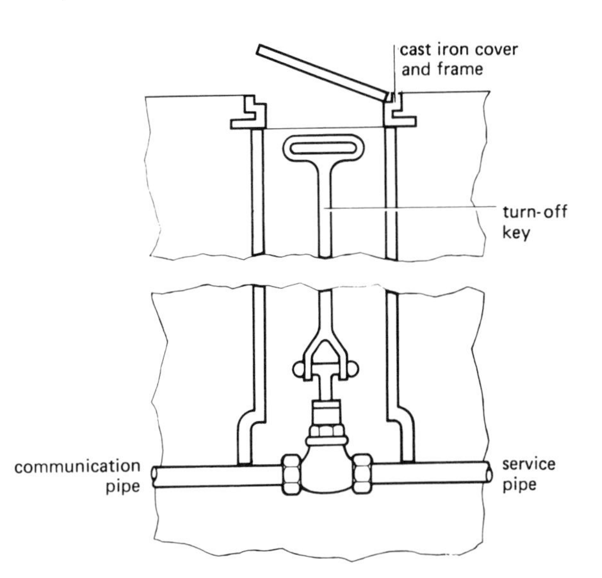

Figure 177 *Fixed at boundary to property*

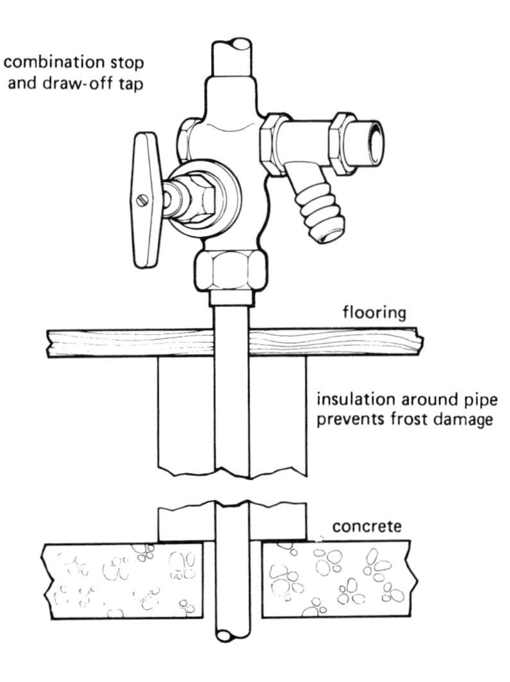

Figure 178 *Fixed on immediate entry to property*

Types of system

Direct system

In districts where the mains supply is capable of delivering adequate quantities of water at good pressure, the water undertaking may permit a direct system of supply to all buildings.

All pipes to the cold draw-off points are taken directly from the rising main or service pipe and are subject to pressure from the main. There is therefore no risk of possible contamination that may occur when water is stored within the premises.

Figure 179 shows a suitable installation for the average dwelling, supplying a sink, WC, wash basin, bath and a cold feed cistern for the domestic hot water supplies.

A drain-off point should be provided on the service pipe where it enters the buildings immediately above the stopcock and this is most conveniently catered for by a combined stopcock and draincock.

Indirect system

In some areas the cold water supply is provided by use of the indirect system (see Figure 180). This means that the service pipe rises through the building to the cold water storage cistern and only one direct draw-off point for drinking purposes is permitted. The remaining cold water draw-off points are supplied from the storage cistern.

Storage capacity is normally based on a full twenty-four hour period, but some undertakings may only require storage for periods of six or twelve hours.

As in the direct system, a drain-off should be provided on the service pipe where it enters the building immediately above the stop tap.

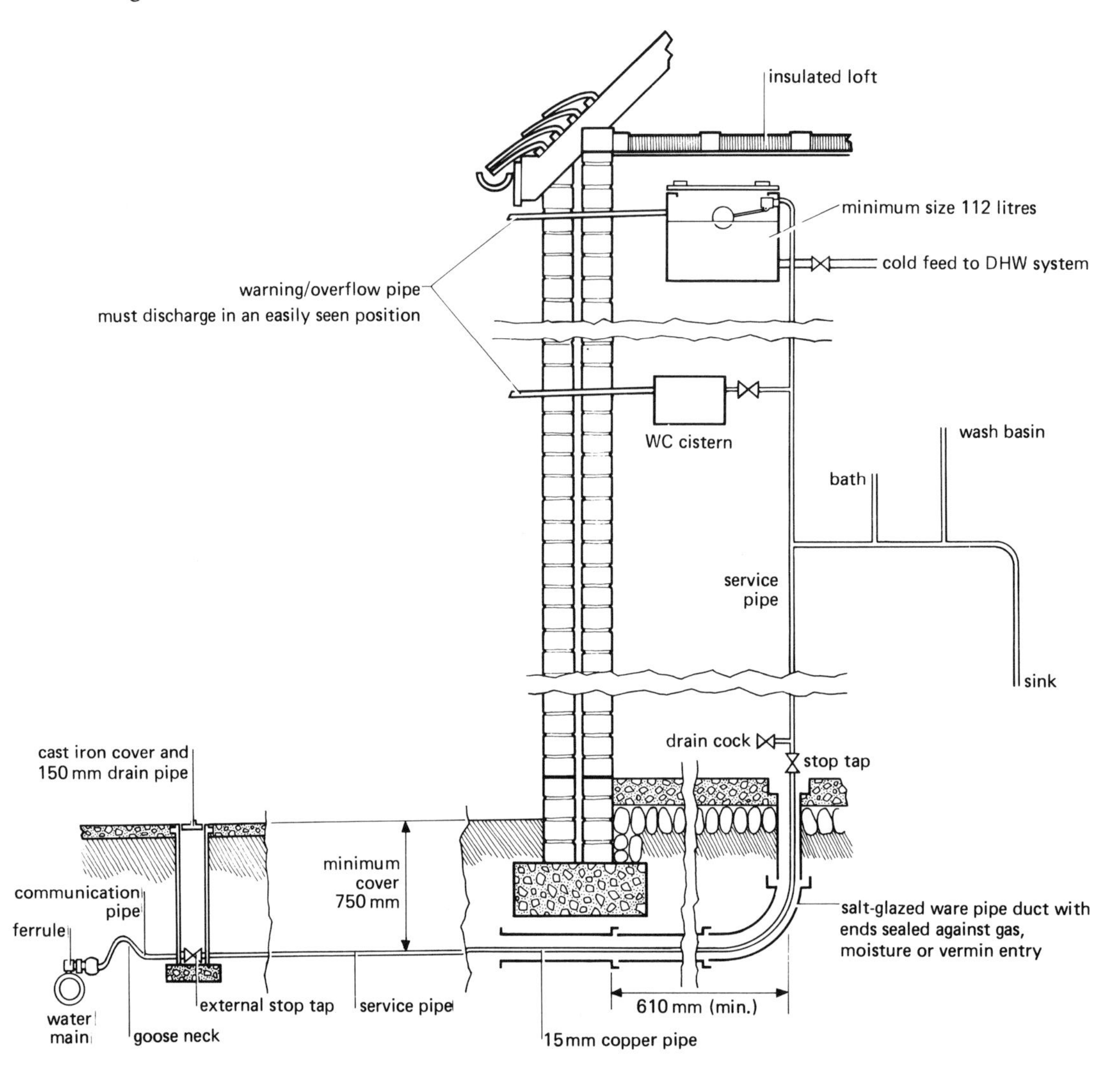

Figure 179 *Direct system of domestic cold water supply*

Cisterns

A cistern is an open topped vessel designed to hold a supply of cold water, which will have a free surface subject only to the pressure of the atmosphere. It should be fixed as high as possible to give adequate pressure flow.

Storage cistern

This is designed to hold a reserve of water to supply cold water to the various appliances fitted to the system.

In districts where all cold water taps and sanitary appliances are supplied from the service pipe, it is not usual to provide a storage cistern as such although it is necessary to have a feed cistern to supply the domestic hot water system.

Feed cistern

This may look exactly like a storage cistern, but it is meant only to hold a reserve of water for the hot water system, and is an essential part of a boiler-cylinder type of hot water system.

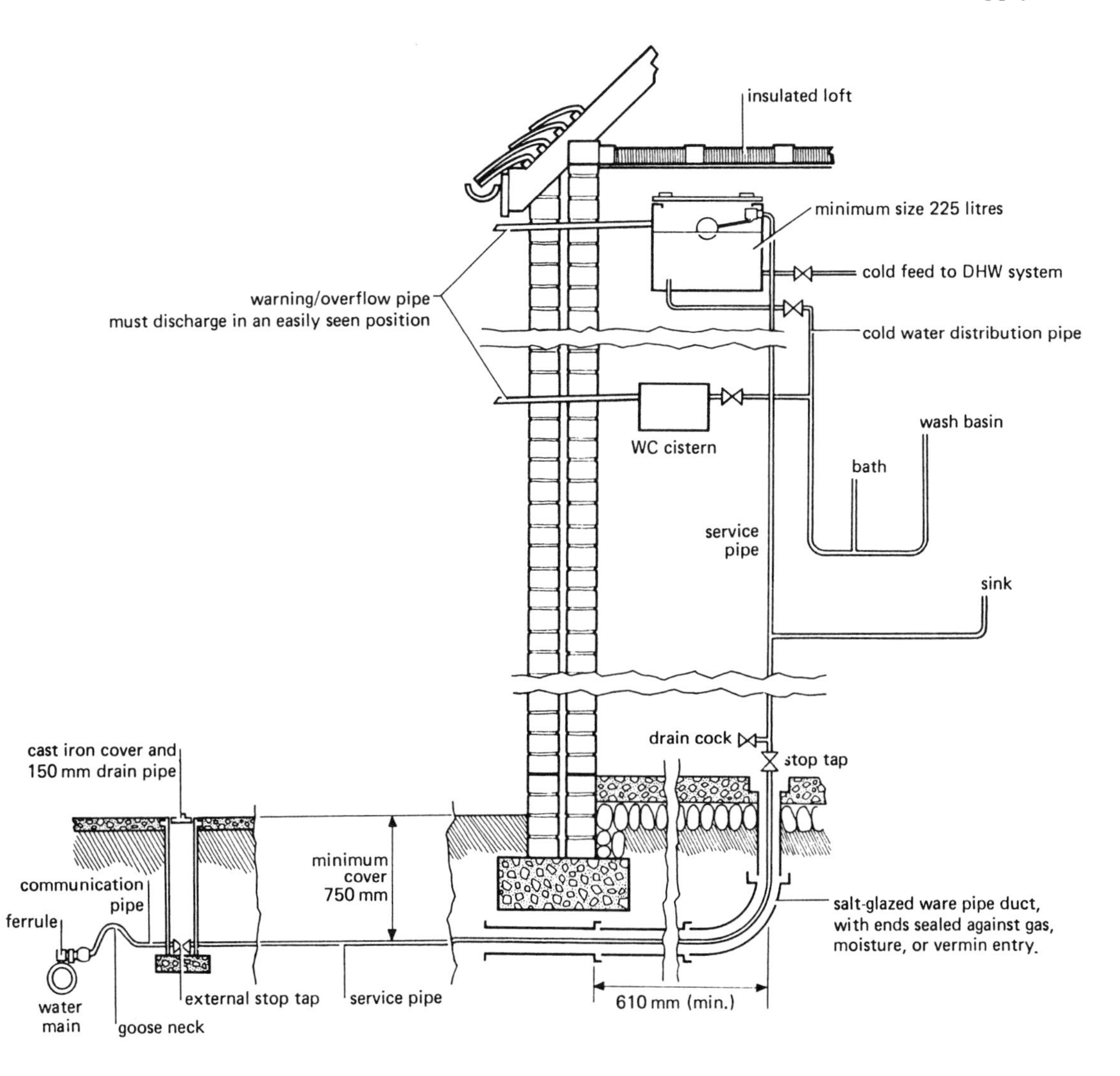

Figure 180 *Indirect system of domestic cold water supply*

Combined storage and feed cistern

This should be large enough to combine both functions and will supply cold water both to appliances and the hot water system. This form of cistern is used where the water undertaking insist that only a drinking supply be taken direct off the company's main.

Capacities and connections

Minimum capacities for cisterns are prescribed by the water undertakings in their bye-laws.

Capacity is defined by the number of litres a cistern will hold, when filled to a level 25 mm, or the internal diameter of the overflow pipe (whichever is the greater) below the invert of the overflow pipe at its connection to the cistern.

Nominal capacity indicates the number of litres a cistern would hold if it were filled to its top edge.

Actual capacity means the number of litres a cistern would hold when filled to its working water line – that is, the level at which a properly adjusted ball valve will shut off and water will cease to feed into the cistern. For cisterns of up to 455 litres capacity the working water level would be about 100 mm below the top edge.

Bye-law 38 of the model water bye-laws 1966 states that where a cistern is used only as a storage cistern, the minimum capacity must be 112 litres. Where used as a storage and feed cistern, its minimum capacity must be 225 litres.

Bye-laws of the Thames Water Authority prescribe a minimum actual capacity of 228 litres for separate feed and storage cisterns, and 364 litres for combined storage and feed cisterns. This is one important example of how the requirements of water undertakings vary, and how important it is, therefore, for you to know the bye-laws of the water undertaking in whose district you work.

If the storage required is more than 4500 litres, it is often advantageous to arrange it in a series of cisterns so interconnected that each cistern can be isolated for cleaning and inspection without interfering with the flow of water to fittings. This can be done by the use of a header pipe of adequate size into which each cistern is connected and from which the distribution pipes branch off. Each branch into and out of the header pipe is provided with a control valve.

Each cistern should have its own float operated valve, and overflow pipe, and a drain tap to facilitate cleaning out. In large storage cisterns, the outlet should be at the end opposite to the inlet to avoid stagnation of the water. If two or more cisterns are coupled together in series without header pipes, the inlet and outlet should be at opposite ends of the series.

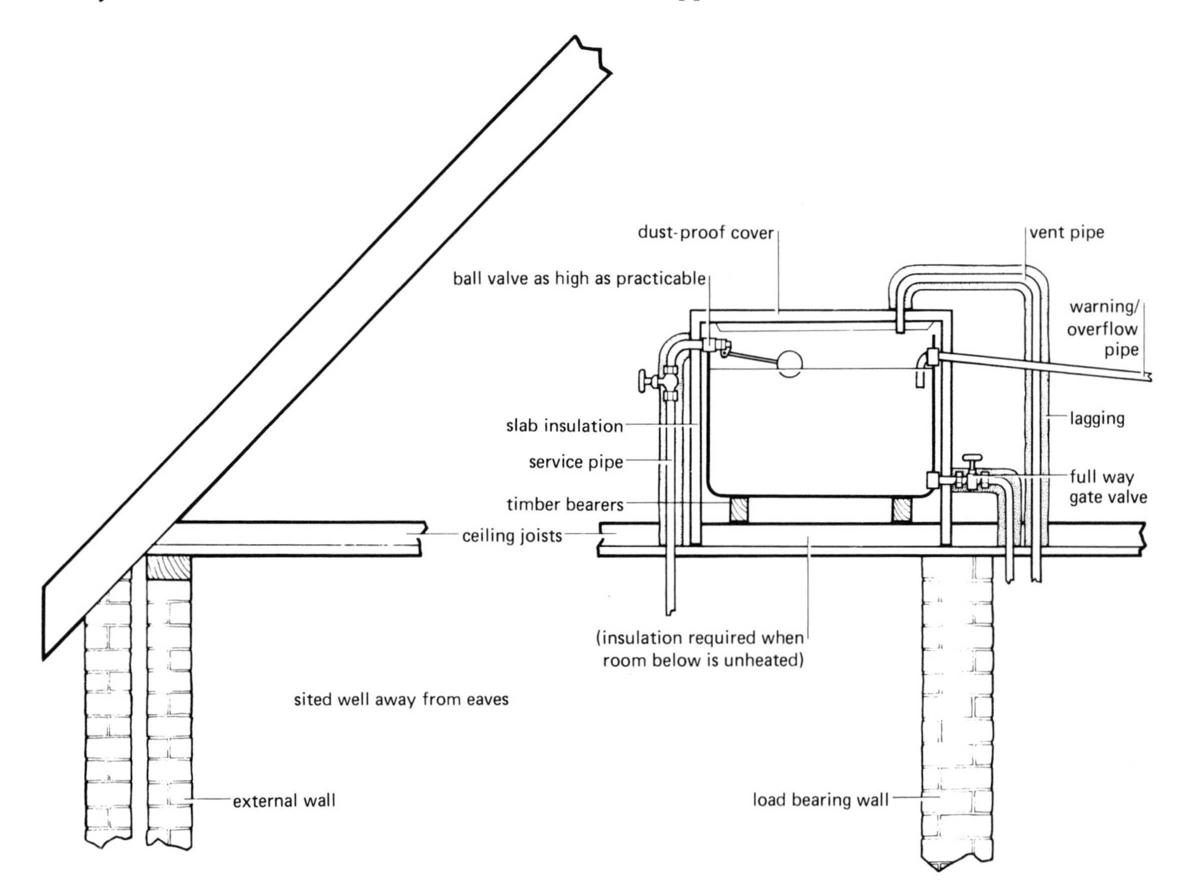

Figure 181 *Siting a storage cistern*

Siting and fixing cisterns

Siting and fixing any cistern will normally be governed by the following factors:

Space available The space available in a loft, roof space, or cistern housing must be adequate to accommodate a cistern.

Head The cistern must be a certain height, or head, above the fittings, in order to provide sufficient water pressure.

Ease of access for maintenance Periodic inspection, cleaning and ball valve adjustment will be necessary.

Temperature The cistern and its contents must not be subjected to extremes of temperature.

Structural tolerances Water is heavy, 1 litre weighs 1 kg. 1 m^3 of water weighs 1000 kg.

Figure 181 shows a good siting of a storage cistern.

It has already been pointed out that the storage cistern, which is fed from the service pipe, must be placed in an elevated position in order to give sufficient pressure at the draw-off points to meet the demand rates. If, however, the storage cistern is placed on the same storey as the draw-offs, larger sized pipes must be installed to compensate for the lack of height.

The distribution pipe, conveying the water from cistern to the draw-off taps etc., should be controlled by a control valve fitted close to the storage vessel. To minimize loss by friction, a full way gate valve is recommended.

Connections for these distribution pipes should be located in the storage cistern in such a way that silt cannot be drawn into pipes. This means that the outlet should be taken from the side, and located at least 25 mm above the bottom of the cistern (see Figure 182). In cases where the outlet is taken from the bottom of the vessel, a suitable connector, providing a 'stand up' above the bottom of the cistern of at least 25 mm, should be used (see Figure 183).

All cisterns used for storing water should be provided with an over-flow pipe. This should be at least one size larger than the incoming supply pipe, and situated 25 mm below it (see Figure 184).

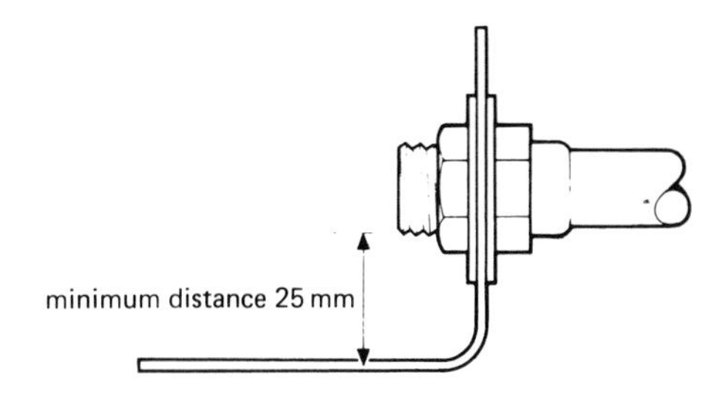

Figure 182 *Outlet from side of storage vessel*

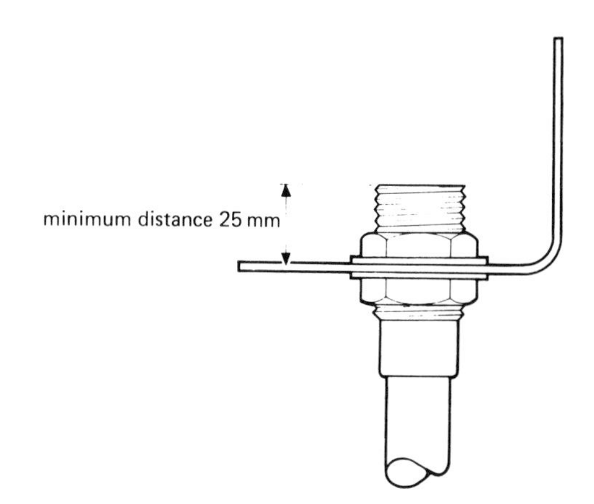

Figure 183 *Outlet from bottom of storage vessel*

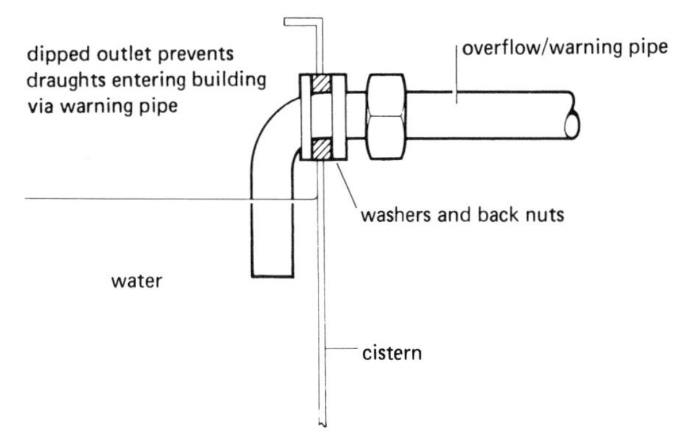

Figure 184 *Dipped outlet*

A dust-tight but not air-tight cover should be placed over the cistern to prevent dust, animals and insects from gaining access and bringing about a risk of contamination.

Materials used in the manufacture of feed and storage cisterns

Galvanized mild steel

Cisterns manufactured in galvanized mild steel to BS 417 have been widely used for many years (see Figure 185). They are obtainable in many thicknesses and sizes. They are formed from black mild steel sheet, and then dipped into baths of acid to remove grease and scale. After this they are dipped into a bath of molten zinc and so coated with a corrosion resistant skin of zinc.

This protective hot-dip galvanizing treatment has been developed from experiments carried out by Galvani, a scientist after whom the process is named.

The cisterns are available in either riveted or welded construction. Their top edges are stiffened with an angled curb which is formed during manufacture and the open top corners are sometimes braced with corner plates which are riveted or welded depending upon the construction.

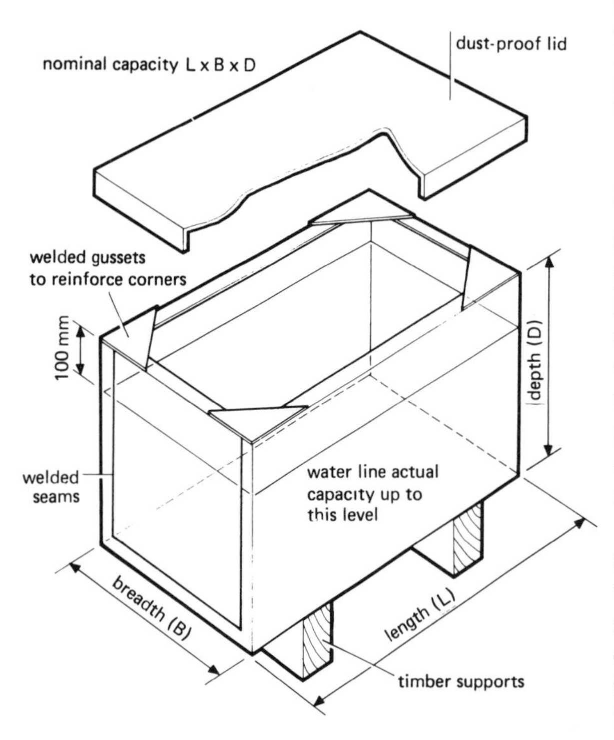

Figure 185 *Cold water cistern*

They are self supporting and can be situated directly on timber bearers. The holes are cut by means of a tank cutter or expanding bit and the jointing done by means of a grummit and jointing paste and back nut. Nylon washers can be used as an alternative. It is good practice to ensure that no cuttings are left in the tank and that the inside is painted with an approved non-toxic bitumastic paint prior to commissioning.

Asbestos cement

Cisterns manufactured from asbestos cement to BS 2777 are completely immune to electrolytic action and the harmful effects of soft and hard waters.

The cisterns are available in a wide range of sizes. The material itself is a poor conductor of heat, and therefore gives some protection against frost damage. This material is virtually everlasting and is not subject to corrosion. It is also self supporting.

The installation of these cisterns is relatively easy. The holes are cut by means of a ring saw and the connection made by means of approved jointing washers and back nuts (hemp grummit and jointing paste or rubber washers).

Plastics materials

Various plastics are now being extensively used for cold water storage cisterns: polyethylene, polyproprylene and polyvinyl chloride (PVC), to mention just a few (see Figure 186). They are strong and resistant to corrosion of all types, virtually everlasting, very hygienic and light in weight. They do not cause, nor are they subject to, electrolytic corrosion, and they have low rates of thermal conductivity – so the stored water retains its heat longer in cold weather.

These cisterns are quieter in filling than a metal cistern, and this is a useful advantage, particularly when the cistern is sited near bedrooms. Furthermore, they are easy to squeeze through small openings, which is an advantage when the cistern is to be placed in an attic or loft (see Figure 187).

Plastic cisterns are manufactured square, rectangular, or circular in shape and are black to

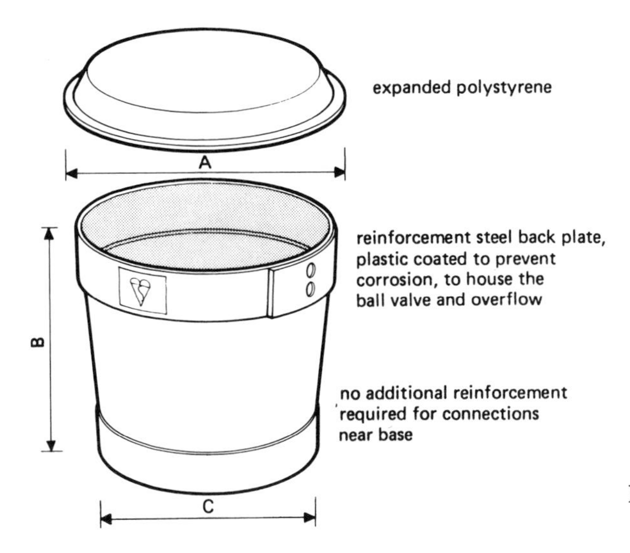

Figure 186 *Low density polyethylene cistern*

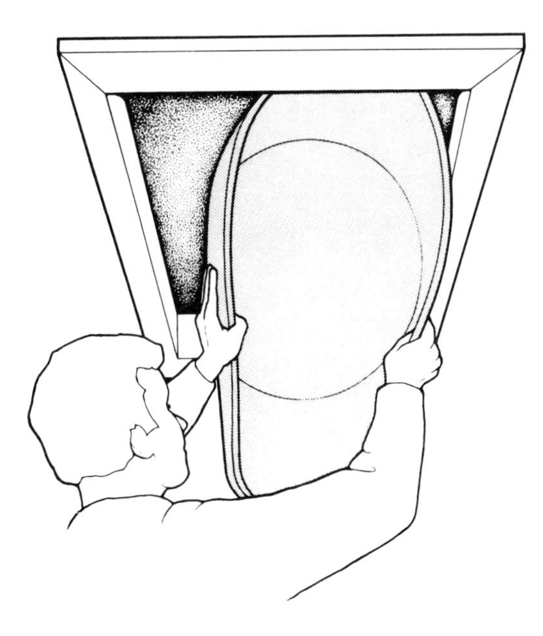

Figure 187 *Flexibility of material enables the cistern to be passed through small openings*

prevent algae growth. They must be fully supported by being placed on a solid decking (see Figure 188).

Holes to enable pipe connections to be made

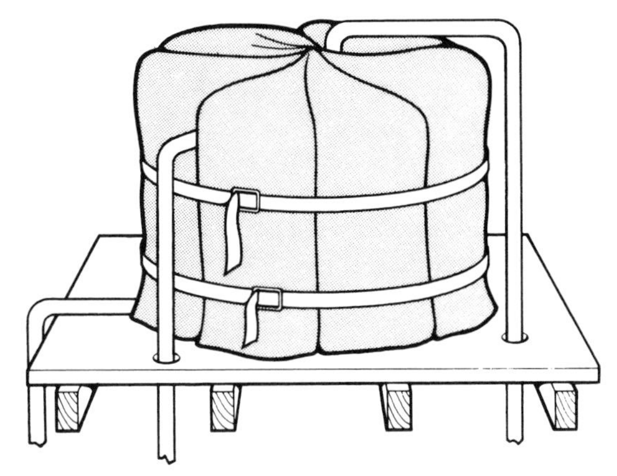

Figure 188 *Base must be fully supported*

are cut by circular saw cutters. The jointing is by means of plastic washers but no oil-based paste of any description must be used, because this softens the material and causes it to break down.

Plastics materials are comparatively soft, so care must be taken in handling and fixing. Sharp instruments and tools can easily cut or puncture the cistern. Naked flames and excessive heat will also damage the material.

Table 16 gives the recommended sizes of cisterns of different materials.

Table 16 *Recommended size of cistern for domestic dwellings*

Material	*Actual capacity*	*Dimensions mm*	*Thickness*
Galvanized mild steel	112 l 225 l	675 × 500 × 500 900 × 600 × 575	16–18 gauge 14–16 gauge
Asbestos cement	112 l 270 l	625 × 475 × 562 775 × 600 × 637	
Plastic	112 l 225 l	650 × 500 × 587 800 × 600 × 737	3–5 mm 3–5 mm

Joints for pipework

The following materials are used to make pipes to convey water in the domestic system.

Lead
This material is no longer used for new installations, but because of its use in the past and the fact that there are still a very large number of houses with lead pipes the modern plumber must be conversant with the jointing of lead pipes. Lead pipes can be joined by means of wiped solder joints, or by lead welding.

Copper
This is by far the most popular metal used today for the manufacture of pipes for domestic use. It has many advantages:
1 It is neat in appearance;
2 It is strong;
3 It is easy to join;
4 It is cheap to install.

Copper can be jointed with compression fittings, capillary fittings, silver soldering, brazing and bronze welding.

Polythene
This material is not very widely used, because it has been superseded by some of the modern plastics such as polyvinyl chloride. It is interesting to note that the methods of jointing are similar to those for copper tube – both manipulative and non-manipulative fittings are used. Fusion welding can also be used.

Steel pipes
Steel pipes are not very popular for domestic work but are very extensively used for industrial work. They form an important part of the plumber's work.

Steel pipes are jointed by the threading (screwing) method and the use of purpose made fittings.

Stainless steel
This extremely strong and attractive looking metal has been able to claim only a very limited share of the market. The method of jointing is by means of compression and capillary fittings as described in the jointing of copper tube.

Jointing lead pipes
Lead pipes are jointed by the skilled manipulation of molten solder into a shaped joint. This is called joint wiping and the necessary skill only comes from practice. The usual method is the 'blow lamp' (gas torch) method. Welding can also be used.

The tools required to prepare and wipe joints on lead pipes are:
Lead saw
Mandril
Bentpin
Pipe opener (auger)
Rasp
Black brush and plumbers' black
Shavehook
Scribing plate and compasses
Wiping cloth
Tallow
Pipe clamps
Dresser
Solder

Underhand wiped joint
Cut the pipes to the required length and square the ends.
Rasp the spigot end for entry into socket, open out socket end with a turnpin or auger and rasp off surplus edge (see Figure 189).
Wire brush socket and spigot ends for 150 mm to remove surface oxides.
Dust over brushed lengths with chalk to degrease and paint with plumbers' black and dry off (see Figure 190).
Measure and clean ends of the pipe with a shavehook to a bright finish, then apply tallow coating (flux) to the shaved ends (see Figure 191).
Fit the spigot end into socket and secure in position.
Using the blow lamp method, heat the pipe and rub the stick of plumbers' solder on to the shaved surfaces to tin them.
Add further heat and solder until the required amount has been deposited.

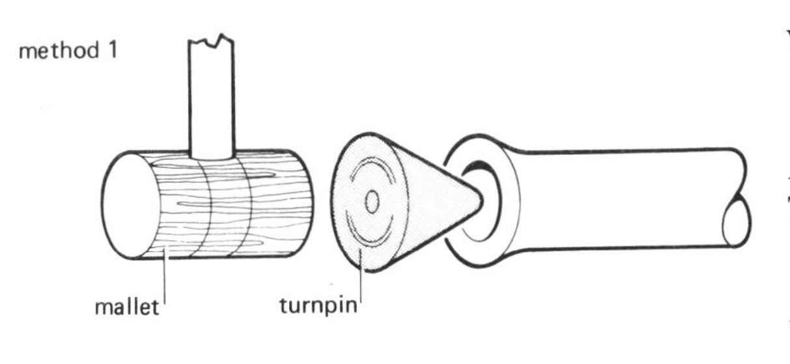

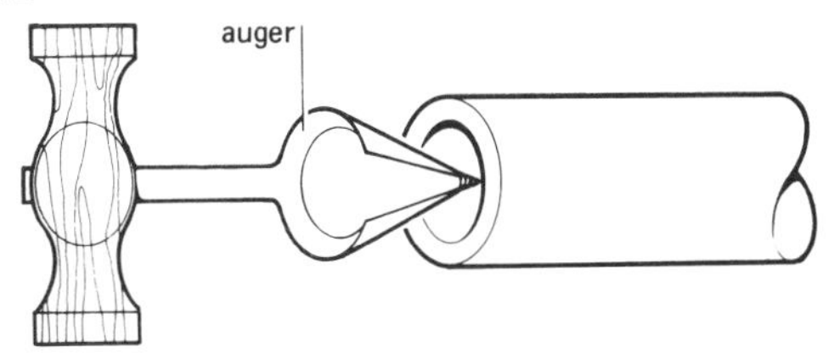

Figure 189 *Method of preparing underhand joint*

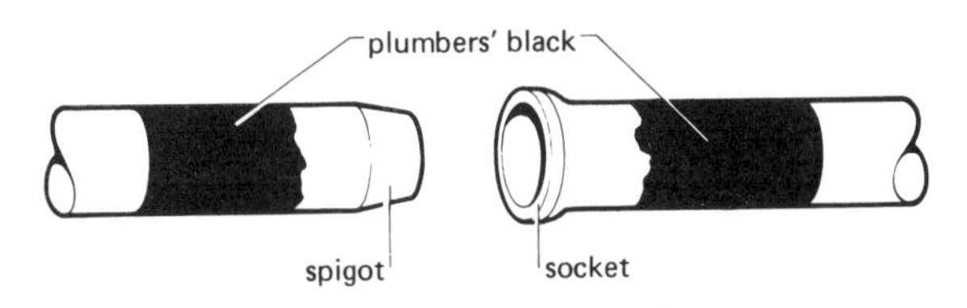

Figure 190 *Preparing ends*

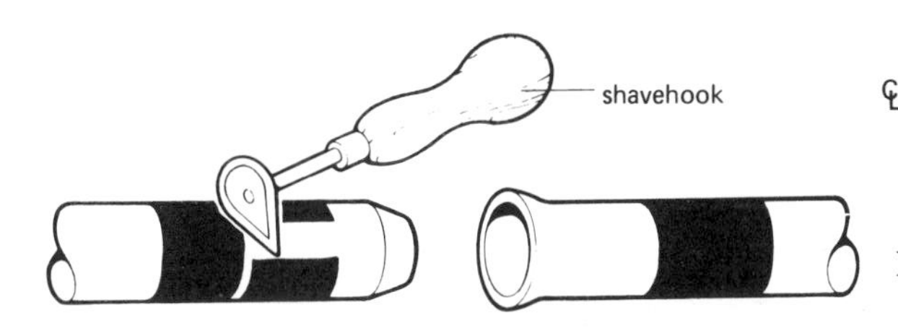

Figure 191 *Removing black*

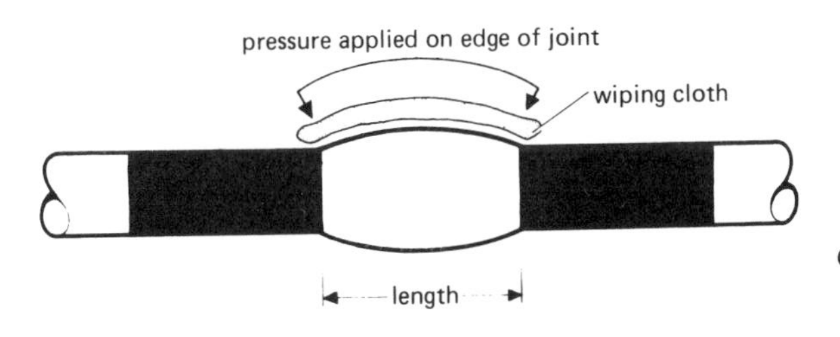

Figure 192 *Wiping*

With a wiping cloth (moleskin) and applying further heat, roughly shape the solder joint (see Figure 192).

Apply further heat, to the edges.

The cloth is rotated round the pipe, pressure is applied to the edges, and a small surplus of solder is carried around the joint.

When the joint is free from cavitations and overwipes, the surplus is dragged off with a piece of cloth and the solder allowed to solidify and cool before unfixing the pipe.

Branch joint

Cut the pipe to the required length.

Straighten the pipe and square the end.

Rasp one end to form a spigot.

Using the pipe opener make a hole in the main pipe (see Figure 193).

Using bent pin and hammer work up the lead to form a socket to receive the spigot (branch) (see Figure 194).

Paint pipes with plumbers' black.

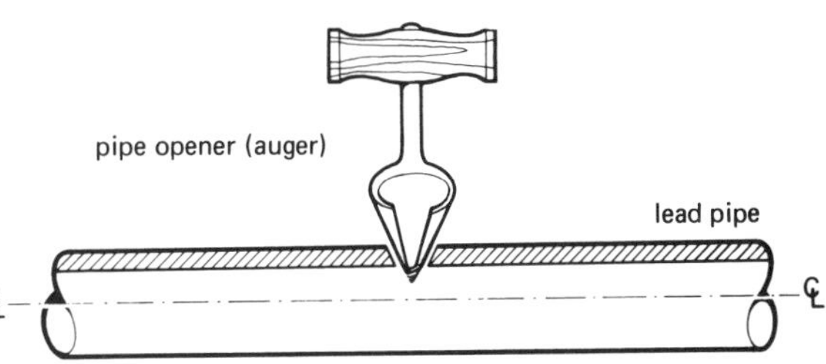

Figure 193 *Preparing branch joint on lead pipe: stage 1*

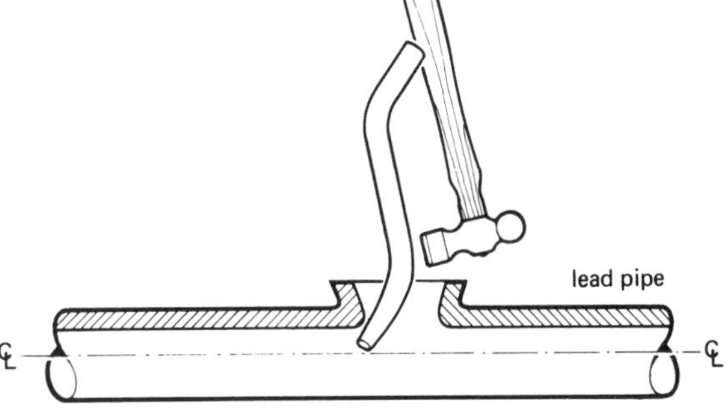

Figure 194 *Preparing branch joint on lead pipe: stage 2*

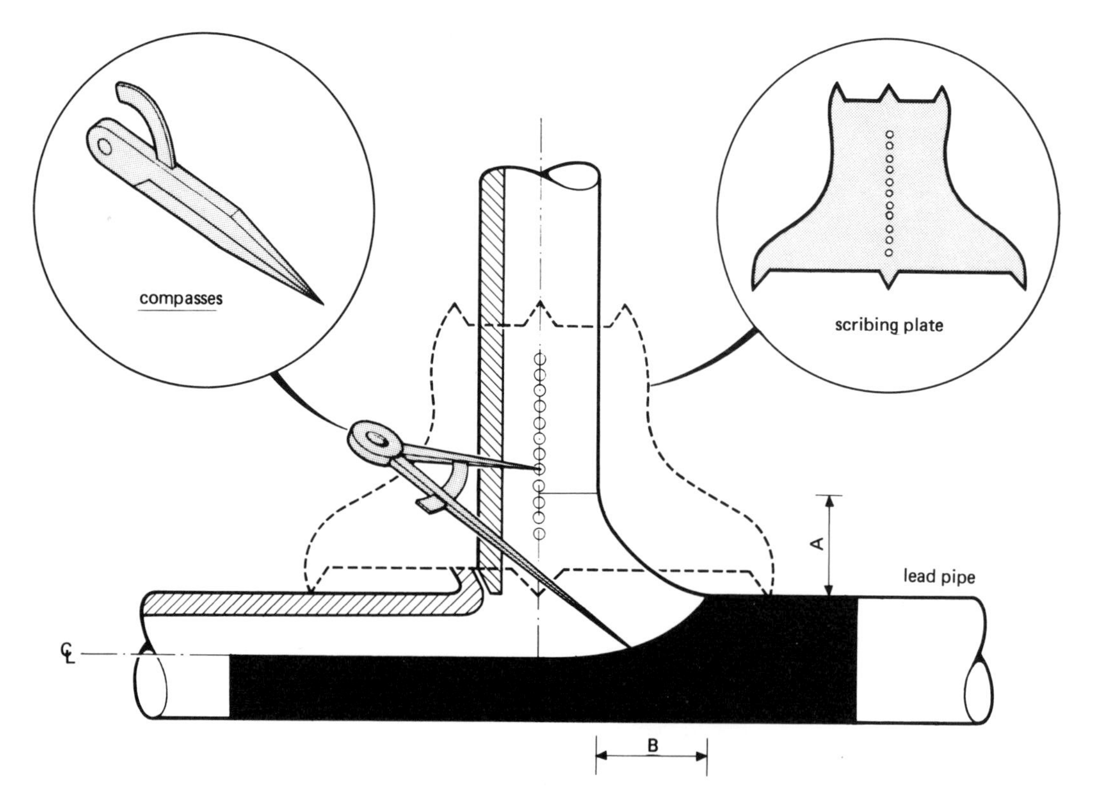

Figure 195 *Preparing branch joint on lead pipe: stage 3*

Scribe length and shape of joint with dividers and scribing plate (see Figure 195).
Thoroughly clean the pipe with the shavehook.
Immediately apply flux (tallow).
Fit the pipes together and hold firm with clamps.
Apply solder and wipe joint to required shape and finish.

Figures 196–200 show a selection of joints used on lead pipework in domestic situations.

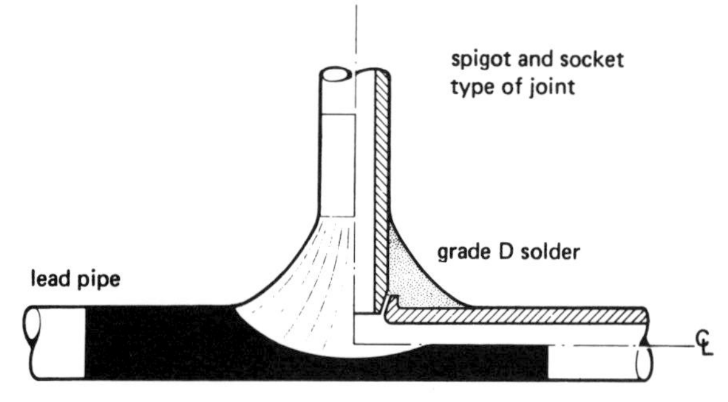

Figure 197 *Wiped branch joint (lead to lead)*

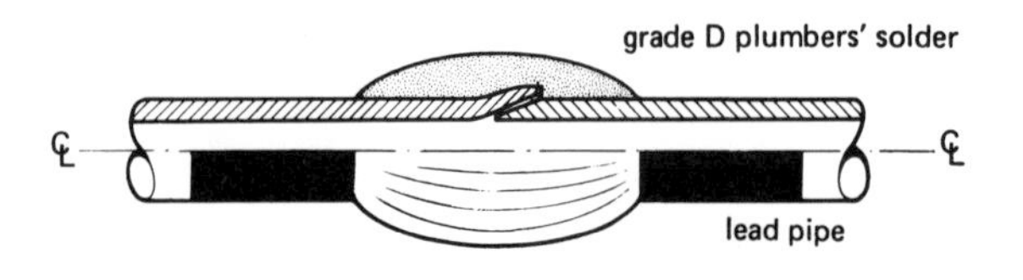

Figure 196 *Wiped underhand joint (lead to lead)*

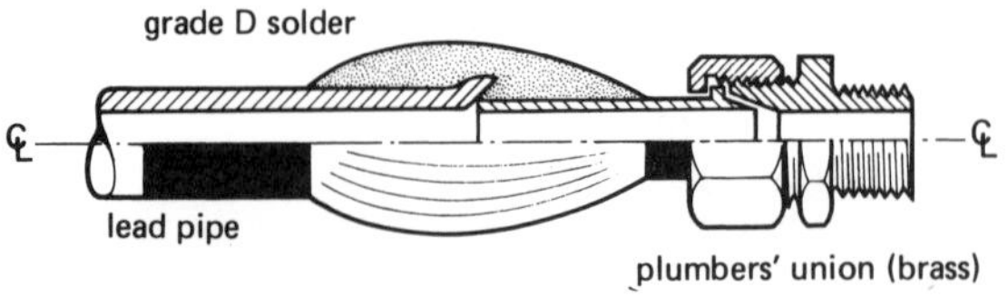

Figure 198 *Wiped underhand joint (lead to iron or copper)*

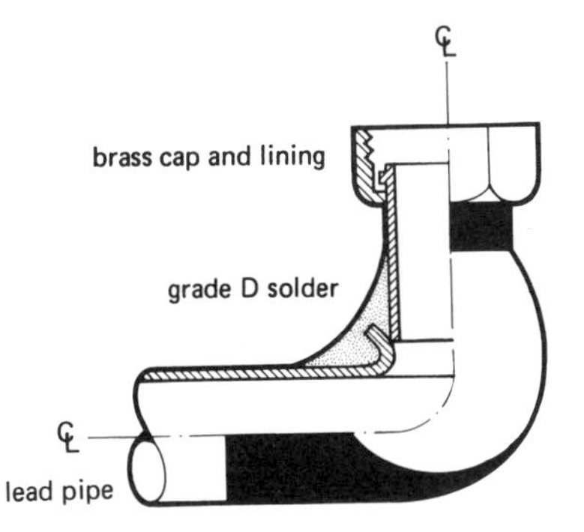

Figure 199 *Wiped knuckle joint (lead to iron)*

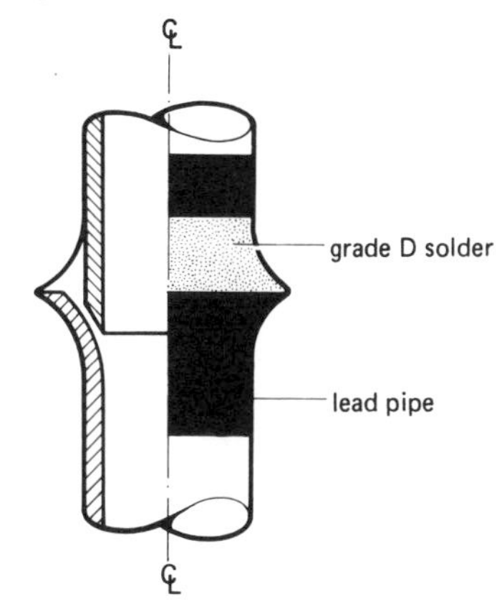

Figure 200 *Wiped taft joint (lead to lead)*

Jointing of copper tube

Copper tube can be jointed in several ways. The most common methods are:

1 Compression fittings
2 Capillary fittings
3 Silver solder
4 Brazing
5 Bronze welding
6 Copper welding

Compression joints

These are divided into two groups:
Manipulative fittings
Non-manipulative fittings

Manipulative fittings

These require the end of the tube to be cut square and to length. The nut is then slipped over the tube end and the tube opened to allow for a brass olive ring to be inserted or a ridge formed with a rolling tool. The nut is then tightened and the copper tube end trapped and squeezed between fitting or olive and nut. This manipulation of the tube end ensures the nut fitting will not pull off. Many water authorities insist on this type of fitting being used on underground services. This type of joint is unaffected by vibration and withstands tensile and other stresses.

There are several types of manipulative fittings on the market. Their common factor is that some form of work (manipulation) is performed on the end of the tube before it is assembled with the fitting. The following examples show two typical fittings.

Kingley fittings In this case a special tool is inserted in the end of the tube which, by means of a small ball bearing, forms a bead on the tube (see Figures 201 and 202).

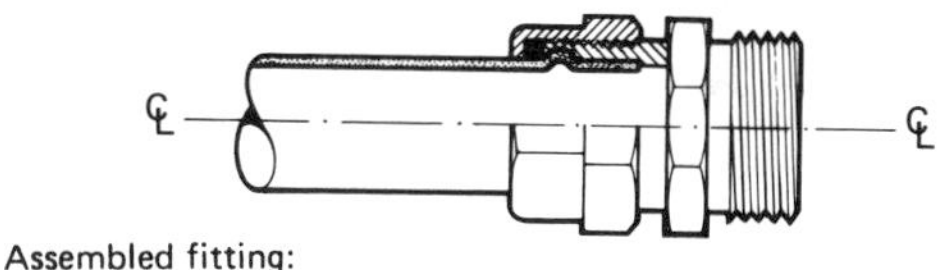

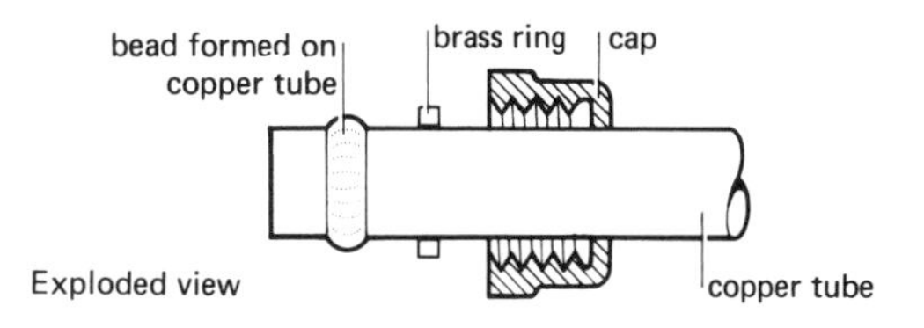

Figure 201 *Kingley joint: manipulative fitting*

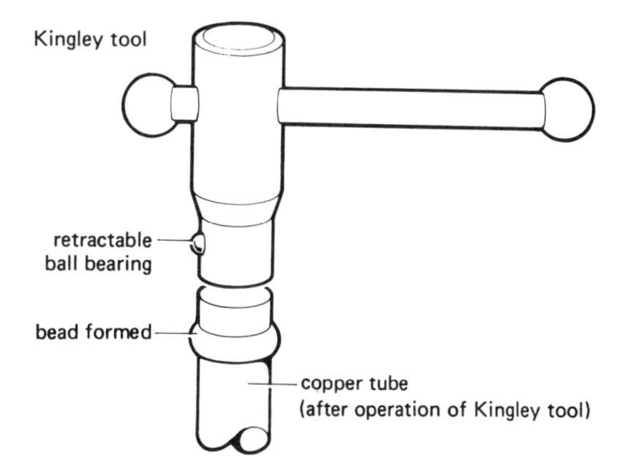

Figure 202 *Copper tube with Kingley tool*

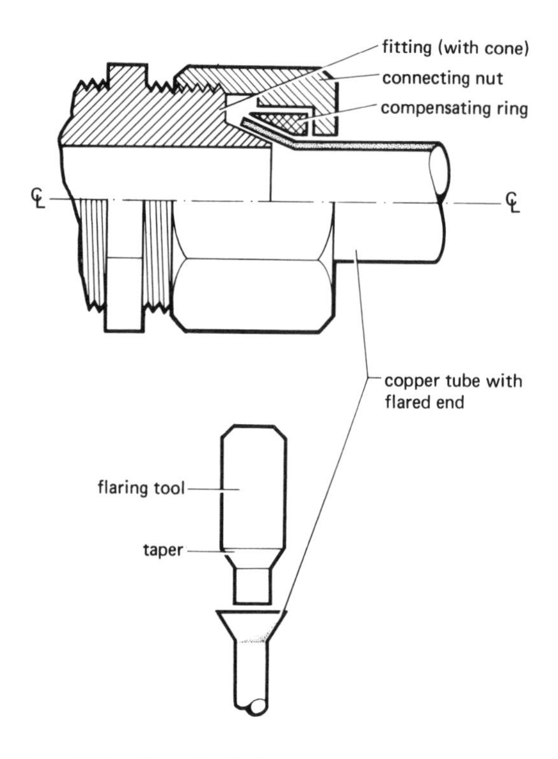

Figure 203 *Kuterlite fitting*

Kuterlite fittings In this case the end of the tube is flared out so that it can be compressed between the inner cone of the fitting and the shaped compensating ring (see Figure 203).

Method of jointing with manipulative fitting

Ensure the tube is cut square and true.

Remove any burr from the pipe.

Place nut and compensating ring washer on to the pipe before any work is done in the preparation of the flared end of the tube.

Using the appropriate tool, i.e. Kingley tool or flaring tool, form the end of the tube as required.

Insert tube and assemble fitting. Slide ring or washer and nut into place.

Tighten nut by hand, ensure that the nut runs freely and is not cross threaded. Then with the aid of a correctly fitting spanner tighten nut half to one revolution.

Note: Do not over-tighten compression joints as this can cause distortion and subsequent leaks.

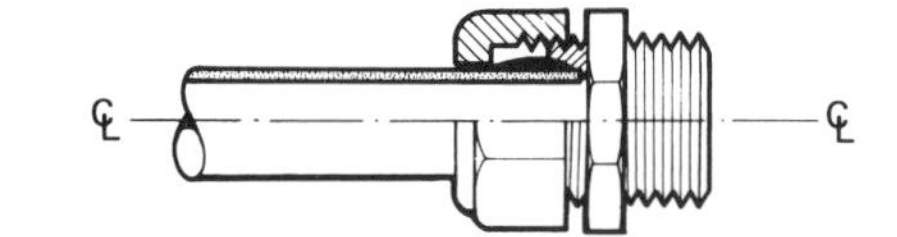

Assembled fitting: half section, half elevation

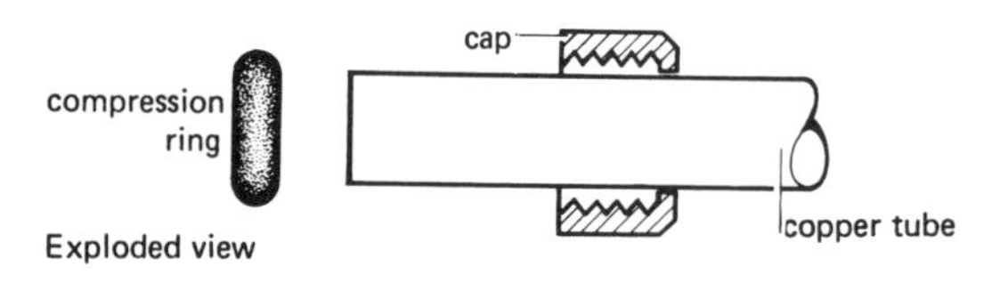

Exploded view

Figure 204 *Compression joint: non-manipulative*

Non-manipulative fittings

These require only the end of the tube to be cut square and to length. The nut is then slipped over the tube end followed by a soft copper wedge shaped ring or brass wedding shaped ring. The tube is then inserted into the fitting socket, the nut tightened and the ring compressed between the inside of the fitting and the outside of the copper pipe (see Figure 204).

This type of fitting is very popular due to its ease of application on both half hard and thin wall copper tube. They can also be used in certain cases on soft copper tubes. It is important to remember that this type of fitting should never be used on copper tube buried in the ground.

Method of jointing with manipulative fittings

Cut the ends of the tube square (by means of fine-toothed hacksaw).

Remove any burr with a fine file.

Place nut and compression ring on pipe and insert pipe in fitting up to the stop (see Figure 205).

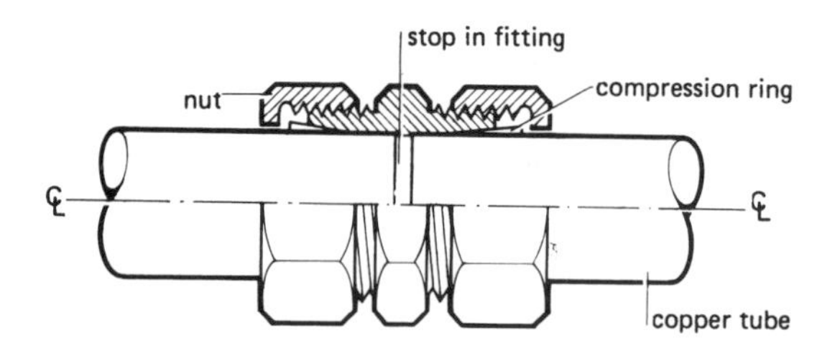

Figure 205 *Compression joint: non-manipulative*

Tighten nut hand-tight. Ensure good fit and then with correctly fitting spanner tighten half to one revolution.

Note: Do not over-tighten; experience will tell you the correct amount.

Capillary joints

Capillary type fittings are also divided into two groups:

Fittings with integral solder rings.

End feed fittings.

Integral solder rings

This type of fitting can be used for both above and below ground work. It can be used extensively in construction, gas, refrigeration, marine and engineering pipelines conveying air, water and oil. It is also without doubt an extremely attractive looking fitting and is also simple to make. The above points make this type of fitting the most popular, and the best known of all capillary fittings. Figure 206 shows an example of this type of fitting.

The joint relies on the phenomenon of capillary attraction in the making of the joint. Each fitting contains the correct amount of solder as an integral part. When heat is applied the solder turns from a solid to a liquid and is then drawn by capillarity around the whole of the joint in the tight space between the outside of the pipe and the inside of the fitting.

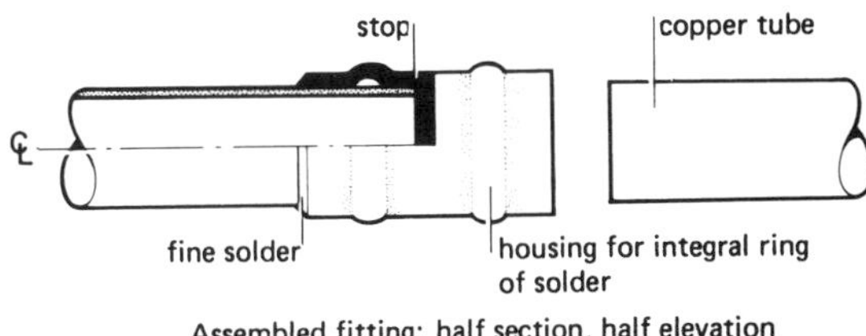

Figure 206 *Capillary joint*

Method of jointing with integral solder ring

Cut the ends of the tube square.

Remove all burrs.

Thoroughly clean outside of tube and the inside of the fitting with steel wool or fine sandpaper.

Lightly coat both cleaned surfaces with a suitable flux.

Insert the tube into the fitting until it reaches the stop.

Apply heat (torch) to the joint until sufficient heat is generated to melt the integral ring of solder. Continue to apply heat until the solder appears at the mouth of the fitting and forms a complete ring.

Allow to cool without disturbance.

Remove flux residue.

Owing to the fact that copper is an excellent conductor, the heat applied at one point is quickly transferred round the whole of the joint, particularly for the smaller size of pipe. It is still advisable to apply the heat wherever possible to all parts of the joint. In some cases where this is not possible, the use of a heat resistant mat around the back of the joint to reflect the heat is recommended. It is worth noting that under heating results in more joint failures than over heating.

End feed fittings

This type of fitting is used in exactly the same circumstances as for the integral solder ring fittings and is identical in all ways except that the solder has to be added to the end of the fitting. The jointing surfaces inside the fitting and outside of the pipe are first cleaned, fluxed and assembled. Heat is applied until the temperature is high enough to melt the solder which is then drawn by capillary attraction into and around the whole joint. All surplus flux residue should be removed, otherwise it will continue to act on the surface of the copper and so leave staining and corrosion marks.

Method of jointing with end feed fittings

Cut the ends of the tube square.

Remove all burrs.

Thoroughly clean outside of the tube and inside of the fitting with steel wool or sandpaper.

Lightly coat both cleaned surfaces with a suitable flux.

Assemble tube and fitting (tube to touch the stop).

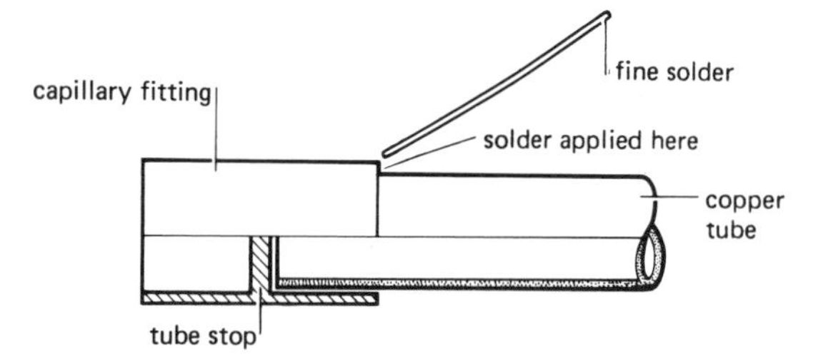

Figure 207 *Capillary soldered joint*

Apply heat until the temperature is high enough to melt a rod of fine solder. Add sufficient solder to fill the joint space (see Figure 207).
Allow to cool without disturbance.
Wash off flux residue.

The Acorn fitting for copper tube
This is a system of plastics push-fit pressure pipe fittings to be used in conjunction with copper pipe to BS 2871 (see Figure 208). There are two sizes – to accept 15 mm and 22 mm copper pipe – and a limited range of fittings available.

They are a technological breakthrough, enabling copper capillary soldered fittings and mechanical compression fittings to be replaced with an all-plastics push-fit system. Jointing can be carried out much more quickly than with conventional methods and the fittings are suitable for domestic hot and cold water supply systems, and in domestic space heating systems. They are not at present recommended for gas distribution work.

The method of jointing is by manually push-fitting the copper pipe into the socket. The assembly force required, which takes the pipe end beyond a resistance within a fitting, to be grabbed in place is well within the physical capabilities of the average installer, and requires no special technique or tools (see Figure 209).

The fittings have a polybutylene body, and incorporated within this is a special stainless steel grab ring. The body also houses a washer and rubber sealing ring. All are retained within the body by a screwed on end cap. The fittings are coloured *brown*.

Polybutylene is an advanced engineering plastics formulated specifically for resistance to high temperatures and pressures. All the materials are

Figure 208 *Exploded view of Acorn fitting*

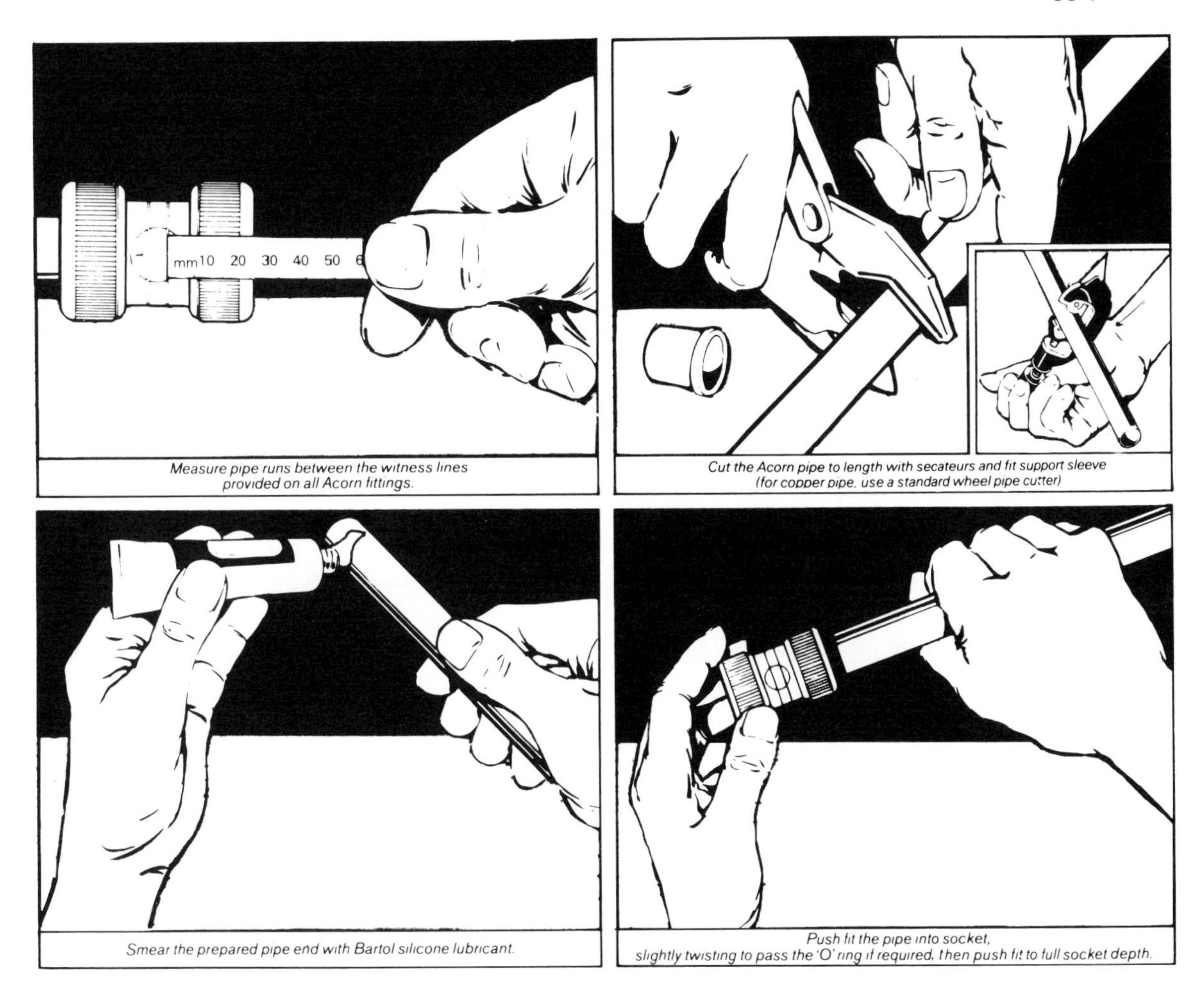

Measure pipe runs between the witness lines provided on all Acorn fittings.

Cut the Acorn pipe to length with secateurs and fit support sleeve (for copper pipe, use a standard wheel pipe cutter)

Smear the prepared pipe end with Bartol silicone lubricant.

Push fit the pipe into socket, slightly twisting to pass the 'O' ring if required, then push fit to full socket depth.

Figure 209 *Method of connecting Acorn fitting*

special in that they have unique properties specific to Acorn and all are non-toxic and accepted by the National Water Council for potable (drinkable) water distribution. The fittings are totally resistant to dezincification and have a high chemical resistance.

Because domestic central heating systems are unpressurized and vented, the same conditions of operation as apply to indirect hot water supply can be used. The fittings have been designed with an integral expansion allowance. They are also flexible giving more allowance for deflection.

The grab ring is stainless steel, specially heat tempered, and is also resistant to corrosion by aggressive water, particularly chloride attack. The grab rings must be renewed if the fitting is to be reused.

Where joints are required to metal threads, especially tapered iron threads, the material strength required is very high. Brass is far more suitable for this type of connection.

Table 17 gives the operating conditions within which Acorn can be used. Acorn fittings are designed to operate for short periods of time above these limits in the event of any control system malfunction.

They are unaffected by proprietary corrosion inhibitors. Anti-freeze (ethylene glycol based)

Table 17 *Operating conditions suitable for Acorn fittings*

Application	*Maximum for continuous use*	
	Temperature	*Pressure*
Cold water supply	20 °C	9.8 bar
Hot water distribution	65 °C	1.5 bar
Central heating	82 °C	1.5 bar

should not be used. Acorn fittings in exposed positions should be insulated as a protection from frost.

The fittings have been accepted by the National Water Council for use in hot water and cold water services and agrément certification has been granted.

To sum up, these fittings have the following beneficial properties:

(a) They are resistant to corrosion
(b) The are non-toxic
(c) They can be easily repaired, replaced and modified
(d) They have no scrap value
(e) They are tough and flexible
(f) No special tools or fitments are required
(g) No flux or torches are required
(h) They are suitable for hot and cold water systems
(i) They have a simple push-in-fit joint for pressure work.

Jointing polythene pipes

Polythene pipes may be joined by compression fittings or polyfusion.

Compression fittings

These require only the end of the tube to be cut square. The nut is passed over the tube end followed by copper wedge ring or brass wedding shaped ring. Into the open end of the tube a copper liner is inserted to strengthen the tube walls and then the tube end is inserted into the socket of fitting. The nut is then tightened and the ring compressed into the polythene tube wall previously strenghtened by the copper insert (see Figure 210). Liquid jointing compound must not be used.

Polyfusion

In this method the end of the tube is cut square and to the required length. The end of the tube and the socket of the polythene fitting is placed inside a heater, until the polythene surfaces melt. The heater is then removed and the tube end (spigot) is pushed into the fitting (socket) and held in place while the melted surfaces fuse (weld) together, thus making the joint (see Figure 211).

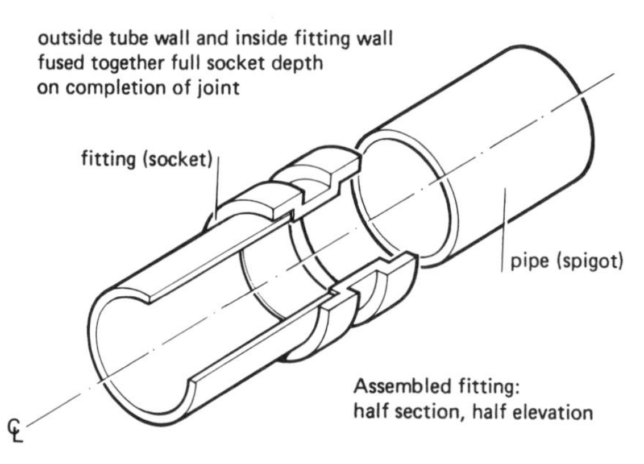

Figure 211 *Fusion fitting*

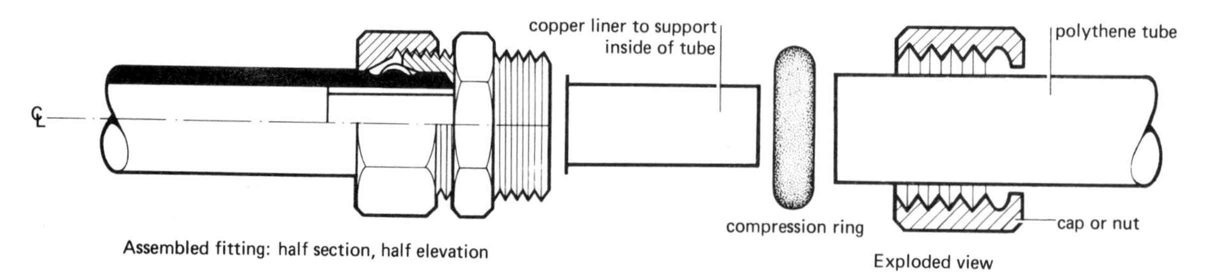

Figure 210 *Non-manipulative compression fitting for polythene pipe*

Manipulative jointing
The ends of the tube are cut square. The nut is then slipped on to the pipe followed by a brass washer. The pipe end is then placed in a special tool which heats and forms a jointing flange which exactly marries up to the inside surface of the fitting. The newly formed flange is now inserted into the socket of the fitting. The brass ring and the nut are then secured into place, making a water-tight joint.

Fusion welding
This term is used for the jointing of plastic by means of a special solution which has the property of melting the surface of the plastic. The tube and fitting must be thoroughly cleaned with spirit. The outside of the tube and the inside of the fitting is then quickly coated with the solution and assembled. The joint should be held together for a short time until the solution hardens. It takes approximately twenty-four hours for full maturity to take place. The two surfaces now form one homogeneous mass.

Jointing steel pipes
It is usual for low carbon steel pipes to be jointed by fittings such as elbows, bends, tees, couplings, flanges etc. All these fittings are threaded with the British Standard Pipe Thread (BSPT) BS 21, thus making all fittings interchangeable irrespective of manufacturer. Steel tubing is usually supplied with a BSPT at each end plus a socket at one end (pipes without threads are also available). The tube is cut square and to the required length while held in a pipe vice. A thread is then cut on the end of the pipe with a tool called stocks and dies (see Figure 212). In one type there are four cutting dies housed in an adjustable stock head. This tool is set on the pipe end and rotated around the pipe. Long handles attached to the stocks head provide the necessary leverage. A lubricant such as oil or grease must be applied to prevent possible thread tearing.

Method of jointing steel pipes
Cut the end of the tube square.
Remove internal burr with the reamer, external burr with a file.

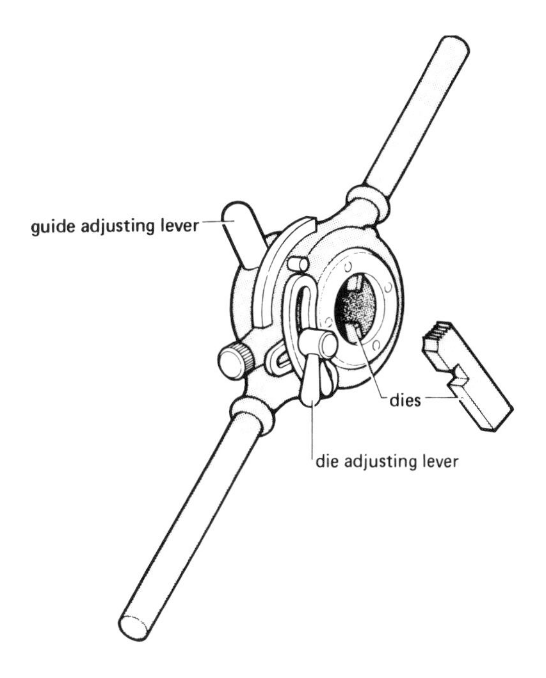

Figure 212 *Hand stocks and dies*

Set stocks and dies to the correct pipe size (for large size pipes it may be necessary to make the thread in two or more cuttings, the required adjustments being made after each cut).
Cut thread to correct length.
Remove the stocks by releasing the dies with the release lever.
Try a fitting to check that the thread is correctly cut.
Wrap the threads with a special plastic tape.
The fitting is now screwed on to the pipe, trapping the tape between the threads and so making a water-tight joint.

Note: Care must be taken not to over-tighten the joint. Fittings can become distorted and can even split if over-tightened, causing leakage.

Taps and valves

The terms 'taps', 'valves' and 'cocks' are used to describe fittings whose purpose is to control the flow of water at draw-off or outlet points or at intermediate positions in the system. Such fittings are usually made of brass or gunmetal with spindles or stems of phosphor bronze or manganese bronze. It is important to know and understand the differences in construction between the different types and to appreciate how these differences affect their use.

Taps

Screw down bib tap (Figure 213)

This is the best known of all water fittings and is in general use as a draw-off tap. Many types have a crutch head handle and an exposed body, but for use indoors a capstan head marked hot or cold, together with an easy-clean cover over the head of the tap gives a much better appearance. A loose jumper may be used if the mains pressure is sufficient to lift the jumper from the seating when the spindle is unscrewed by the operation of the crutch head. In situations where mains pressure is not available as in cases of distribution supplies drawn from a storage cistern, the jumper must be fixed (pinned), i.e. secured to the spindle so that it is lifted off its seating as the tap is opened. Although the jumper is fixed in the worm (threaded spindle) it must be done in such a way to allow the jumper to rotate. This prevents the washer being worn away quickly by the turning action of the spindle. Washers are usually made of rubber, leather or nylon for cold water and fibre for hot water. There is a composition washer now available suitable for both hot and cold water. Figure 214 shows an exploded view of a bib tap.

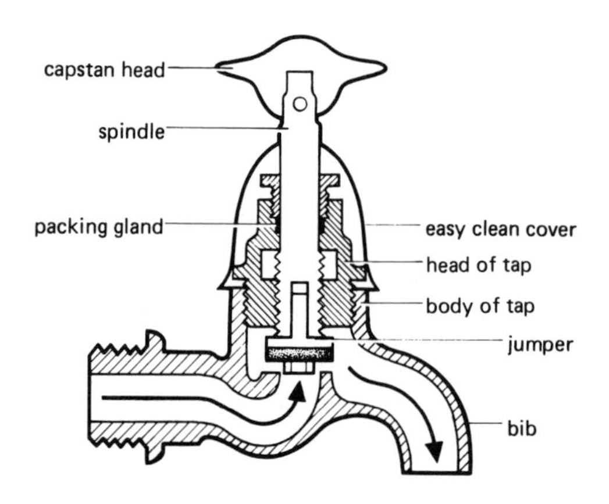

Figure 213 *Bib tap*

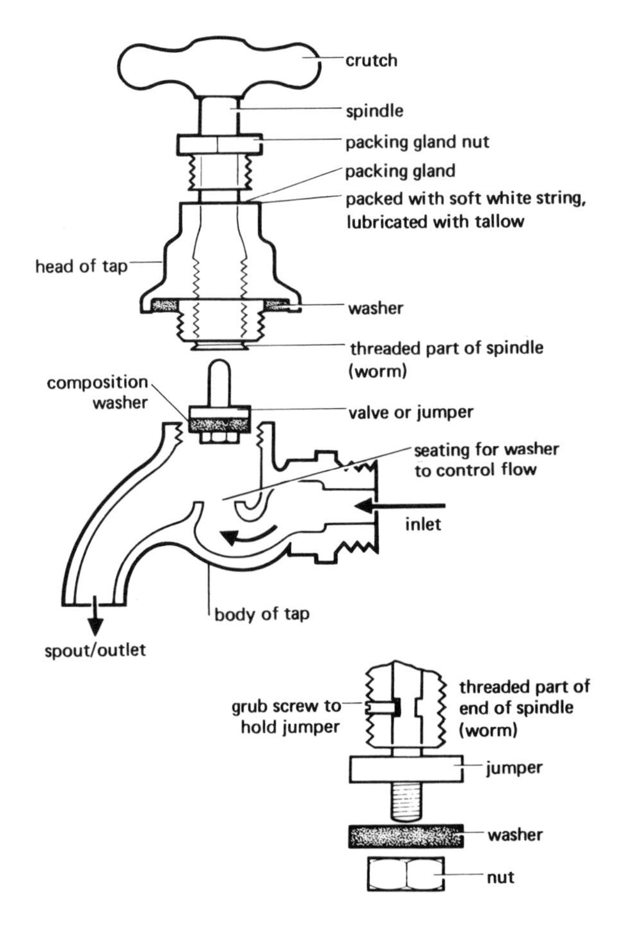

Figure 214 *Exploded view of bib tap*

Pillar tap (Figure 215)

This variation of the ordinary bib is fitted to baths, basins and sink units. Although it is fixed in the appliance, it is so designed that the nozzle outlet must always terminate above its flood level.

Supatap (Figure 216)

This incorporates an automatic closing device for use when the tap is being rewashered. This device enables rewashering to be completed without turning off the supply flow of water.

The washer and jumper holds a check valve off its seating in both the closed and fully open positions. If however the nozzle is unscrewed beyond the full flow position (this can only happen if the gland nut is first removed) the check

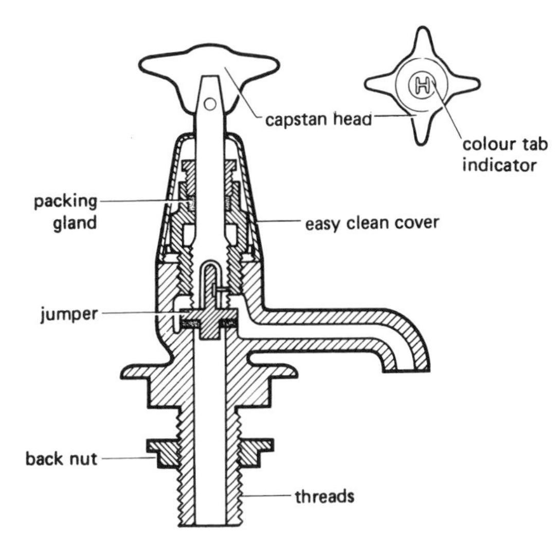

Figure 215 *Pillar tap*

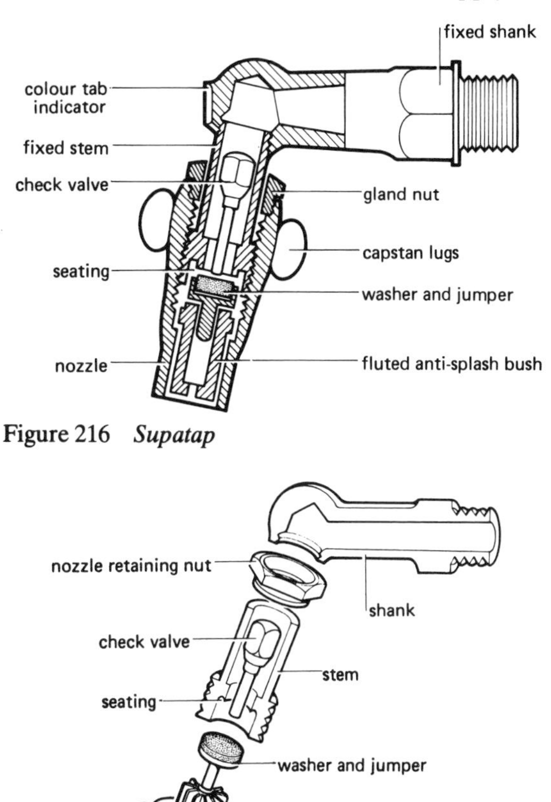

Figure 216 *Supatap*

Figure 217 *Exploded view of supatap*

valve drops on to its seating to shut off the flow of water. The nozzle, complete with anti-splash device, and the valve may then be removed. The valve may then be rewashered and the tap reassembled, all without the water supply being affected in any way. This type of tap is available for use in all domestic appliances, i.e. baths, basins, and sinks. Figure 217 shows an exploded view of the supatap.

Spring loaded tap (Figure 218)

This type of tap might be considered a special variety and would not normally be used on domestic work. Its use would be recommended for places where a large number of people are using the facilities, such as factories or public places. They might also find a use in hotels, hospitals and homes for children and mentally handicapped, where quite often taps are left running. This results in tremendous waste of both water and in the case of hot water, fuel also.

The taps are operated by exerting a downward pressure on the head of the tap which in turn depresses a spring holding the main valve on its seating. Water now flows through the tap, and will continue as long as this pressure is maintained. As soon as this pressure is released, the spring loaded valve is returned to its seating and the flow of water ceases.

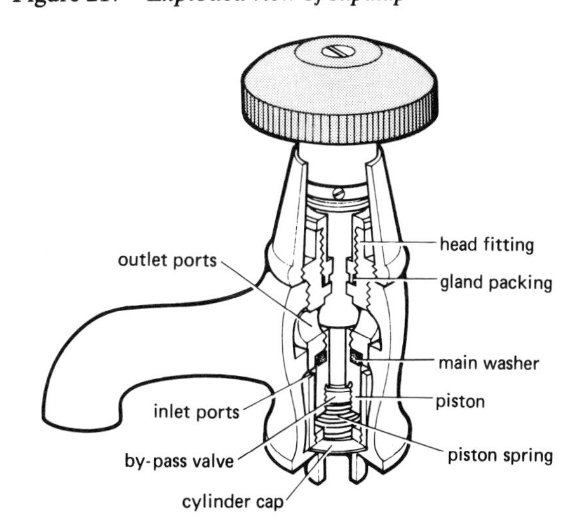

Figure 218 *Spring loaded tap*

This type of tap can cause water hammer in areas of high pressure unless they are of the non-concussive kind. They are also more prone to problems and more expensive than ordinary bib or pillar taps.

Stop tap (Figure 219)
The mechanism of this type of tap is the same as for that of the bib tap, but has a different body shape to suit its particular function which is to control the flow of water in a section of pipework. These taps have inlets and outlets to suit connection to lead, steel, copper and plastic. The word inlet, or alternatively a direction arrow, must be stamped on the body of the valve. This is most important, for should the tap be fixed the wrong way round no water will be able to pass through. In all cases a stop valve must be provided just inside the boundary of the premises in a convenient and accessible position, usually in the drive or path and invariably the service pipe follows near to the line of drains. Because the stop tap will be at least 760 mm below the ground level, a loose key is advisable to facilitate its operation. Stop taps are recommended inside the property, the number being governed by the size and complexity of the building.

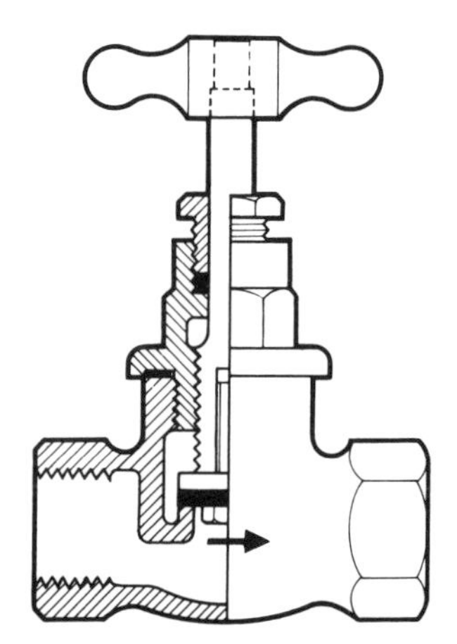

Figure 219 *Stop tap*

Stop and draw off tap (Figure 220)
Some stop taps have incorporated in the design an additional small tap or drain tap to enable the householder to drain the system of water. This combination of stop and draw off should be fitted at the entry of the service pipe into the dwelling and is particularly useful in frosty conditions or when the occupants are going away. The drain cock is sometimes a small bib tap with identical principles or alternatively it could be in the form shown with a hose pipe connection.

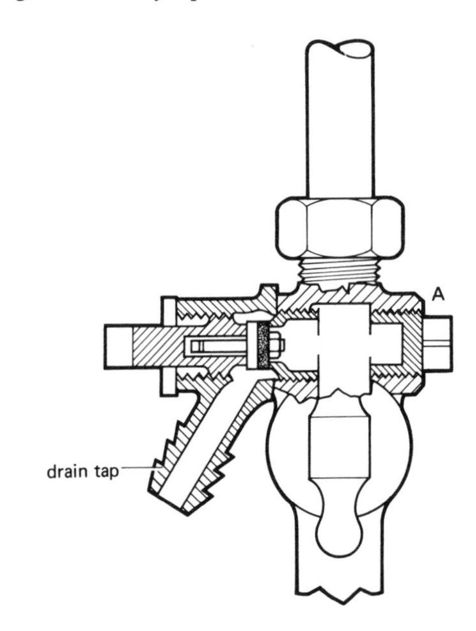

Figure 220 *Stop and draw off tap*

Mixer taps
Many combination hot and cold mixer sets are available for all sanitary appliances, in which the hot and cold supplies are separately controlled. The outlet appears to be just one pipe but in fact it contains two separate pipes. Figures 221, 222 and 223 show three different types of mixer tap, while Figure 224 shows an internal view.

Cocks

Draincock (Figure 225)
These draw off taps are generally fitted to drain off the water from boilers, cylinders and even sections of pipework. They are available with either loose valves and/or fixed valves according to the requirement. The tap is operated by a purpose made loose key which fits on the hexagonal head. The key should not be fixed to the tap permanently for general use, only sited near for use in emergency or for repair work.

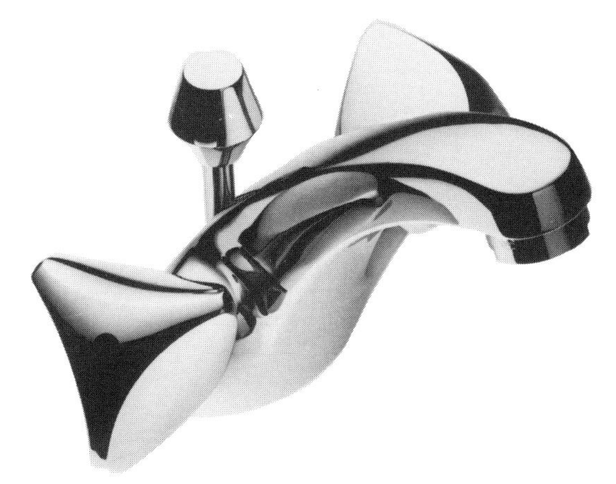

Figure 221 *Basin mixer*

Figure 222 *Sink mixer*

Figure 223 *Pillar bath mixer*

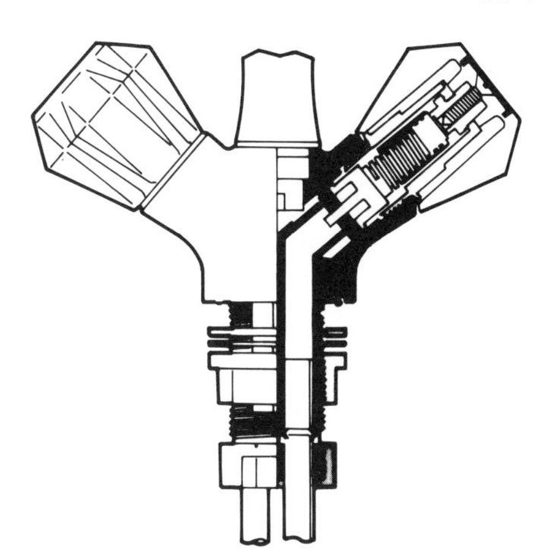

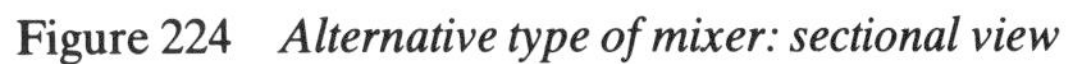

Figure 224 *Alternative type of mixer: sectional view*

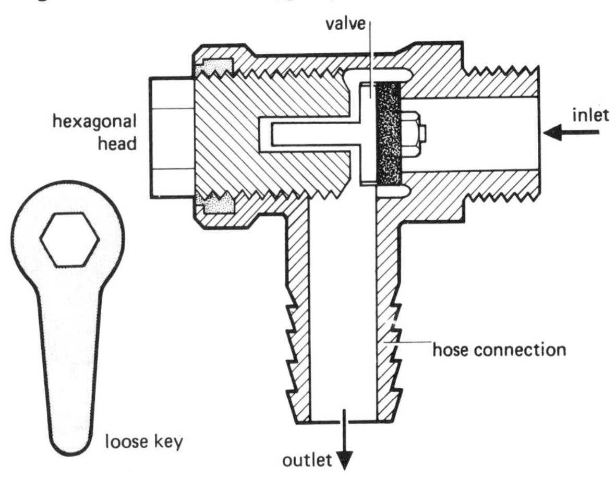

Figure 225 *Draincock*

Plug cocks (Figure 226)
These cocks were the earliest form of control on water and gas supply. They are now used only on gas supply and as a drain off cock for hot water systems. The reason they proved unsuccessful for cold water systems is due to the sudden stopping of the flow of water by a simple quarter turn of the key so setting up serious and damaging water hammer.

Valves
Valves used to control the flow of water along a pipeline are either globe valves or gate valves. The type chosen is governed by the pressure in the system. Both kinds of valve operate on the principle of the screw closing slowly so avoiding water hammer.

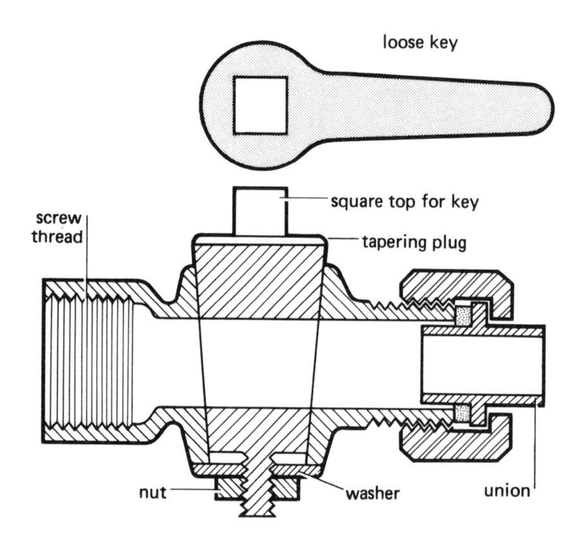

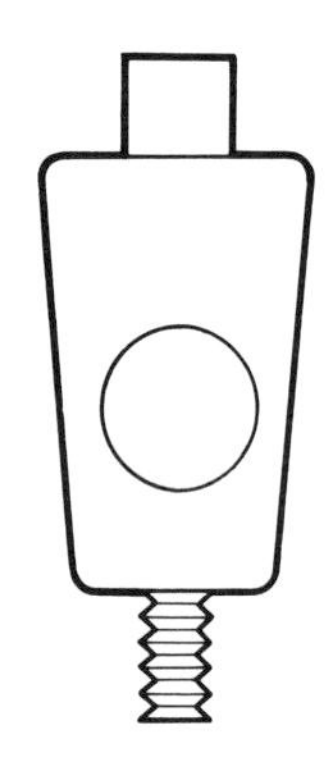

Figure 226 *Plug cock*

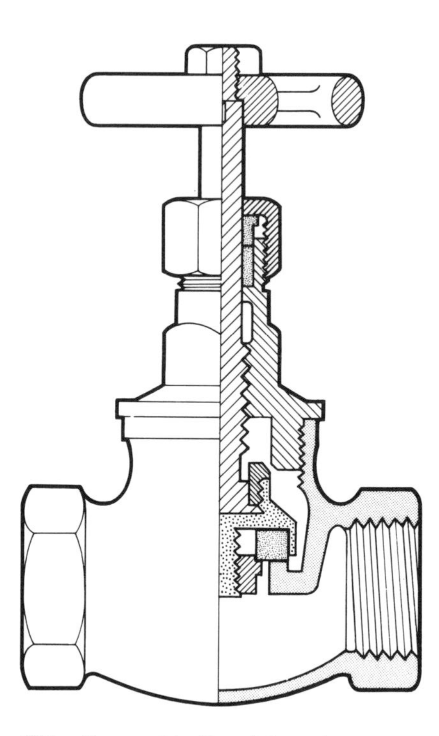

Figure 227 *Renewable disc globe valve*

Globe valve (Figure 227)
These types is used on high pressure systems. There are several types, metal-metal valve and seating for heating systems or composition washer for high pressure cold water systems. In this case the jumper could be of the loose fitting variety.

Gate valves (Figure 228)
These are known as fullway valves due to the fact that when they are fully opened there is no restriction to the flow of water through them. These types of valve are recommended where it is important that there should be the least possible restriction to the flow, as with low pressure distribution from storage cisterns. A gate valve is operated by a screw spindle generally of the non-rising type. In this, instead of rising out of the fitting when the valve is opened, it is rotated without rising and the wedge shaped gate climbs up the spindle screw thread.

Ball valves
A ball valve is simply a control actuated by a lever arm and a float which closes off the water supply

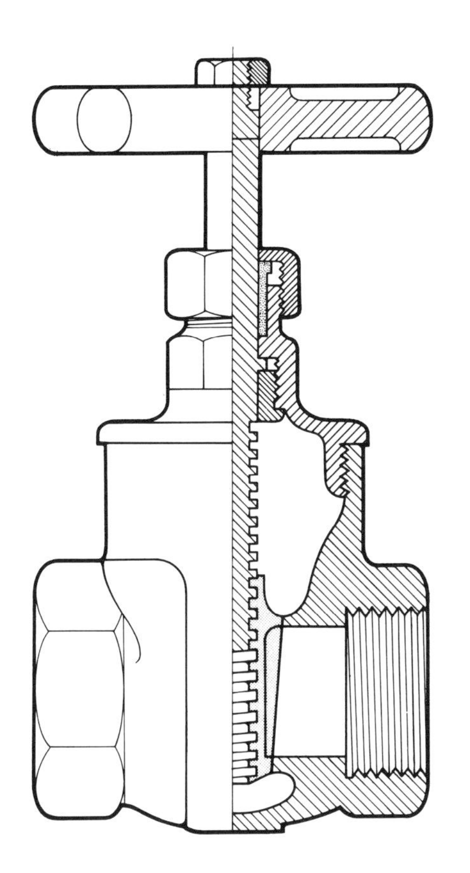

Figure 228 *Fullway gate valve*

when a predetermined level of water has been reached. There are many different types and it is important to be able to recognize each to know when and where each should be fitted.

All ball valves must be manufactured from non-corrosive material. The two used today are brass and plastic. The brass valve is made from a hot pressing which is then drilled, threaded and dressed to the dimensions required. The nozzle seating is a replaceable nylon unit and the washer must be a good quality rubber 3 mm thick. The float which is attached to a 6 mm brass rod is generally made of plastic which has taken over from copper which for many years was the accepted material.

Ball valves are classified as follows:

High pressure ball valves These must be capable of closing against pressures of *1380 kN/m²*.

Medium pressure ball valves These must be capable of closing against pressures of *690 kN/m²*.

Low pressure ball valves These must be capable of closing against pressures of *276 kN/m²*.

The visual difference between the three types of valve is in the size of the orifice (the hole through the nozzle). For high pressure water supply the orifice must be small, for medium pressure the size of the orifice will be a little larger, while for low pressure the orifice is larger still, almost full way.

Croydon ball valve (Figure 229)
This was widely used at one time but has been generally superseded by the others. The movement of the piston is in a vertical direction and tends to have a rather jerky and sluggish action.

Portsmouth ball valve BS 1212 (Figure 230)
This has the advantage of being easily dismantled or removed from the cistern in the case of malfunction, without having to undo the backnuts to remove the body. The parts are renewable and readily accessible. This is particularly important with regard to the nylon seating which can be

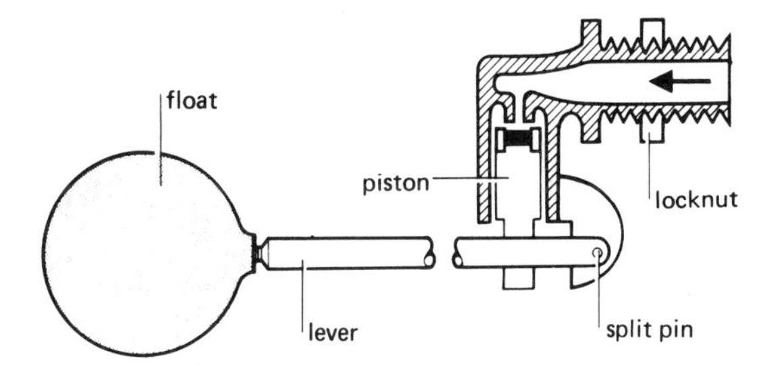

Figure 229 *Croydon pattern ball valve*

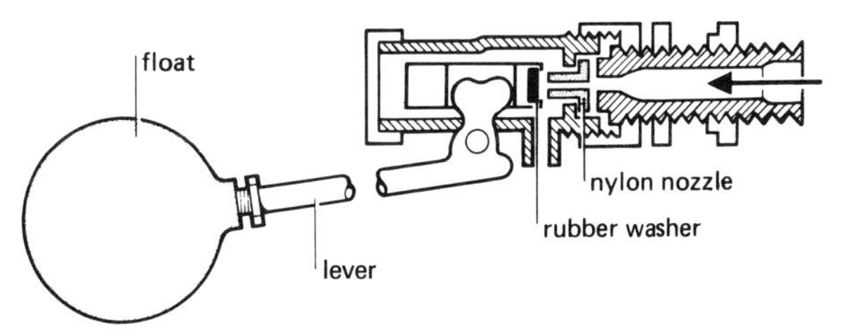

Figure 230 *BS 1212 Portsmouth ball valve*

suitable for low, medium or high pressure, thus making this type of valve universal.

Nylon is a chemically inert plastics material and so has a high resistance to mechanical wear and is now extensively used in place of metal seatings.

The movement of the piston in this type of valve is in a horizontal direction and has a much smoother and better action.

Figure 231 is an exploded view showing all the component parts. It will be noted that the outlet is not threaded to receive a silencing pipe. It is now against regulations to fit these, as they may cause back syphonage. This type is being superseded by valves with a top outlet.

Diaphragm valve (Figure 232)

This type of valve is similar in some respects to the BS 1212 Portsmouth. It resembles the British

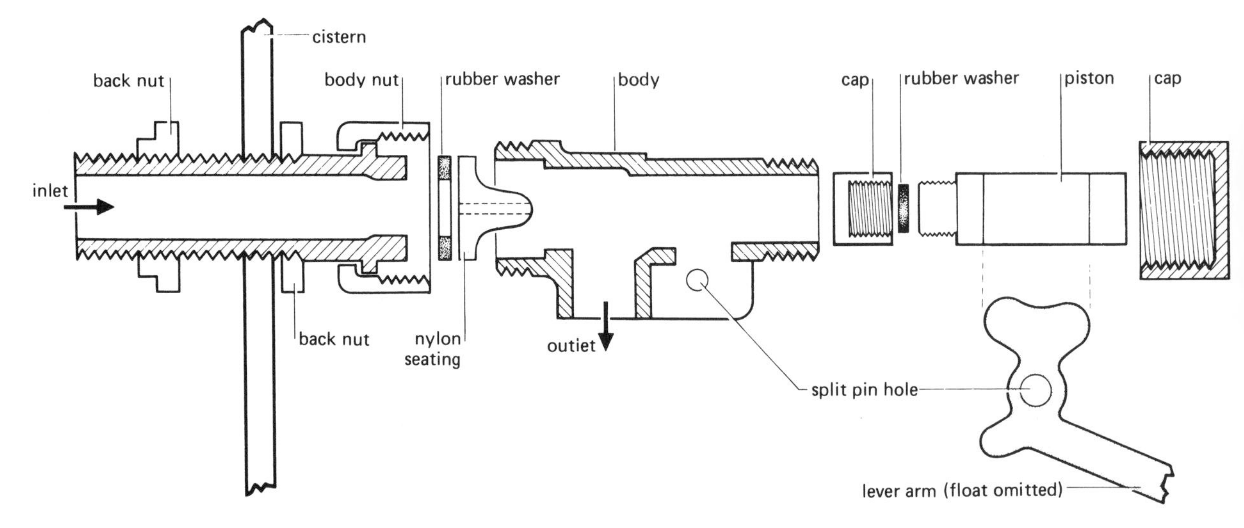

Figure 231 *Exploded view of Portsmouth ball valve*

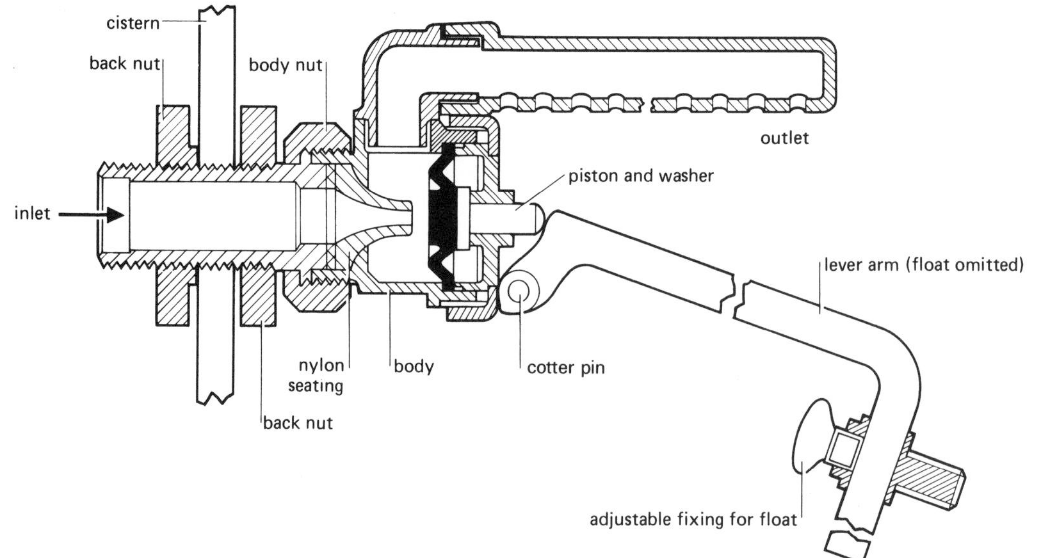

Figure 232 *Diaphragm valve*

Research Station's diaphragm valve in some features, but differs from it in that the design of the outlet completely overcomes the problem of back syphonage.

Although small holes in the outlets were at one time accepted as a prevention to back-syphonage, this is no longer the case. Several methods have been tried, including a very thin plastic tube which collapsed when the supply was subjected to negative pressure. There are now several variations to the above with regard to the outlet. The one illustrated delivers the water through a number of fine streams, which is comparatively quiet and completely overcomes back syphonage. This type of valve is covered by BS 858.

Building Research Station diaphragm ball valve (Figure 233)

The Government's Building Research Station (BRS) design team set about designing a valve free from corrosion problems and from the nuisance caused by lime deposits, both of which interfere with the proper working of the valve.

The advantages of this valve are:

1 No water touches the working part, thus overcoming the corrosion and lime problems.
2 The large inlet chamber breaks the speed with which the water enters and so reduces wear and noise.
3 A simple and convenient method of adjustment for the float.

Rewashering is a simple operation and is achieved by unscrewing the large nut which secures the ball valve head to the body (see exploded view in Figure 234). In this respect the BRS diaphragm valve resembles the BS 1212 ball valve.

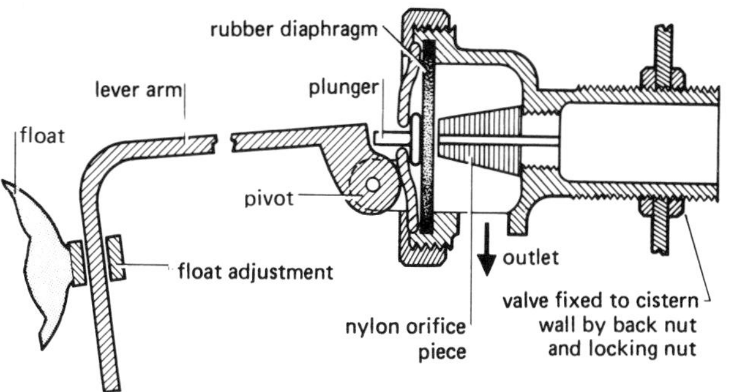

Figure 233 *British Research Station diaphragm ball valve*

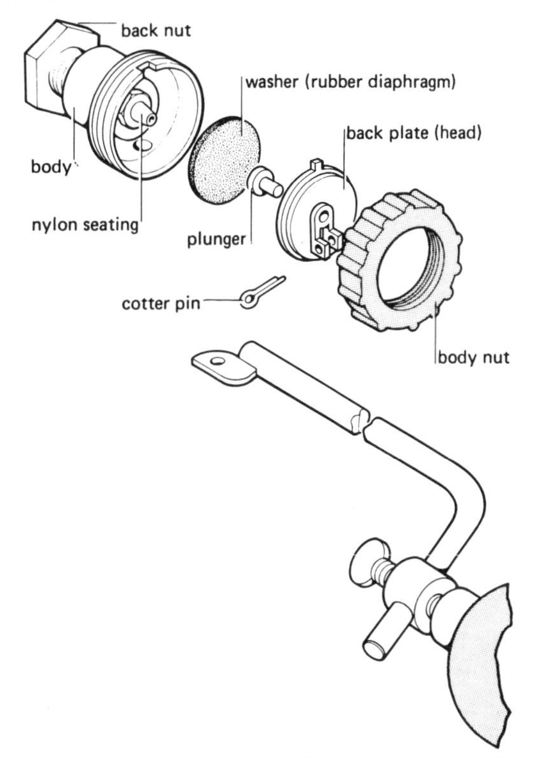

Figure 234 *Exploded view of BRS diaphragm valve*

Self-assessment questions

1 The minimum depth of cover required for a cold water service pipe is:
(a) 600 mm
(b) 750 mm
(c) 900 mm
(d) 975 mm

2 The type of control valve fitted to a cold distribution pipe from a cold water cistern to allow a full-bore flow of water is:
(a) ball valve
(b) gate valve
(c) pillar valve
(d) safety valve

3 To avoid undue weakening of the joints when notching for pipe runs, the notching should be:
(a) on the underside in the centre of the span
(b) on the underside of the span one third of the distance from the support
(c) on the top close to the supported end
(d) on the top in the centre of the span

4 On pipework which is subject to condensation the most suitable type of pipe clip is a:
(a) saddle band clip
(b) spacing clip
(c) pipeboard
(d) continuous support

5 Which type of ball valve has a large circular washer which also acts as a water seal to prevent corrosion of the working parts of the ball valve?
(a) Croydon
(b) diaphragm
(c) Portsmouth
(d) equilibrium

6 Which of the following are suitable materials for connecting a pipe to a polythene cold water storage cistern?
(a) lead washers and putty
(b) hemp and oil based jointing compound
(c) metal washers and jointing paste
(d) plastic washers

7 To avoid damage to the ferrule in the water main due to settlement the communication pipe is:
(a) given a swan neck bend
(b) wrapped in greased tape
(c) surrounded in concrete
(d) carried through an earthenware drain

8 In a cold water system the distribution pipe differs from the service pipe in that:
(a) part of it is supplied by the water authority
(b) it is not subject to mains pressure
(c) it is subject to mains pressure
(d) it is placed at least 750 mm below ground

9 Corrosion to a galvanized cold water storage cistern before fixing can be minimized by:
(a) applying two coats of non-toxic bitumastic solution
(b) lining it with copper
(c) applying two coats of white lead paint
(d) softening the water

10 Underground copper service pipes are required by water authorities to be joined by means of:
(a) bronze welding
(b) wiped soldered joints
(c) manipulative compression fittings
(d) non-manipulative compression fittings

6 Hot water supply

After reading this chapter you should be able to:

1 Recognize and name materials and components of the system.

2 Select materials for given jobs and state reasons for selection.

3 Recognize and name each type of hot water system.

4 State the functional requirements and working principles of each system.

5 Identify pipelines in accordance with BS 1710.

6 Demonstrate knowledge of heat transfer.

7 Demonstrate knowledge of the principle and methods of providing for expansion.

8 Use different methods of pipe fixing.

Introduction

An efficient and adequate hot water system is an essential part of any home. The effectiveness of any system will be judged by its ability to produce an ample supply of hot water at any draw-off point simultaneously.

The plumber will sometimes be asked to advise a customer on his or her choice of a system most suited to his or her dwelling and to the finance available. The following points must be taken into account.

Type of fuel (gas, oil, solid fuel or electricity)
Type of system (storage, non-storage, instantaneous)
Quantity of water required
Temperature of water
Heating and recovery time
Nature of water in the area
Smokeless zone regulations
Heat losses
Cost of installation and running costs.

Design and pipe sizing will be dealt with in more detail later.

Once the choice of system has been made it is the plumber's responsibility to install the component parts and pipework according to the local water authority's regulations and the respective codes of practice, with all of which he or she must be familiar.

The domestic hot water system has three main components:

The boiler or heating element
The hot storage tank (usually a cylinder)
The cold storage and feed cistern

In addition, there is all the relevant pipework.

Figure 235 shows a typical direct system of hot water supply in isometric projection.

Note: The sizes of components and pipes given in Figure 235 are the minimum allowed by regulations and codes of practice and will be found adequate for most ordinary dwellings. It is a

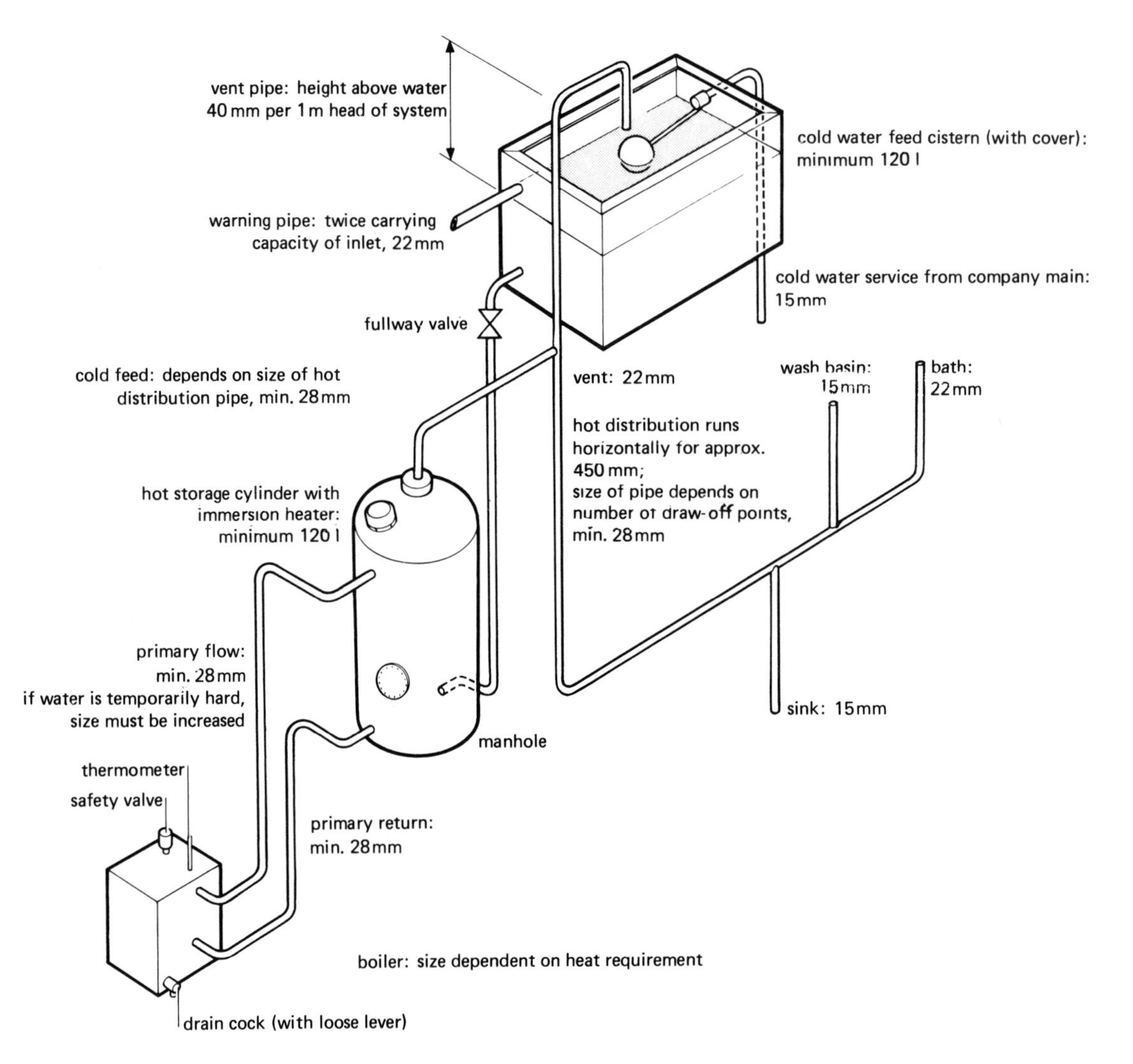

Figure 235 *Direct system of hot water supply*

common mistake to undersize the pipework which could result in an inefficient system. It is therefore of paramount importance to consider all the factors involved in each particular case before deciding on pipe sizes; if in doubt choose the larger diameter pipe.

Boiler

There are many different types, shapes and sizes of boiler, including back boilers, high output back boilers and independent boilers.

Back boilers

In many houses throughout the country back boilers in one form or another are the main source of heat. They are cheap to purchase, install and run, because they form an integral part of the normal fireplace or cooking range and share available heat.

High output back boiler

This type of boiler forms part of the normal fireplace. The difference between this and the ordinary back boiler is one of size: the high

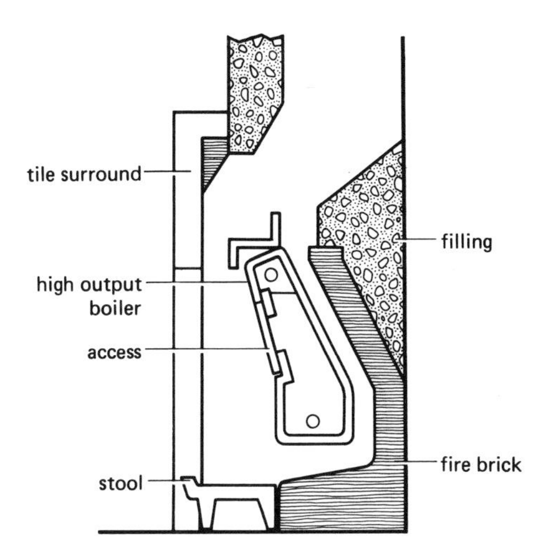

Figure 236 *High output back boiler*

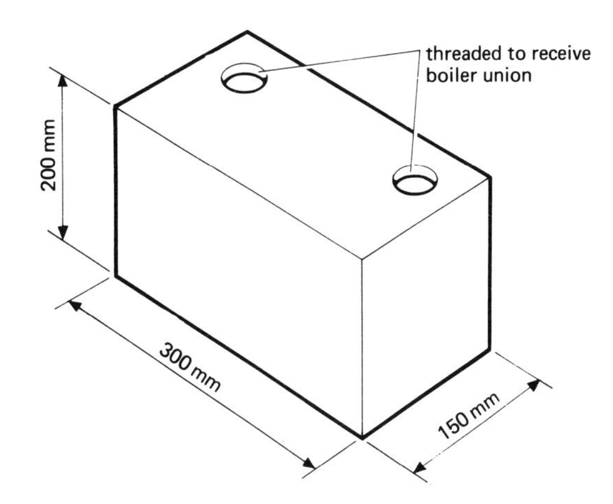

Figure 237 *Block boiler*

output boiler is much bigger and presents a much larger surface area to the flames and flue gases, enabling a greater heat value (see Figure 236). Some boilers form the actual flue and so have an even greater heating surface.

Independent boiler
An independent boiler is so-called because it is independent of any other heating service, although some independent boilers would impart a quantity of heat to the room in which they are installed as well as performing the main function of heating the domestic hot water.

Boilers are made generally from mild steel or cast iron but in special cases stainless steel or copper may be used. At an additional cost the boilers can be 'bower barffed' during manufacture to prevent rusting. Some boilers are fitted with a manhole to allow for periodical cleaning out of scale.

Figures 237–242 illustrate some of the points to be borne in mind when fitting a back boiler.

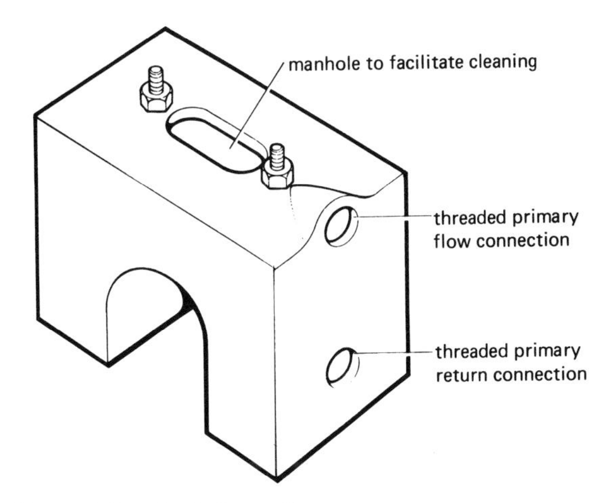

Figure 238 *Arch flue boiler*

Cold feed cistern

Cisterns are often misnamed tanks: cisterns are open to the atmosphere; tanks are closed vessels. Nowadays cisterns are usually manufactured from galvanized steel or plastics. They should be sited as high as practicable to give the required head

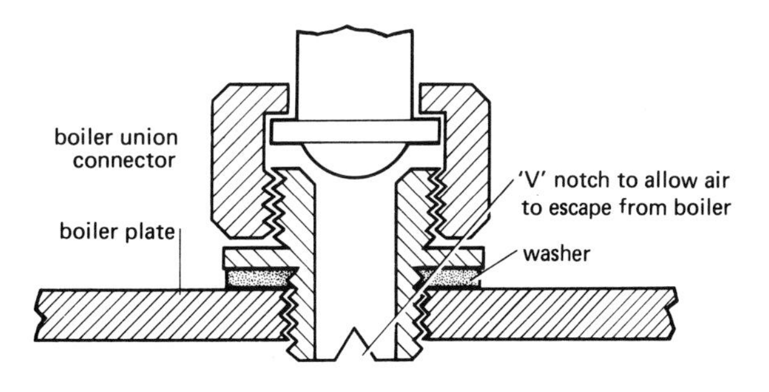

Figure 239 *Detail of flow connection to top entry boiler*

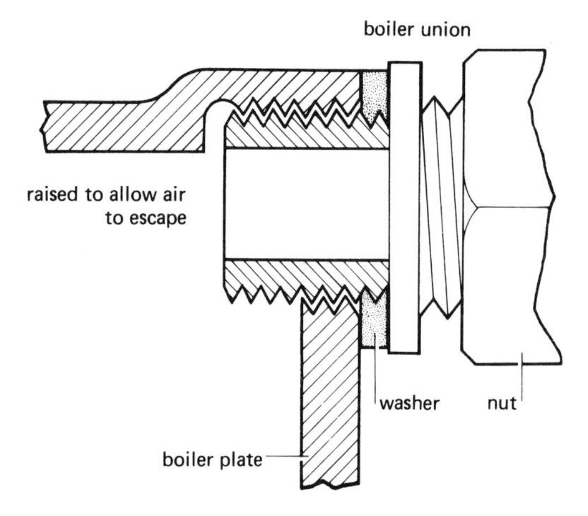

Figure 240 *Detail of flow connection to side entry boiler*

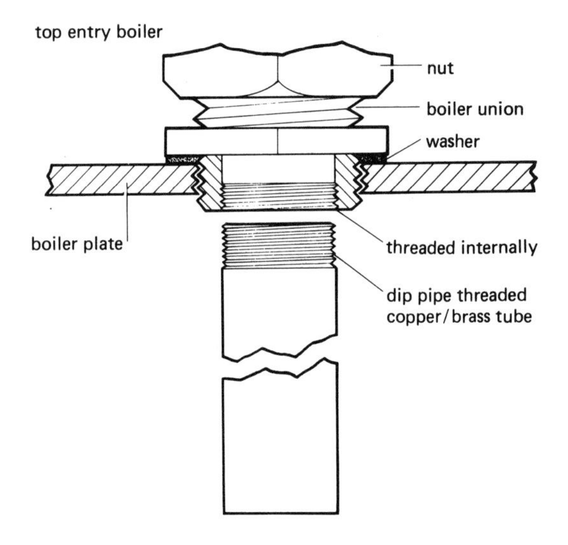

Figure 241 *Detail of return connection with dip pipe*

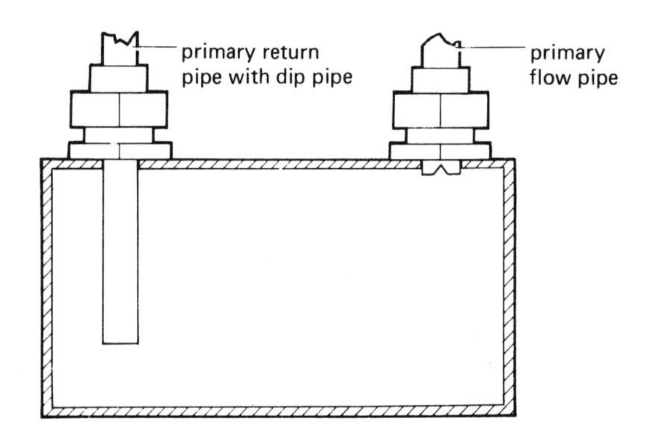

Figure 242 *Block boiler*

(see Chapter 5). The function of a cistern fed directly from the authority's main is to store a minimum of 112 litres of water and to feed the hot storage vessel. It also acts as an expansion vessel when the system is heated.

Hot storage tank

These vessels are manufactured from many different materials; for instance, copper, galvanized steel and plastics, and in various sizes and shapes. The most commonly used is a copper cylindrical tank 900 mm high × 450 mm diameter, with a minimum capacity of approximately 112 litres.

Their function is to store the hot water before it is drawn off at the appliances. The replacement (cold) water is supplied from a cold feed cistern.

It is recommended that hot water storage cylinders should be fixed in the vertical position rather than the horizontal position, as this aids stratification of the water: the taller the cylinder the better. However, in some cases, it may be advantageous to fix the cylinder in a horizontal position.

Primary flow pipe

The heated water rises in the boiler and is conveyed by this pipe to the hot store (cylinder). The flow pipe is connected at the top of the boiler to a connection on the hot storage cylinder above the primary return pipe. The best position is approximately two thirds the height of the cylinder. This gives a supply of hot water quickly. The size of the pipe must be the same as for the primary return pipe. There must *not* be any control valve fitted to this pipe.

Primary return pipe

The water leaves the hot store by this pipe and is conveyed to the boiler. The connection is near the base in each case. The size of this pipe will depend on many factors; size of system, length of run and nature of water (temporary hardness), but for ordinary domestic systems a minimum of a 19 mm bore tube should be used. There must *not* be any control valves fitted to this pipe.

Vent

This pipe is connected to the top of the cylinder and is carried up (after travelling in a horizontal direction for approximately 450 mm) to terminate over the cold feed cistern. The function of this pipe is three-fold:

1 To allow air to escape from the system during the initial filling;
2 To permit air to enter the system and facilitate the draining of water from the system;
3 To allow gases liberated from the heated water to rise and escape from the system.

Hot water supply

A hot water supply is taken to the appliances by means of a pipe connected to the vent. The height of the vent pipe above the water in the cistern must be sufficient to allow the water to expand in the system without escaping. The recommended height is 75 mm per 1 metre head of system (static head).

The hot water supply pipe for ordinary dwellings usually terminates as a dead leg at the appliance. However, for large installations where the length of dead leg exceeds that laid down by water authority regulation the hot supply becomes a secondary flow and return. The supply pipe returns and connects back into the hot storage vessel approximately one-third the distance from the top.

Cold feed
This pipe conveys cold water from the cold feed cistern to a connection near the base of the hot storage vessel (tank). It must be large enough to allow water to enter the hot store as fast as it is required to supply the flow to the appliances simultaneously. For domestic use, the minimum recommended is 19 mm bore tube. The cold feed pipe must be a direct feed from cistern to tank, with no other branches taken from it, and it must be controlled by a fullway value. The connection to the cistern should be approximately 50 mm above the base to prevent sediment from the cistern entering the hot water system.

Overflow and warning pipe
Both these functions are sometimes fulfilled by the same pipe, but in other cases two different pipes are necessary. Their definitions are as follows:

Overflow pipe
This is part of an appliance fixed to prevent overflowing. It is often an integral part of the appliance, and allows water to enter the waste pipe between the plug and the trap. In certain cases for large cisterns, an overflow pipe could be connected directly to a discharge system provided that a separate small warning pipe is connected to discharge in a prominent position.

Warning pipe
This is a pipe connected to a cistern above the water line, but under the ball valve. This pipe, which should have twice the carrying capacity of the inlet, should be fixed so that if the ball valve should malfunction and the cistern fill to above that set, the warning pipe will discharge the additional water to a prominent position and so give warning to the householder that attention is required.

Expansion of water

Before the student can understand the circulation of water he or she must first understand and appreciate the effect of adding heat to a volume of water.

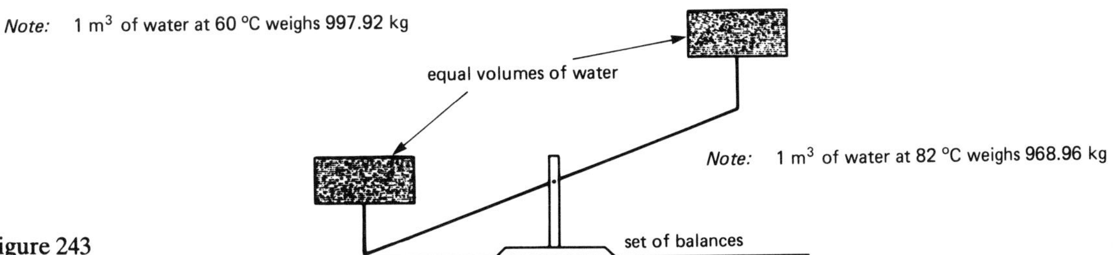

Figure 243

Water, when heated, becomes lighter volume for volume than colder water as shown in Figure 243.

This can be easily understood and demonstrated by the following simple experiment.

Two containers are filled with equal volumes of water at the same temperature and placed on a scale (Figure 244).
Heat one container. The water will expand and overflow (Figure 245).
Although both containers are full of water (equal volumes) the cold water is heavier and this container drops (Figure 246).

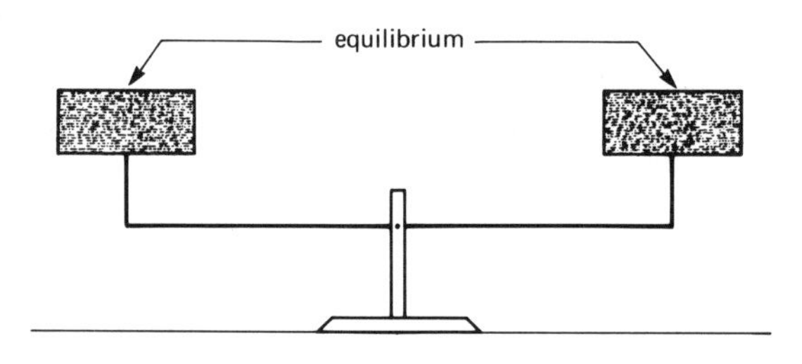

Figure 244

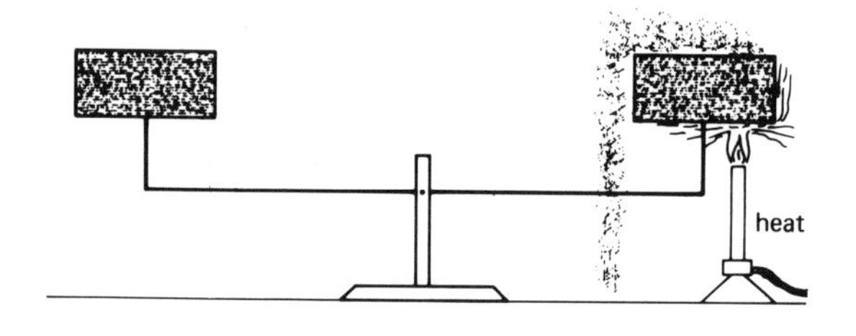

Figure 245

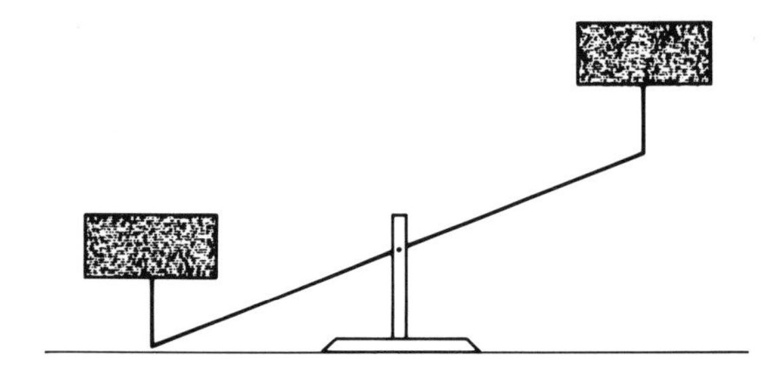

Figure 246

The following points are important in understanding circulation:

1 Water expands on being heated (increased volume). A volume of cold water is heavier than the same volume of hotter water (cold is more dense).
2 A volume of cold water, when heated, requires a larger storage space (expansion allowance).

Circulation of water

The factors that influence the circulation of the water are:

1 Temperature difference of flow and return water
2 Length of circulation pipes
3 Diameter of circulation pipes
4 Dips or bends in circulation pipes.

Water circulates or moves when heated or cooled through convection currents. This movement will continue so long as there is a difference of temperature, and is increased as the temperature difference is increased.

Heat movement

There are three forms of heat movement:

1 Conduction
2 Convection
3 Radiation

Conduction

This is the movement of heat in a solid. The heat passes from one particle to another without the particles themselves moving.

Convection

This is the movement of heat which takes place in liquids and gases. The particles themselves actually change places.

Radiation

This is the movement of heat in direct rays from the source without heating the intervening space.

Note: Heat travels in direct rays. It can be reflected and in many ways behaves similarly to light.

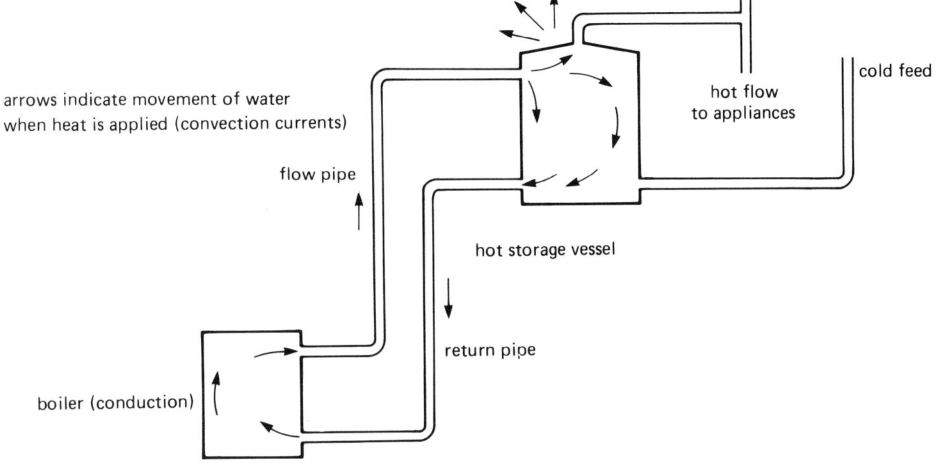

Figure 247 *Circulation of water*

The three forms of heat movement are put to use in the domestic hot water system (see Figure 247). Conducted heat takes place between the source of heat, i.e. solid fuel, gas or oil flame, and the boiler plate, which in turn passes the heat to the water inside the boiler by conduction. The water when heated begins to move (circulation). This is due to *convection*. The source of heat, pipes and hot storage vessel, all transmit *radiated heat*.

Circulation pressure

Circulation pressure is the difference in density between the vertical flow and return columns of water. For example, if the flow temperature is at 80 °C and the return temperature is at 50 °C, then the circulating pressure will be the difference in density (weight) of those two columns of water. The vertical circulating height is taken from a centre line through the boiler to a centre line through the hot storage vessel, shown as X in Figure 248.

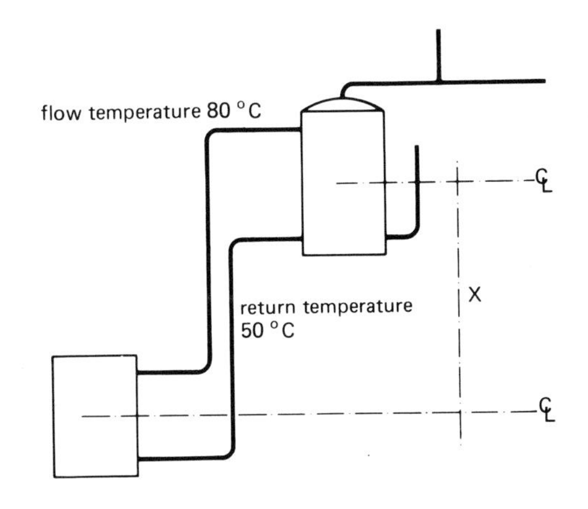

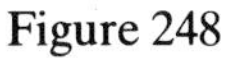

Figure 248

Simple heating circuit

Figure 249 shows a gravity system of heating which operates on the difference in density of the flow and return columns of water. The primary flow pipe should be carried direct to the highest point terminating as a vent over the feed and storage cistern. The primary return starts at a convenient high level and returns with a fall from this point back to the source of heat. Any dip or reverse fall would mean that the flow of the circulation of the water would be reduced.

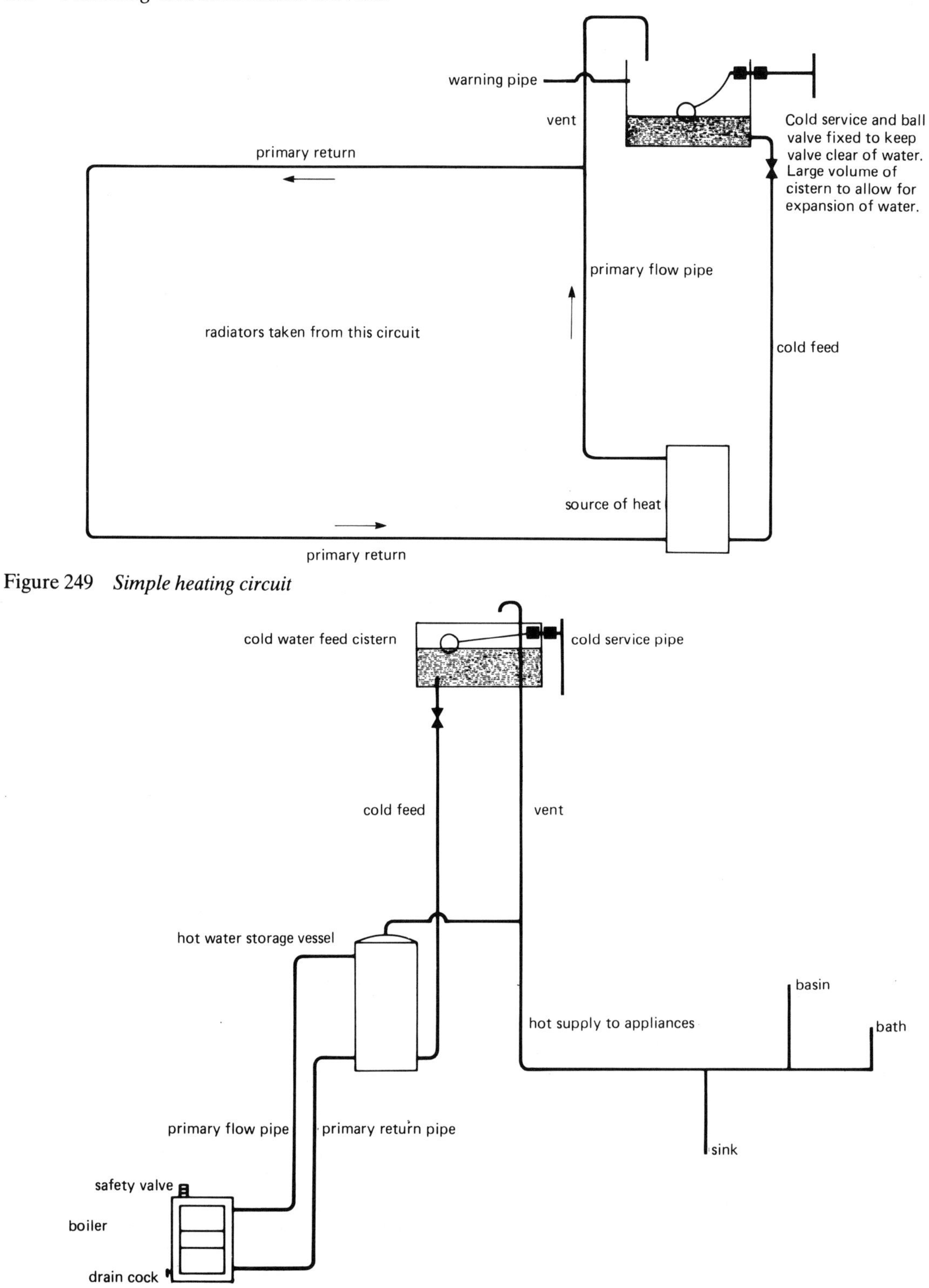

Figure 249 *Simple heating circuit*

Figure 250 *Cylinder system of domestic hot water supply (direct)*

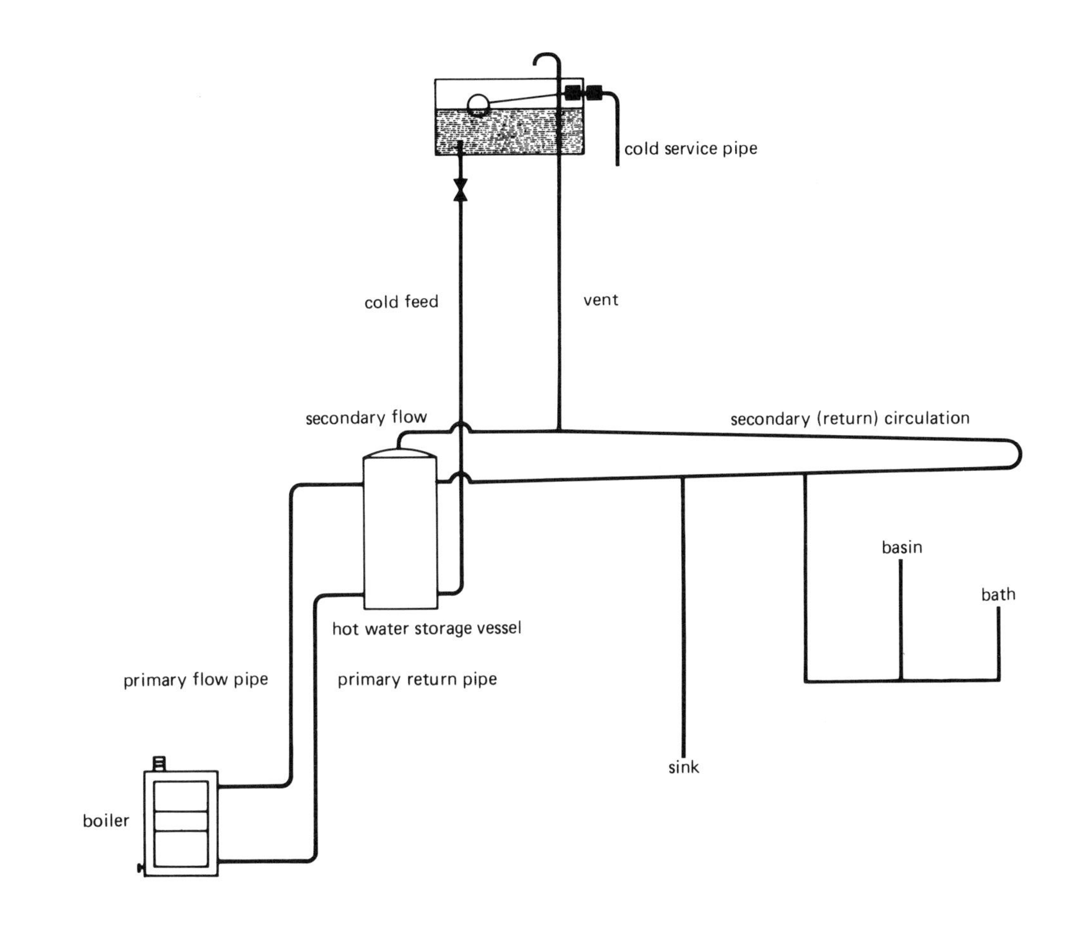

Figure 251 *Direct cylinder system with secondary return*

Direct cylinder system

The system shown in Figure 250 is a *direct* domestic hot water system. In this type of system the water which is heated in the boiler is drawn off at the appliances, so there is a continual change of water taking place.

Direct cylinder system with secondary return

Figure 251 illustrates the conventional domestic hot water system with the inclusion of a secondary (return) circulation. A secondary circulation is used in all cases where the hot storage vessel cannot be sited near the appliances, resulting in long lengths of pipes (dead legs). Hot water supplies to appliances are taken from this secondary circulation.

Combined cylinder and tank system (conventional layout)

This system (see Figure 252) is not used in ordinary dwelling houses, but can be used to advantage in the following cases:

1 Where there is very little head above appliances fitted high up in a building.
2 Where the supply pipes, i.e. cold feed and hot water supply pipe, are too small to supply all appliances.
3 In tall buildings.

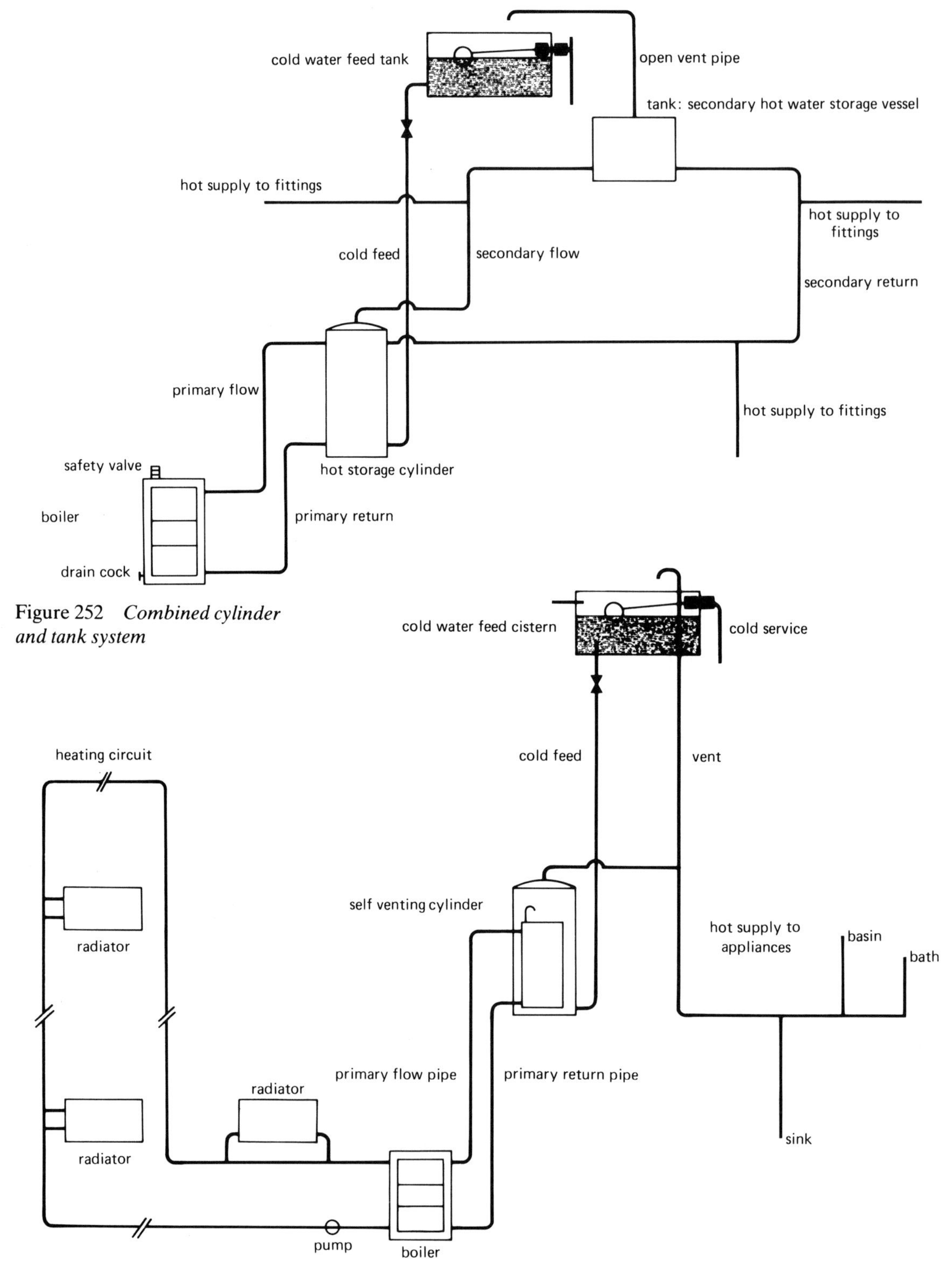

Figure 252 *Combined cylinder and tank system*

Figure 253 *Indirect cylinder system*

Indirect cylinder system
Figure 253 illustrates how an ordinary direct domestic hot water system can be adapted to incorporate a heating system by means of a single feed/single vent cylinder. The system is thus converted into an indirect system, which is required when radiators are fed from the domestic hot water system boiler. There are many different types of self-venting cylinders available which work on the same principle of forming an air lock in the inner cylinder to prevent mixing of the waters. It is advisable to check the manufacturer's literature when selecting and installing these self-venting cylinders.

In a conventional indirect system of hot water supply, there are two cold feed cisterns; one supplying the domestic hot water system, and the other supplying the heating circuit (see Figure 254). The two systems are completely separate. The domestic hot water is heated indirectly by means of the calorifier (cylinder within a cylinder) – hence the name of the system. Single feed systems are shown in Figure 253.

This type of system is used where:

1 Domestic hot water and heating is fed from the same boiler.
2 The water is of a temporarily hard nature.

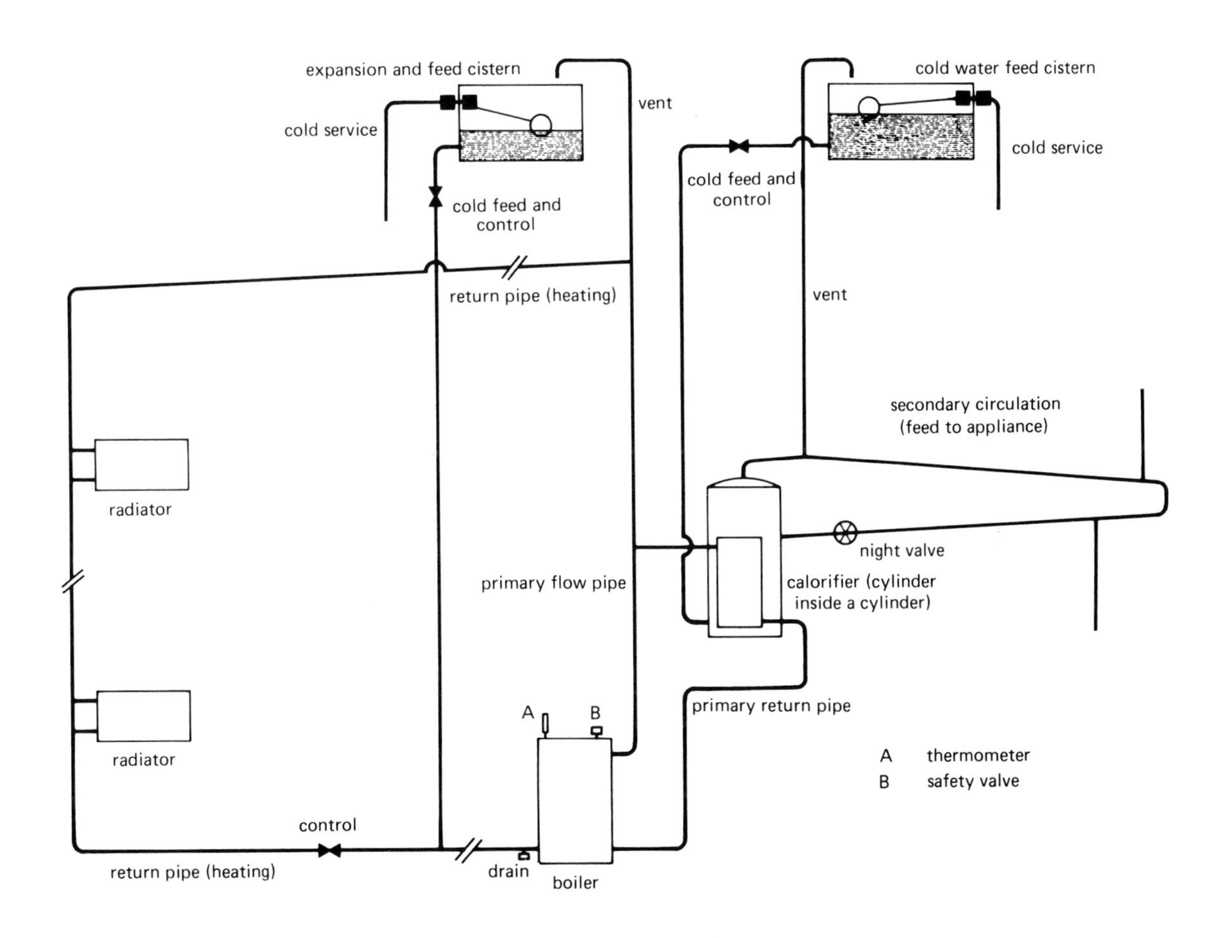

Figure 254 *Indirect cylinder system*

Table 18 *Requirements of indirect hot water cylinders (Table 2: BS 1566 Part 1, 1972)*

BS type ref.	External diameter A mm	External height B mm	Capacity litres	Heating surface m^2	Thickness of copper sheet – Grade 2, Test pressure 2.20 bar. Max. wkg. hd. 15m. – Body and top mm	Grade 2 Bottom mm	Grade 3, Test pressure 1.45 bar. Max. wkg. hd. 10m. – Body and top mm	Grade 3 Bottom mm	Coil test pressure 7.00 bar max. wkg pressure 3.50 bar Tube size mm	Height of screwed connections H mm	J mm	L mm	M mm	P mm	Screwed conns. Internal threads	Primary conns. External threads
0	300	1600	96	0.35	0.9	1.4	0.7	1.2	28	1250	100	100	540	150	G1	G1
1	350	900	72	0.27	0.9	1.4	0.7	1.2	28	700	100	100	400	150	G1	G1
2	400	900	96	0.35	0.9	1.6	0.7	1.2	28	700	100	100	400	150	G1	G1
3	400	1050	114	0.42	0.9	1.6	0.7	1.2	28	800	100	100	470	150	G1	G1
4	450	675	84	0.31	1.0	1.6	0.7	1.2	28	450	100	100	300	150	G1	G1
5	450	750	95	0.35	1.0	1.6	0.7	1.2	28	550	100	100	340	150	G1	G1
6	450	825	106	0.40	1.0	1.6	0.7	1.2	28	625	100	100	370	150	G1	G1
7	450	900	117	0.44	1.0	1.6	0.7	1.2	28	700	100	100	400	150	G1	G1
8	450	1050	140	0.52	1.0	1.6	0.7	1.2	28	800	100	100	470	150	G1¼	G1
9	450	1200	162	0.61	1.0	1.6	0.7	1.2	28	950	100	100	540	150	G1¼	G1
10	500	1200	190	0.75	1.2	1.8	0.9	1.6	35	950	150	150	540	200	G1½	G1¼
11	500	1500	245	0.87	1.2	1.8	0.9	1.6	35	1200	150	150	670	200	G1½	G1¼
12	600	1200	280	1.10	1.4	2.5	1.2	2.0	42	950	150	150	540	200	G2	G1½
13	600	1500	360	1.40	1.4	2.5	1.2	2.0	42	1200	150	150	670	200	G2	G1½
14	600	1800	440	1.70	1.4	2.5	1.2	2.0	42	1350	150	150	800	200	G2	G1½

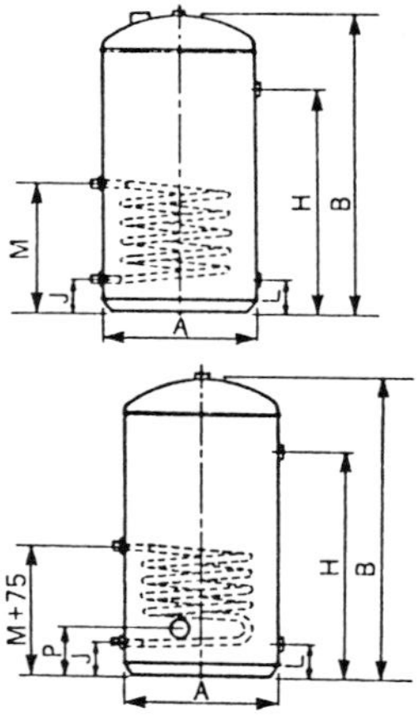

*Fifth connection on sizes 0–9 and immersion heater connection on all sizes not fitted unless requested.

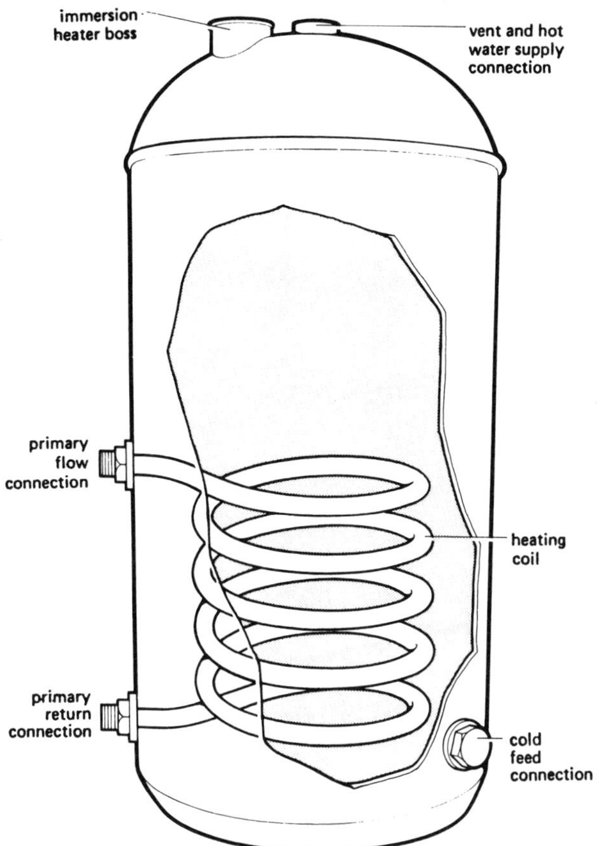

Figure 255 *Cylinder (indirect)*

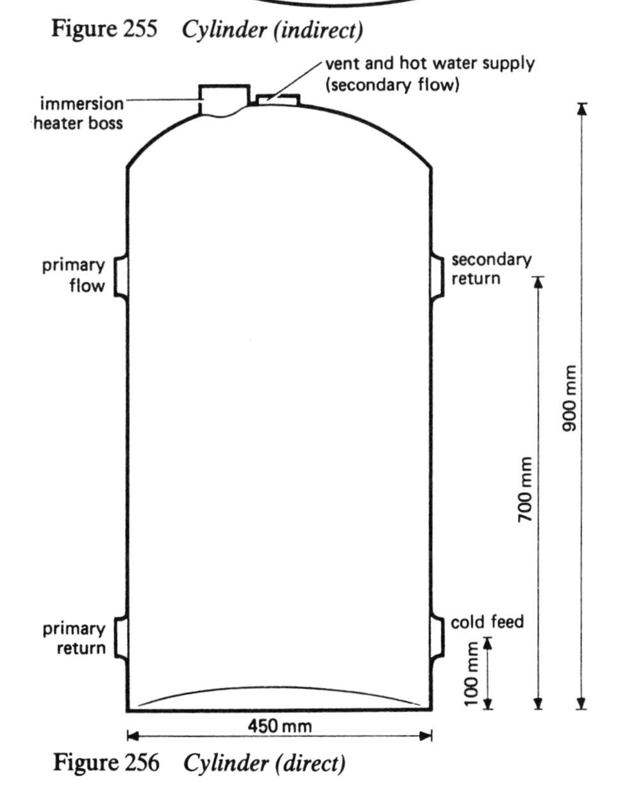

Figure 256 *Cylinder (direct)*

Hot water cylinders

The bye-laws of all water authorities require that indirect hot water cylinders shall comply with BS 1566. This is also a requirement of the National House Builders Council, British Gas Corporation, and indeed most local authorities. Table 18 gives these requirements in detail.

Figure 255 shows an indirect hot water cylinder. Figure 256 shows a copper direct hot water cylinder with a capacity of approximately 112 litres.

Pipe fixing

All pipework whether conveying hot or cold water should be correctly supported and fixed. Unless this is done the result could be mechanical damage and/or unnecessary pipe movement noises.

There are many types of clip, bracket and fixing some of which are illustrated in Figures 257–265.

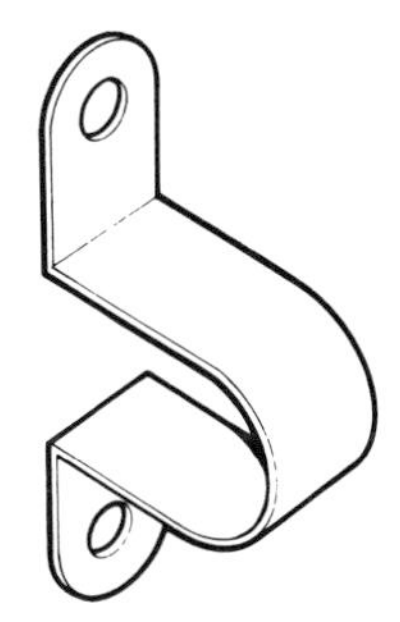

Figure 257 *Saddle clip*

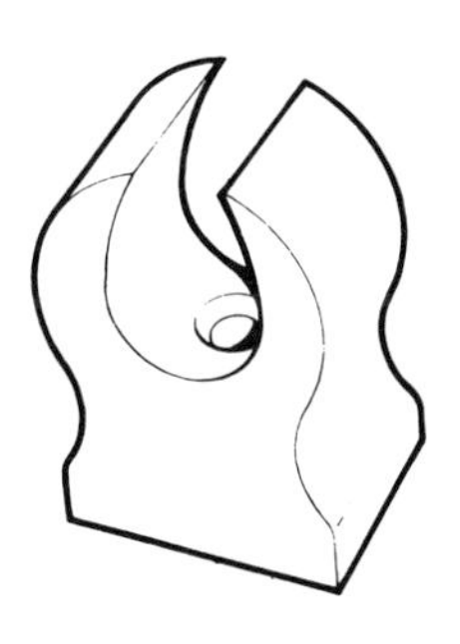

Figure 258 *Snap action PVC spacing clip*

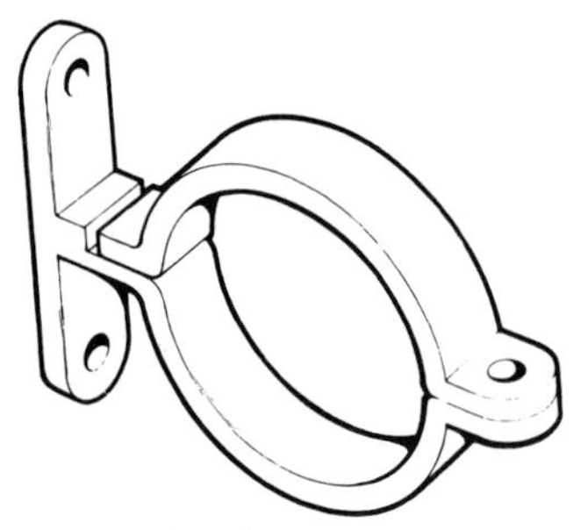

Figure 259 *Screw-on bracket*

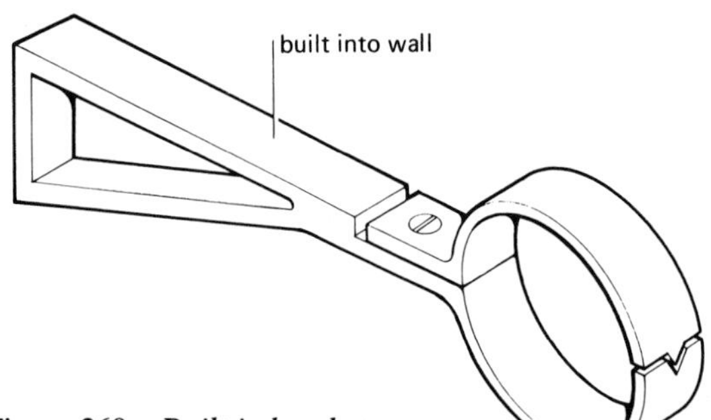

Figure 260 *Built-in bracket*

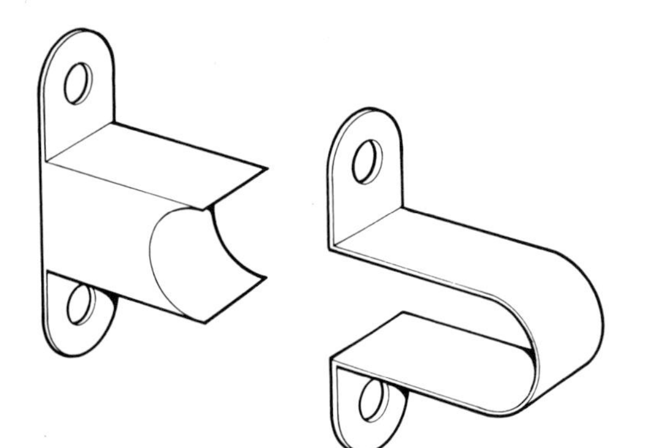

Figure 261 *Two piece spacing clip*

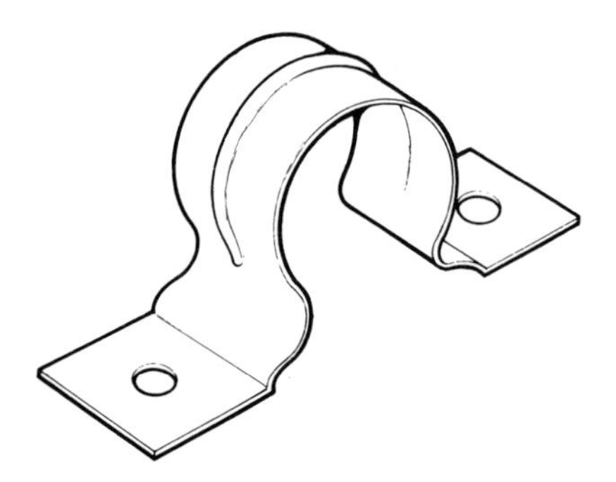

Figure 262 *Single spacing clip*

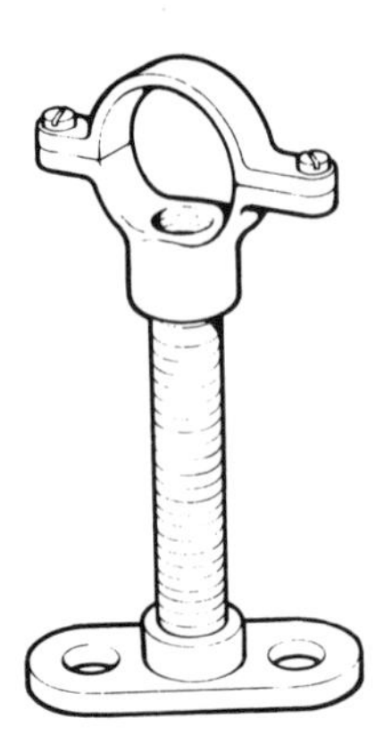

Figure 263 *Two piece pipe ring with extension rod and back plate*

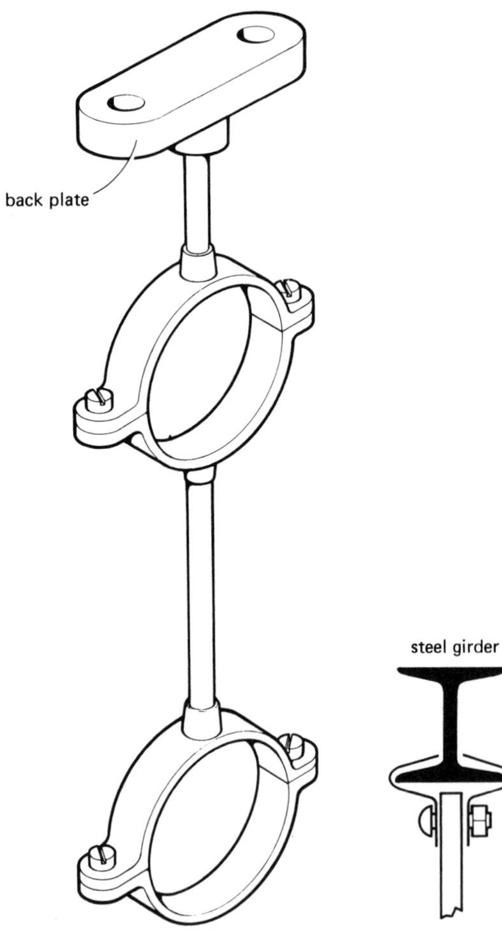

Figure 264 *Variation of two piece pipe ring multiple carrying arrangement*

Figure 265 *Alternative method of fixing two piece pipe ring*

Table 19 *Recommended fixing distances (CP 310) (nominal pipe sizes)*

Material	*Fixing position*	*Diameter in millimetres (nominal)*								
		13	*19*	*25*	*32*	*38*	*50*	*63*	*75*	*100*
Low carbon steel	Horizontal	1.8 m	2.4 m	2.4 m	2.7 m	3.0 m	3.0 m	3.6 m	3.6 m	3.6 m
	Vertical	2.4 m	3.0 m	3.0 m	3.0 m	3.6 m	3.6 m	4.5 m	4.5 m	4.5 m
Copper	Horizontal	1.2 m	1.8 m	1.8 m	2.4 m	2.4 m	2.7 m	3.0 m	3.0 m	3.0 m
	Vertical	1.8 m	2.4 m	2.4 m	3.0 m	3.0 m	3.0 m	3.6 m	3.6 m	3.6 m
Lead	Horizontal	0.6 m	0.6 m	0.6 m	0.6 m	0.6 m	0.76 m	0.76 m	0.76 m	0.76 m
	Vertical	0.76 m	0.76 m	0.76 m	0.76 m	0.76 m	← 1.0 m–1.2 m →			

The type chosen will be governed by several factors such as:

(a) The type and size of water pipe
(b) Whether conveying hot or cold water
(c) The type of background, i.e. brickwork, timber etc.
(d) Whether the pipes are to be visible or hidden
(e) Whether the pipes project from the surface or not, i.e. spacing clips or saddle.

Fixing lead pipe

Lead pipe is heavy and relatively soft, particularly when heated. It is therefore prone to sag if not well supported. One acceptable method is to secure by means of galvanized steel clips at regular intervals: i.e. 610 mm for horizontal work (Figure 266) and 760 mm for vertical work (Figure 267).

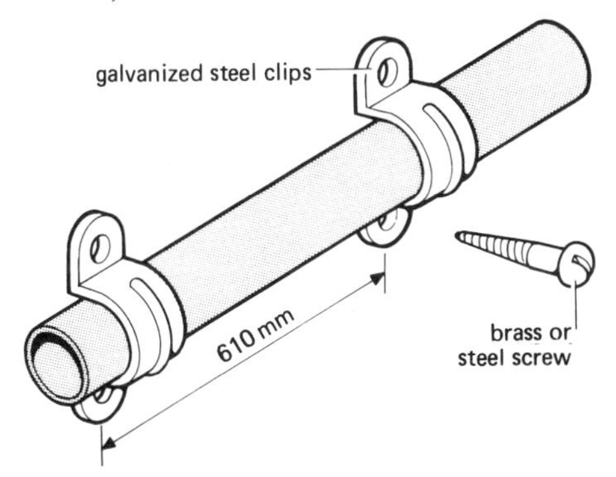

Figure 266 *Horizontally fixed pipes*

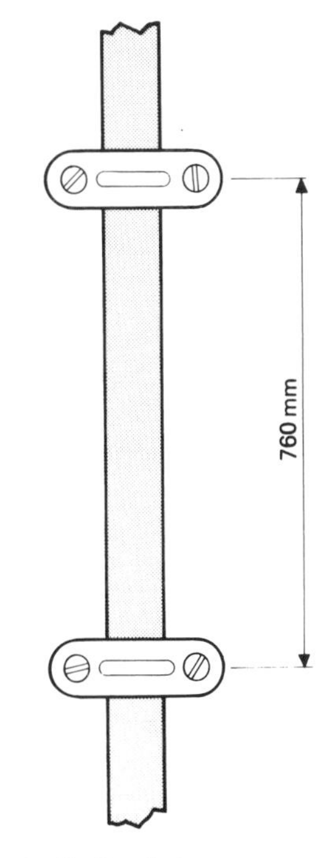

Figure 267 *Vertically fixed pipes*

Figure 268 shows an alternative method of fixing lead pipe, using a wrought iron pipe hook.

Lead pipe or other non-rigid pipes can be held in a horizontal position using a continuous support. Figure 269 shows a continuous support of a decorative form where appearance is important. The pipe is held by the recess.

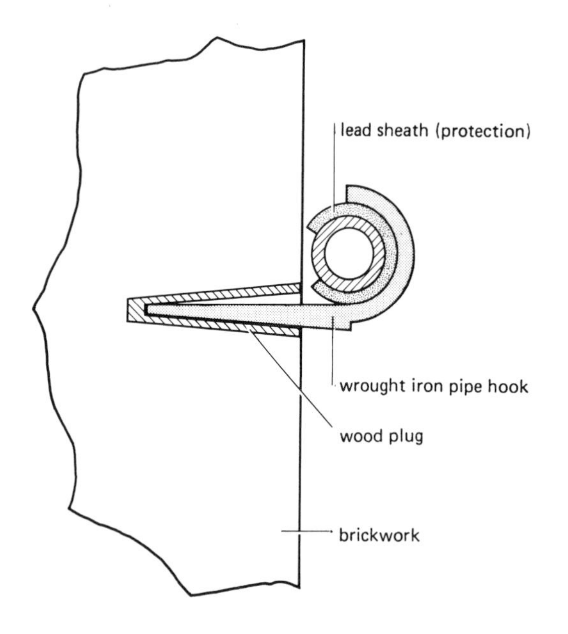

Figure 268 *Alternative method of fixing lead pipe*

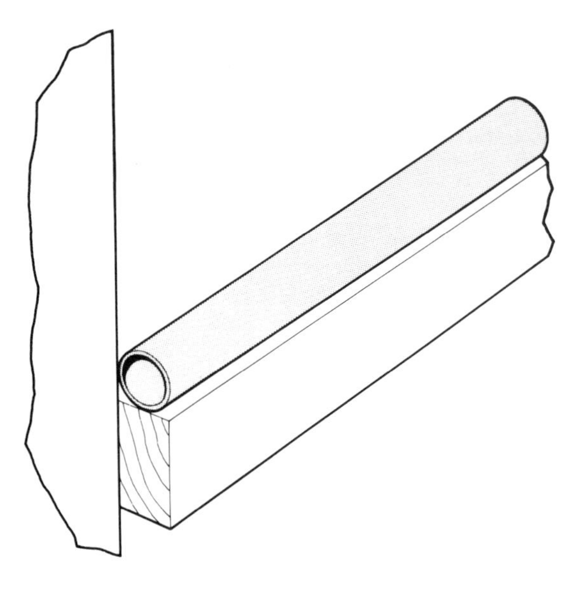

Figure 270 *Timber continuous support for pipe*

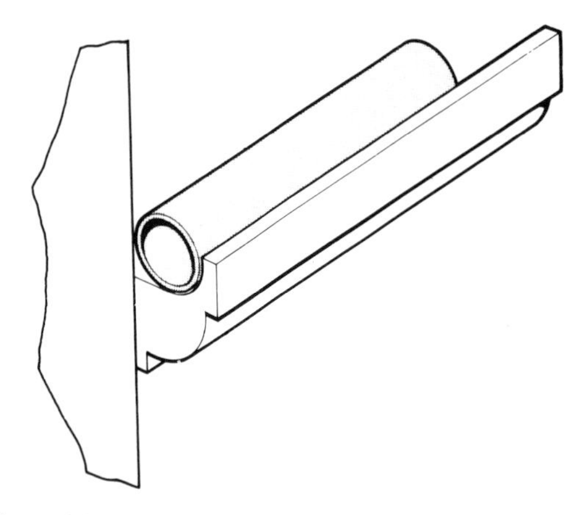

Figure 269 *Continuous support for pipe*

Where appearance is not a dominant factor a plain piece of timber can be used as a continuous support, as shown in Figure 270. The pipe would need to be held in place by hooks or clips.

Pipe fixing in floor spaces
Figure 271 shows the method used for first floor joists and Figure 272 shows the method used in

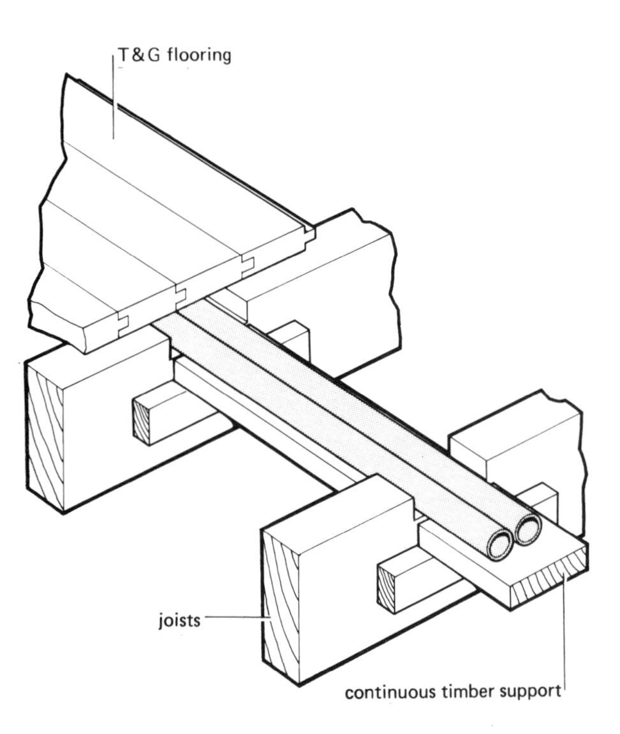

Figure 271 *Pipe fixing for first floor joists*

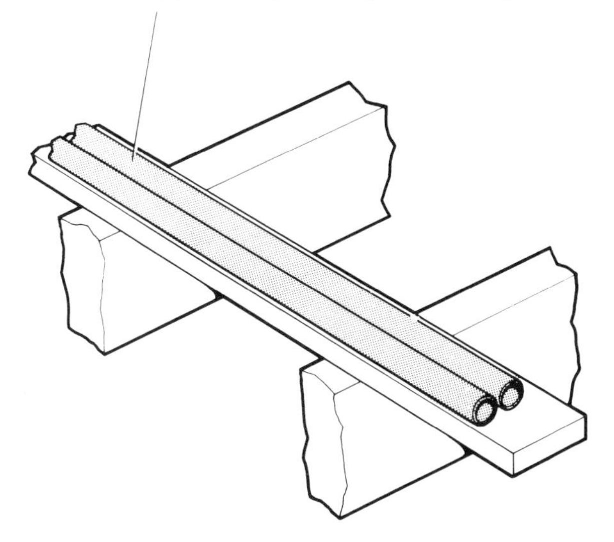

Figure 272 *Pipe fixing in roof spaces*

roof spaces. The pipes should be insulated in roof spaces.

A method of fixing pipes to be run under the ground floor is to secure them to the underside of the floor joists as shown in Figure 273. The space between the site concrete and the ground floor must be ventilated by means of fresh air inlets. This means that this space is very cold and the pipes should therefore be well insulated.

Where timber joists have to be notched to receive the pipes, the amount removed should not be more than is necessary to house the pipes plus a small addition which prevents the pipes being trapped (see Figure 274). Excessive cutting will greatly weaken the joist. It is also advisable to cut the joist as near to the wall as possible, as in Figure 275. It is a better method to drill the joist, but it is difficult to insert the pipes. Hot water (copper) pipes expand quite considerably when hot water is passed through them. This expansion causes the pipe to move and if it is trapped in the joist or by the floorboard, creaking and cracking noises will result.

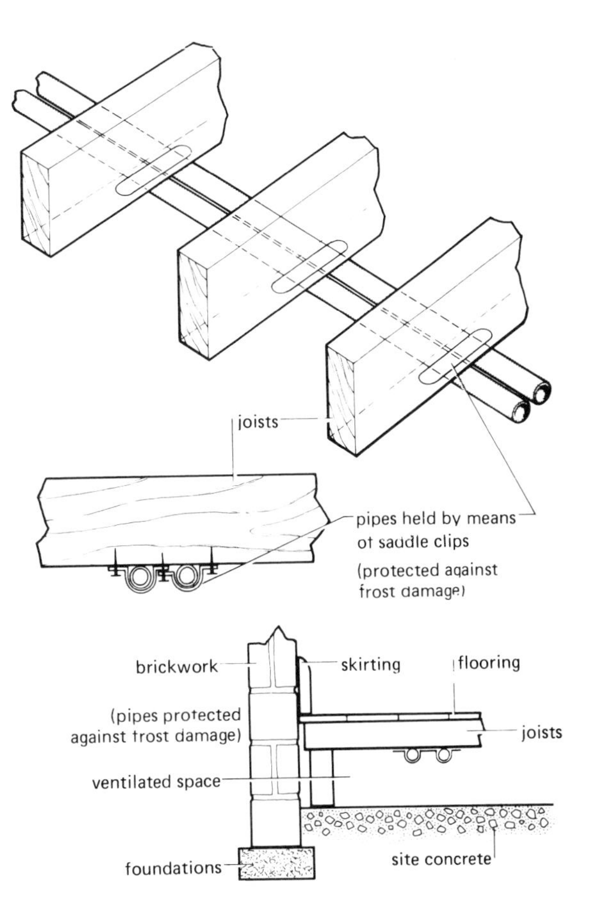

Figure 273 *Pipe fixing under the ground floor*

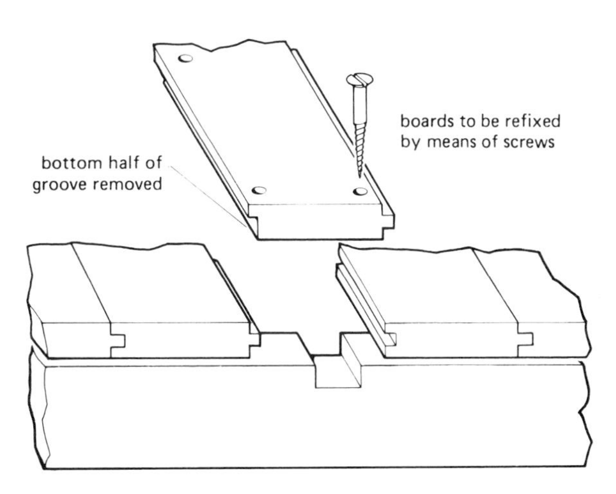

Figure 274 *Refixing floorboards*

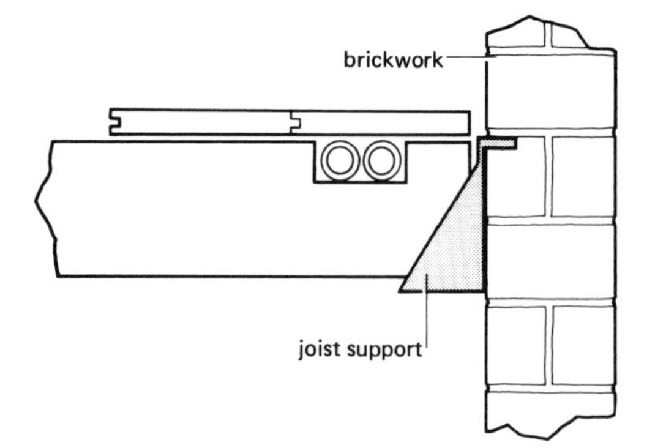

Figure 275 *Joist cut near the wall*

Pipe fixing in brickwork

Sometimes it is desirable to have all the pipework hidden from view. There are several acceptable ways of doing this, of which three are described below.

Method A

This is the easiest and cheapest way to bury the pipe in plaster or cement. However, it is *not approved by many authorities.*

A chase is cut in the brickwork large enough to take the pipe or pipes (see Figure 276). The pipe is wrapped with a protective wrapping such as building paper or thin felt to prevent detrimental action taking place between the pipe and either the cement or plaster. The pipe is then secured in an approved manner by clips or hooks and finally covered by cement or plaster in the normal way.

The disadvantages are:

1 Damage can be done to the pipe by the householder by, for instance, knocking nails etc, in the walls.

Figure 276 *Chase cut in brickwork to house pipe and plastered over*

2 Should there be a leak it is more difficult to carry out the repair.
3 Leaks can occur and considerable damage take place before it is discovered.
4 Pipes are more prone to freezing.

Method B (Figure 277)

This is a more acceptable method, although it must be noted that it is not completely satisfactory. In some ways it is very similar to method A, the pipe being housed in a chase in the brickwork. In some cases the pipe is not secured to the brickwork and the cavity is not filled with plaster. A piece of timber or other suitable facing material is recessed into the plaster and secured by screws counter-sunk into wooden plugs set in brickwork.

The advantages of this method are:

1 The pipes are readily accessible.
2 Minimal damage to decoration should repairs be necessary.

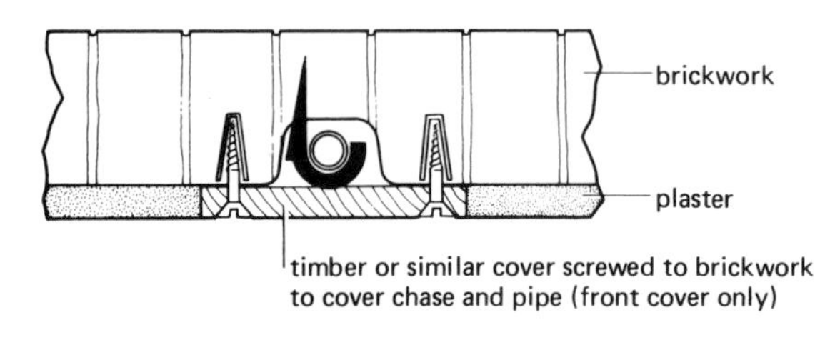

Figure 277 *Pipe covered by timber*

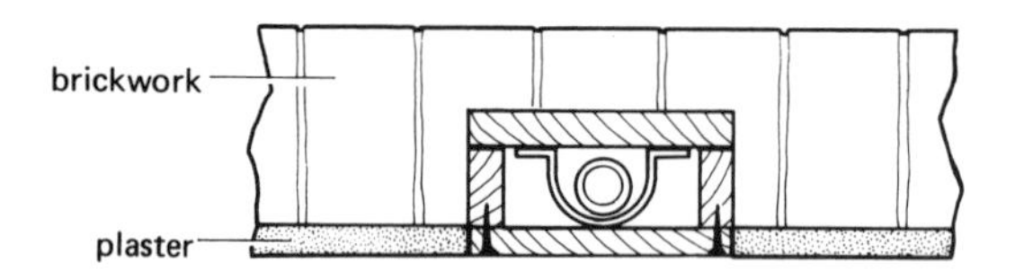

Figure 278 *Complete timber boxing to house pipes*

Method C (Figure 278)

This is by far the best method, although it involves more work and is therefore more costly. In this method a much larger chase is required to house a complete timber lining to accommodate the pipes. The pipes are secured to the back board by clips. It is also normal practice to fill the box with some suitable insulator. The front of the boxing forms a flush finish with the plaster work and is secured by screws.

The advantages of this method are:

1 Neat and tidy cover to unsightly pipework.
2 Pipework readily accessible.
3 Pipework insulated by means of complete timber boxing.
4 Minimal damage to decorations should repairs be necessary.

Self-assessment questions

1 An indirect system of hot water may be installed to prevent:
(a) long runs of pipe
(b) using too much water
(c) the furring up of the pipes
(d) electrolytic corrosion

2 In which of the following places should pipes *not* be situated to reduce the risk of frost attack?
(a) partition walls, roof spaces, solid floors
(b) internal ducts, hollow first floors, internal partitions
(c) skirting boards, suspended ceilings, chases in brickwork
(d) hollow ground floors, eaves, external walls

3 Failure to fit a dip pipe to a back boiler with top connections will result in:
(a) the water boiling
(b) the possibility of reversed circulation
(c) water not heating
(d) the collection of air in the boiler

4 When replacing a floorboard in an upstairs bathroom it is good practice to:
(a) nail it with floor brads
(b) screw it down
(c) leave it loose
(d) replace the old nails

5 What is the minimum size of cold water feed cistern allowed?
(a) 112 litre
(b) 150 litre
(c) 200 litre
(d) 240 litre

6 It is recommended that hot storage cylinders be fixed in a vertical position to:
(a) assist circulation of water
(b) aid stratification
(c) allow free movement of air
(d) assist the filling

7 The circulation of hot water in a domestic hot water system is caused by:
(a) conduction
(b) radiation
(c) stratification
(d) convection

8 The purpose of fixing a secondary circulation to a hot water system is to:
(a) prevent long dead legs
(b) assist the circulation of water
(c) aid the heating process
(d) act as an auxiliary

9 In an indirect system of hot water supply a suitable position for the safety valve is on the:
(a) primary return
(b) secondary return
(c) secondary flow
(d) cold feed

10 In a domestic hot water and heating system the radiators would be connected to:
(a) secondary flow
(b) primary or heating flow
(c) secondary return
(d) hot distribution pipe

7 Sanitary appliances

After reading this chapter you should be able to:

1 Name common materials used for the manufacture of sanitary appliances.

2 Recognize different types of domestic sanitary appliances.

3 Select particular sanitary appliances for stated locations.

4 Understand the basic design criteria applied to sanitaryware.

5 Describe the working principles of domestic sanitary appliances.

6 State the recommended fixing heights of sanitary appliances.

Introduction

Sanitary fittings are usually divided into two groups:

1 Those which are intended to receive waste products from the human body.
2 Those required for dirty, soapy or greasy water.

The first group, which can be referred to as 'soil fitments', includes water closets and urinals, while the second group, often referred to as 'ablutionary' or 'waste fitments', includes baths, wash basins, showers, sinks and bidets.

The general design of sanitary fitments has reached a high standard. For domestic work a great diversity now exists in style and pattern, from costly coloured suites to the more orthodox fitments complying with the basic requirements of the appropriate British Standards.

A good sanitary fitment should be of the simplest possible design, constructed so as to be self-cleansing, and, as far as possible, free from any moving working parts. It should be capable of being easily connected to the appropriate waste pipe or drain, be completely accessible and free from all insanitary casings. The outlet should be simple and capable of permitting the rapid emptying of the fitment. The overflow, where provided, should be efficient and accessible for cleansing purposes.

Sanitary fitments should be constructed of hard, smooth, non-absorbent and incorrodible materials. For domestic use sanitary appliances are usually made from one of the following materials:

Vitreous china
Earthenware
Cast iron
Pressed steel
Plastics (perspex, glass-reinforced polyester)
Fireclay

Table 20 gives the British Standards relevant to sanitary appliances of various materials.

Table 20 *British Standards relevant to sanitary appliances*

Fireclay sinks	BS 1206	1974
Acrylic sinks	BS 4315	1972
Metal sinks	BS 1244 Part 2	1972
Ceramic wash basins	BS 1188	1965
Metal wash basins	BS 1329	1956
Cast iron baths	BS 1189	1972
Steel baths	BS 1390	1972
Acrylic baths	BS 4305	1972
Ceramic washdown WC pans	BS 1213	1945
WC flushing cisterns	BS 1125	1973
Flush pipes	BS 1125	1973
WC seats (plastic)	BS 1254	1971
Quality of vitreous china	BS 1254	1971
Wastes for sanitary appliances	BS 3380	
Glossary of sanitation terms	BS 4118	1981
Sanitary pipework	BS 5572	1978

Sinks

Sinks for domestic purposes are available in a range of sizes and are manufactured from the following materials:
Ceramic glazed fireclay or earthenware
Vitreous enamelled steel
Stainless steel
Fibreglass
Cast iron

Ceramic sinks are generally available in two patterns:
1 London pattern.
2 Belfast pattern.

The London pattern sink is different to the Belfast pattern in that it does not have an overflow. The Belfast sink (see Figure 279) has emerged as the most popular of the ceramic sinks and is available with depths of up to 255 mm. These sinks are fitted with an integral weir type overflow (see Figure 280). Ceramic sinks should be supported on cast iron cantilever type brackets. Fixing heights for sinks may vary according to particular requirements, but are generally fitted 900 mm from floor level to the top front edge of the sink.

Figure 279 *Belfast sink*

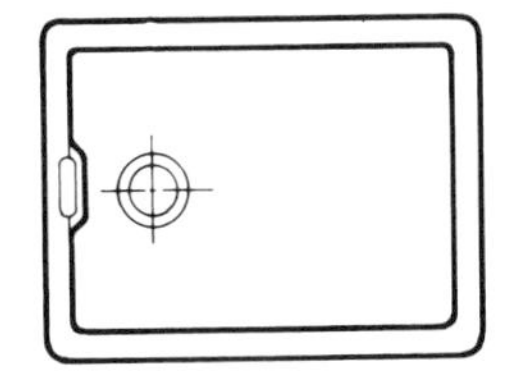

Figure 280 *Belfast sink*

The water supply to ceramic sinks is usually via bib taps fixed to the wall above the sink, but if the sink has been incorporated into a unit or worktop, the supply may be from pillar standard taps fitted to the worktop or unit. Taps fitted above the sink should be at a height to enable a bucket to be filled.

The outlets to ceramic sinks may be rebated or bevelled. The waste fitting will need to be slotted if the sink has an overflow. The fitting itself is made watertight with the use of either putty or mastic compound and metal (usually lead) washers, or rubber and plastic washers. Alternatively, a 'skeleton' type waste fitting may be used (see Figure 281). This employs a long threaded bolt which is tightened from within the sink and pulls the two halves of the waste fitting together on to the outlet section of the sink.

Vitreous enamelled steel and stainless steel are popular materials for sinks. Patterns with single and double drainers and single or double sinks can be obtained. Metal sinks can be supported on cast iron cantilever brackets, a wooden base unit or inset into a worktop. Most metal sinks are provided with tap holes for 12 mm diameter pillar standard taps, or a mixer valve with a swivel arm. Additional fittings such as flexible hand sprays to assist with washing-up are available.

Kitchen sinks are supplied with cold water direct from the service pipe and where mixer taps are fitted they must be of the biflow pattern which does not allow the cold and hot water to mix within the valve. Mixing takes place after the water has left the swivel arm or spout outlet.

Most metal sinks have an overflow combined with waste outlet.

Wastes for domestic sinks should not be less than 38 mm diameter. Control of the outflow is by plug and chain.

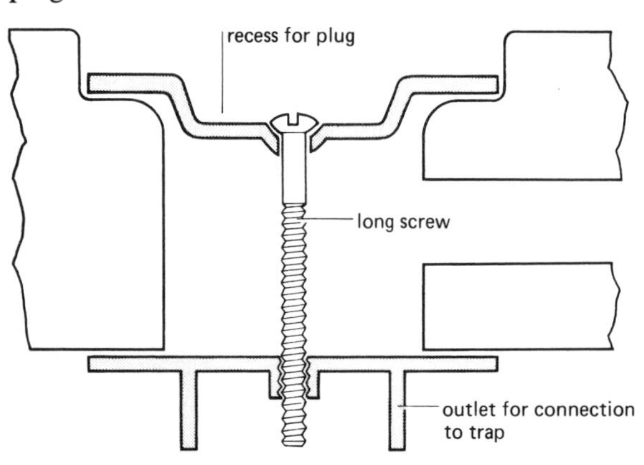

Figure 281 *Skeleton waste fitting*

Showers

A shower is considered to be more hygienic than a bath because the used water is continually washed away. They are particularly well suited for cleaning a number of people simultaneously or in quick succession. A shower occupies less floor space than a bath, and consumes less water.

Shower trays (Figures 282, 283 and 284) are manufactured from:

Acrylic
Glass-reinforced polyester
Ceramic glazed fireclay

Shower trays are available in sizes varying between 700 mm × 700 mm and 950 mm × 950 mm with depths of up to 300 mm.

When the appliance is used only as a shower tray, it will not require a plug and chain and the waste fitting can be flush topped (see Figure 285),

Figure 282 *Ceramic shower tray*

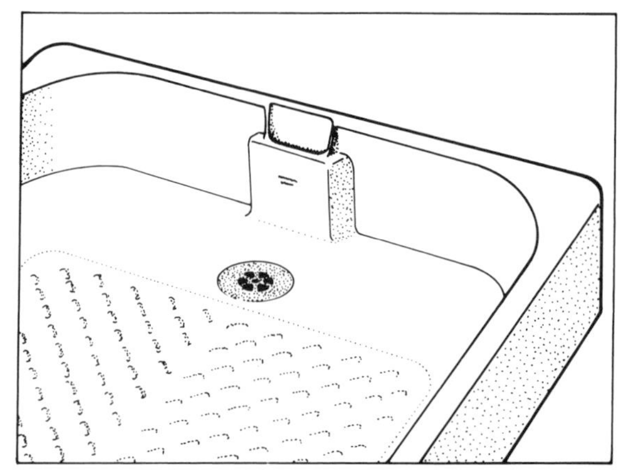

Figure 283 *Ceramic shower tray with integral overflow*

Figure 284 *Acrylic shower tray*

Figure 285 *Flush top waste fitting*

Figure 286 *Recessed waste fitting*

but when the tray is also used as a foot bath the waste fitting must be recessed to receive a plug (see Figure 286). The tray must also be fitted with an integral overflow.

Prefabricated shower cabinets are available complete with all fittings, and when assembled only require connection to water supply and waste services to be ready for use.

Wash basins

Wash basins are made in a wide variety of shapes and sizes, ranging from small cloakroom basins to large bathroom fitments. Each type offers different refinements to the basic functional requirements of a wash basin.

For identification purposes wash basins may be classified into three groups:

1 Wall basins (Figure 287).
2 Pedestal basins (Figure 288).
3 Vanity or inset basins (Figure 289).

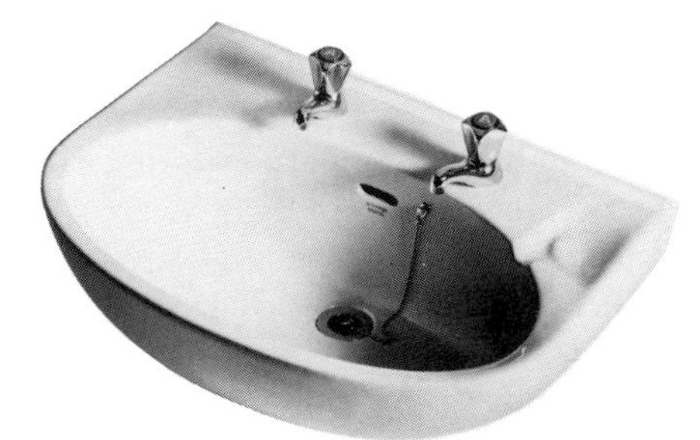

Figure 287 *Vitreous china wall hung basin*

Figure 288 *Vitreous china pedestal basin with integral back overflow*

Figure 289 *Vitreous china inset basin with integral front overflow, concealed mixer valve and pop-up waste*

Wall basins

These are available with various means of support. The fixing may be concealed, employing the use of secret brackets, or cantilevered from the wall by means of a corbel which is part of the basin. This corbel is built into the wall with cement/sand mortar. Traditional bracket fixing involves the use of cast iron support brackets which may be built in (Figure 290) or screwed to the supporting wall (Figure 291).

The main advantage with wall fixed basins is that the floor beneath is left clear.

Wall basins are manufactured to fix to a flat wall surface or to be contained within a 90° corner. This type of basin is called an angle basin (Figure 292).

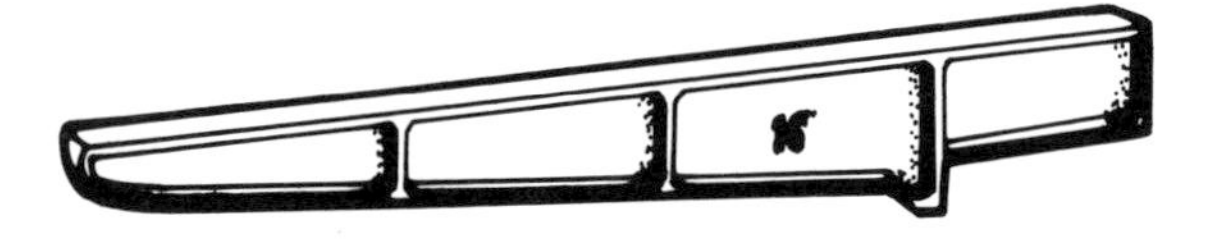

Figure 290 *Built-in cantilever bracket*

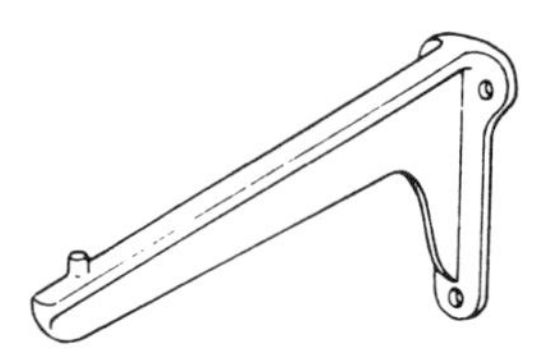

Figure 291 *Screw to wall bracket*

Figure 292 *Vitreous china wall hung angle basin*

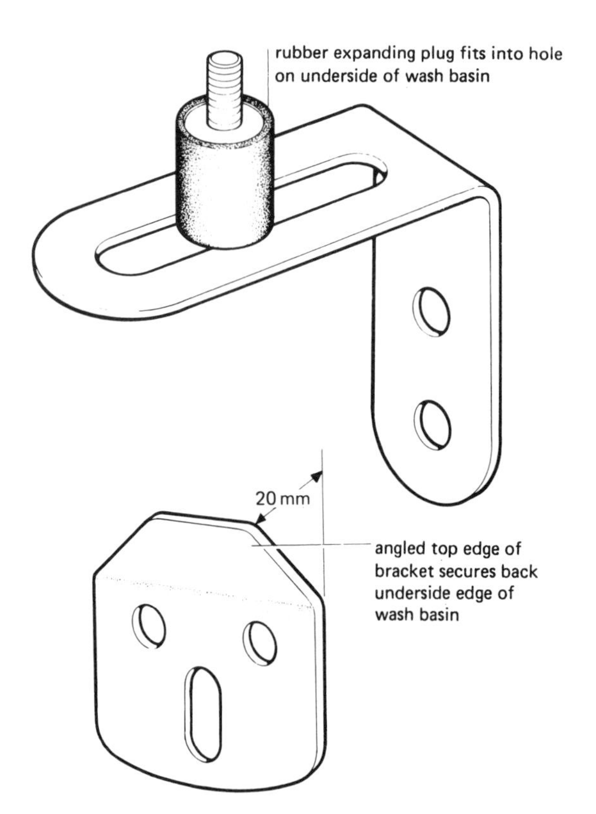

Figure 293 *Wash basin safety brackets*

Pedestal basins

The pedestal provides a means of support for the wash basin and should be large enough to conceal the waste services and supplies. The basin should also be secured to the back wall. This is usually done by screw fixing through holes provided under the back edge of the basin or by using safety brackets (Figure 293). Many manufacturers provide a bracket to secure the basin to the pedestal. The pedestal itself should also be secured to the floor.

Vanity basins

These are manufactured to be inset into a cabinet, worktop or shelf unit. Some types of basin have a metal trim to make the seal between basin and surrounding surface (Figure 294), others are self-rimming (Figure 295).

These basins can be used singly or in range

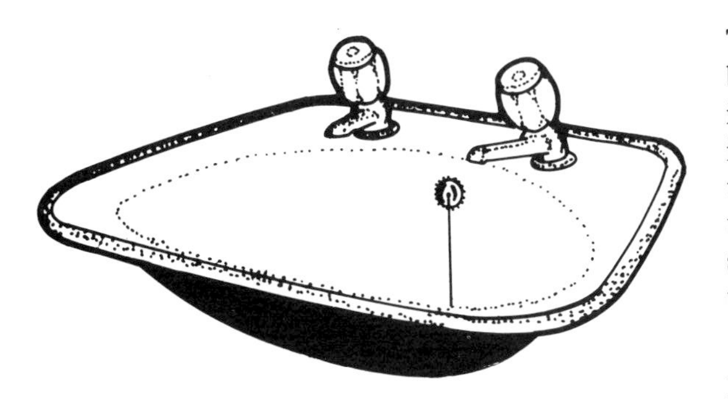

Figure 294 *Pressed steel inset basin with stainless steel trim, combined waste and overflow*

Figure 295 *Vitreous china inset basin with integral back overflow and recessed slotted waste fitting*

form, often saving valuable space. They may also be used to provide supplementary facilities in bedrooms, dressing rooms and cloakrooms.

Wall basins and pedestal basins are usually manufactured from vitreous china or ceramic glazed fireclay. Vanity basins may also be made from these materials and pressed steel, stainless steel and plastic.

The control of the water supply into a wash basin is usually by 12 mm pillar valves which are connected to the hot and cold service pipes. Several manufacturers supply spray taps and mixer valves suitable for wash basins.

The taps or valves discharging water to a wash basin must have their outlets above the flood level of the appliance to obviate the risk of siphonage or contamination.

The waste outlet from a wash basin should never be less than 32 mm in diameter and should incorporate a slot to accommodate the overflow integral to the basin. Control of the outlet is usually by plug and chain, although pop-up waste units are available. Self draining recesses to hold soap are formed within the horizontal top edge of the basin.

Baths

Baths designed for domestic use vary little in basic shape whether made from cast iron, pressed steel or acrylic materials (see Figure 296). However, refinements in design are available beyond the standard rectangular shape: handgrips, recesses for soap, non-slip bottoms and drop fronts to make access easier and safer are features now incorporated by most manufacturers.

Also available are corner baths which use up more floor space and are generally more expensive than a rectangular bath, and 'sitz' baths

Figure 296 *Typical domestic bath complete with shower, waste overflow, handgrips, side and end panels*

which are short and deep and incorporate a seat. This type of bath may be used where floor space is limited and is also suitable for elderly people, since the user maintains a normal sitting position.

The following materials are used for the manufacture of baths:

Vitreous enamelled cast iron
Vitreous enamelled pressed steel
Acrylic plastic (perspex)
Glass-reinforced polyester

The control of the water supply into a bath is usually by 19 mm pillar valves or via a mixer valve which may also incorporate a diverter to direct the flow of water to a shower rose.

The waste outlet from a bath should never be less than 38 mm in diameter. The waste fitting may incorporate a slot to receive any water which leaves the bath via the overflow. Alternatively the overflow may connect into the trap itself (if permitted by the water authority), or discharge through the external wall.

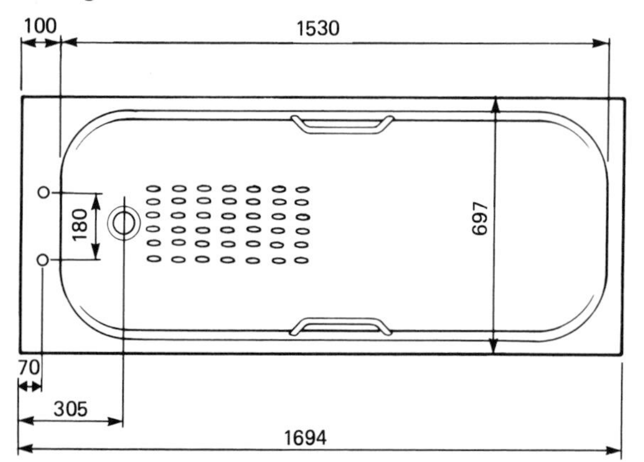

Figure 297 *Plan of acrylic bath with approximate dimensions in millimetres*

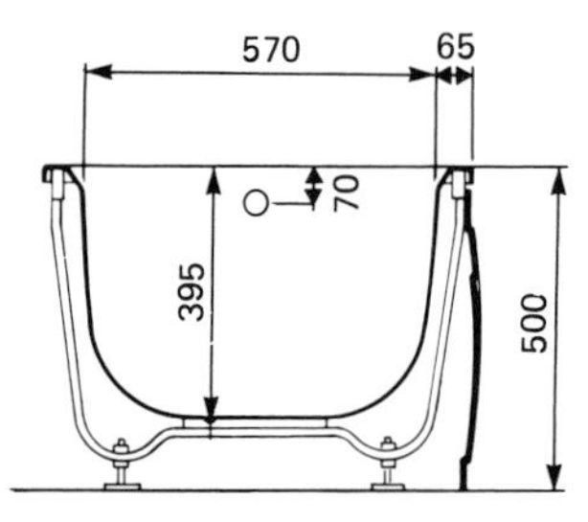

Figure 298 *Section through acrylic bath showing approximate dimensions and typical support detail*

Rectangular baths are manufactured in standard sizes ranging from 1.5 m to 1.85 m in length and 0.7 m to 0.85 m in width (see Figure 297). Most baths have four or five feet which are adjustable to gain the required height and level position (see Figure 298). Baths are fitted level as the necessary fall is provided in the bottom by the manufacturer. The height at which a bath is fitted may be governed by several factors which may include:

1 Structural limitations.
2 Height of bath panel (if moulded).
3 Customer requirements.
4 Sufficient space to allow waste pipe connection to be made.

Generally baths are fitted as low as is possible for ease of access, safety and good appearance. When fixing it must be remembered that using a bath exerts considerable force on its feet and if the feet are resting on a timber floor settlement may take place. This can be eliminated by fixing steel plates underneath the feet to distribute the load and building in the side and end sections of the bath to their adjoining walls. Some manufacturers provide brackets for securing the bath to the wall.

There are several types of leg, cradle, or support available for baths and when fixing any bath it is most important to adhere to the manufacturer's fitting instructions.

Water closets

A water closet consists of a pan containing a quantity of water and a vessel or device capable of providing a flush of water to remove excreta, wash any soiled surfaces clean and re-seal the trap in the pan.

Water closet pans are manufactured from:

Vitreous china
Glazed earthenware
Stainless steel

Flushing cisterns are manufactured from:

Vitreous china
Glazed earthenware
Stainless steel
Pressed steel
Cast iron

Galvanized mild steel
Plastics materials

Water closet pans may be divided into *two* categories: *washdown* (Figure 299) and *siphonic* (Figure 300).

Within these categories, there are several types of each, determined by details such as whether it is floor supported (pedestal) or wall supported (corbel), and the type and position of outlet: horizontal, 'P' or 'S', straight, or turned right or left handed.

Washdown pans

Washdown pans are the most common form of water closet, and the modern pedestal washdown pan has been developed from earlier short and long hopper types.

Corbelled washdown closets are used in situations where hygiene is particularly important or where a clear floor area is required. Two types are available, one requiring a firm solid back wall into which the corbel extension of the pan is built, and the other which rests on a metal frame fixed to the floor and the back wall.

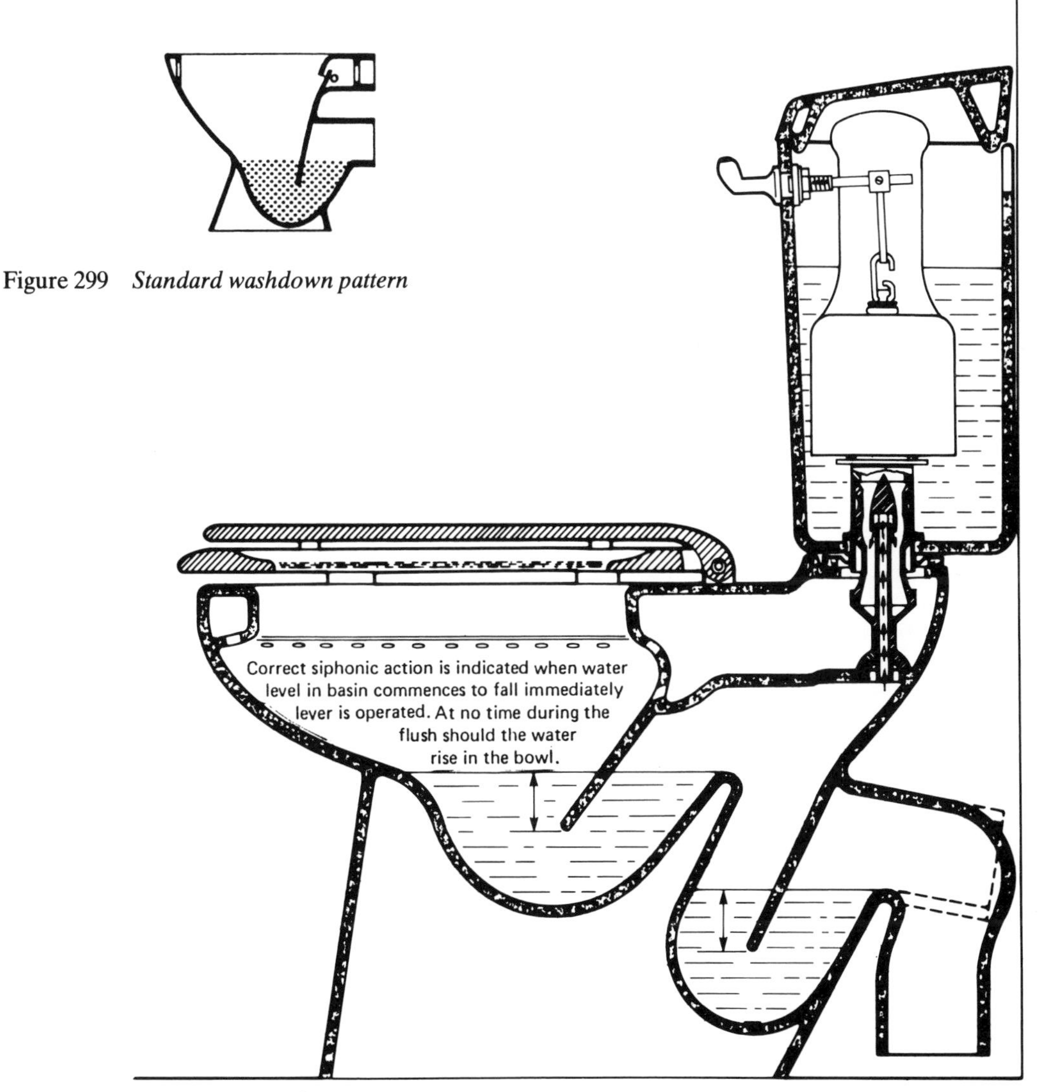

Figure 299 *Standard washdown pattern*

Figure 300 *Siphonic water closet suite*

Siphonic pans

Siphonic water closets are more positive and silent in operation than are washdown closets, and rely on siphonic action to remove the contents from the pan. Siphonic closets are either single trap or double trap (as in Figure 300)

Single trap

Single trap siphonic pans are designed so that the bowl of the pan and its outlet form the short and long leg of a siphon. The outlet or long leg is shaped so as to restrict the outlet flow of water.

The operation is as follows. The cistern is flushed and water passes via the open flushing rim into the pan and through to the long leg of the siphon. Due to the partial restriction in the outlet, this momentarily fills with water which, as it runs away, causes siphonic action which removes most of the water and contents of the trap. The trap is resealed with water from a special 'retention' or 'after flush' chamber built into the pan itself.

Double trap

Some double trap siphonic pans (see Figure 300) use a pressure reducing valve (A in Figure 300) which is located between the outlet leg of the cistern siphon and the air chamber between the two traps. The operation of this type of suite is as follows.

The cistern is flushed, and water flows past the pressure reducing valve A. Air is drawn through the valve from chamber B so creating a drop in air pressure in chamber B. This causes the water level in the pan to drop and siphonic action commences. The water from the cistern enters the pan via the perforated flushing rim, assisting with the siphonic action, and washing clean the pan surfaces. As the siphon ceases, this water reseals the traps in the pan.

Pedestal type pans are manufactured with two or four holes through their base to provide screw fixing to the floor beneath. Ideally, a pan should be fixed so that it can be disconnected at a later date if necessary. The pan must be securely fixed to the floor, eliminating the possibility of movement.

The method of jointing the pan outlet to the drain/discharge pipe will depend upon factors such as:

1 Whether the supporting floor is solid (concrete) or suspended (timber).
2 The material from which the drain/discharge pipe is manufactured.
3 Whether the pipe connection is spigot or socket/collar.

When a WC pan is supported by and fixed to a solid floor, the method of jointing may be rigid or flexible. If the floor supporting the WC pan is suspended and movement or settlement may occur, the jointing method should provide flexibility to allow for any subsequent structure movement (see Figure 301).

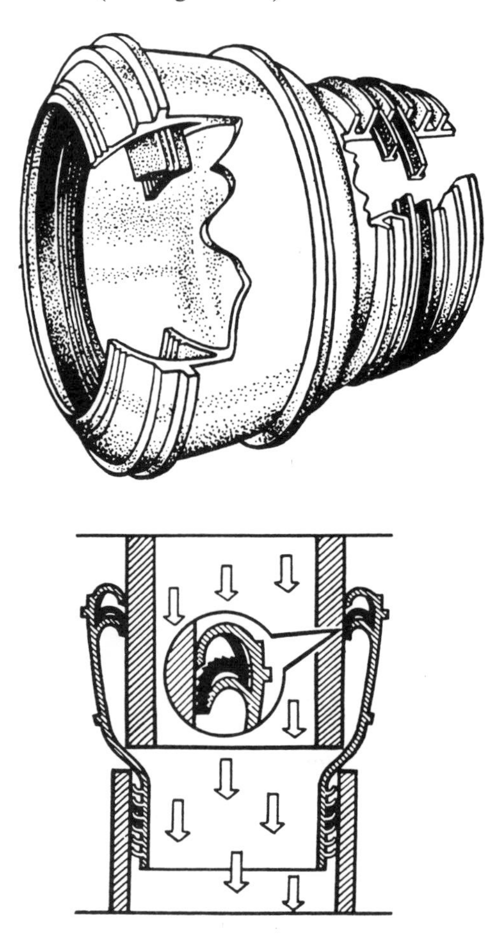

Figure 301

Flushing cisterns

Modern types of flushing cisterns are also known as water waste preventors (WWPs) because, under normal conditions, water cannot leave the cistern other than by siphonic action. There are two main types of WC flushing cistern:
1 Plunger or piston (Figure 302).
2 Bell (Burlington) (Figure 303).

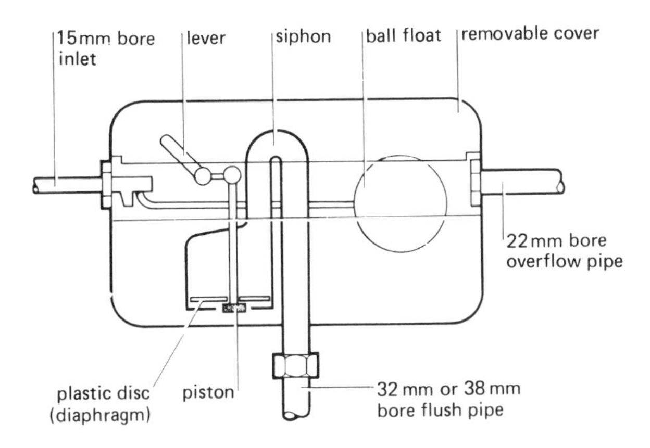

Figure 302 *Plunger or piston flushing cistern*

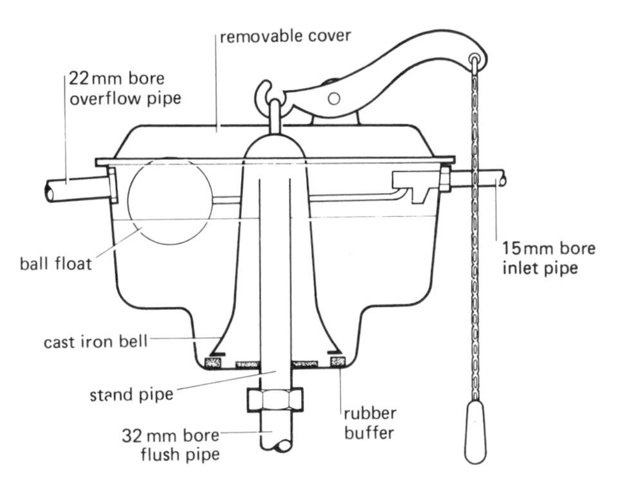

Figure 303 *Bell (Burlington) cistern*

Both types employ siphonic action to remove the water from the cistern to the flush pipe/WC pan. The capacity of the cistern is governed by the water supply authority, most of whom stipulate a 9 litres flush of water.

The cisterns operate as follows.

Plunger or piston cisterns

When the lever handle or chain pull is operated, the diaphragm plunger is lifted. As it rises, it displaces water over the crown of the siphon. This water falls down the long leg of the siphon taking some air with it which creates a reduction in air pressure in the siphon. The greater air pressure acting upon the surface of water in the cistern forces water past the diaphragm and through the siphon until the water level is low enough in the cistern to permit air to enter the siphon. This breaks the siphonic action and causes the flush to cease. The cistern refills via the ball valve to its working water level.

Some manufacturers produce a dual flush cistern. These incorporate an air relief tube (Figure 304) in the diaphragm chamber which breaks the siphonic action when the cistern is half emptied so reducing water consumption. The dual flush operates as follows.

When the lever handle is depressed and immediately released, the siphonic action commences and empties the cistern until air is drawn in through the air relief tube. This breaks the siphonic action at this level, so that only 4.5 litres of water is used.

When a full 9 litre flush is required, the lever handle is depressed and held in this position. This causes the diaphragm to seal the air relief tube and a normal 9 litre flush takes place, the handle being released when the siphonic action ceases.

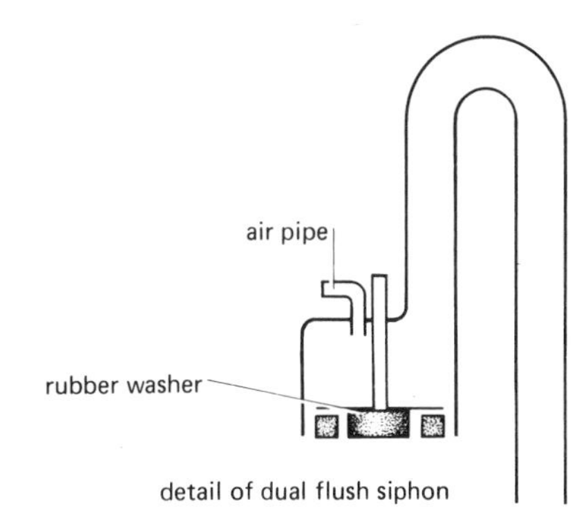

Figure 304 *Dual-flush siphon incorporating air relief tube*

Bell flushing cistern
A pull action on the chain causes the lever arm to move on the fulcrum and raise the bell inside the cistern. When the chain is released, the bell drops and displaces water which flows down the funnelled standpipe into the flush pipe. As this water drops, it takes air with it creating a reduction of air pressure inside the bell. Atmospheric pressure acting upon the surface of the water in the cistern forces this water under the lower edge of the bell through the bell and down the standpipe. This siphonic action ceases when the water level in the cistern is low enough to permit air to enter the bell and equalize pressure.

Fixing of cisterns
Cisterns may be fixed at either high level, low level, or close coupled to the WC pan. The high level position is most efficient for washdown WC pans and should be fixed 1.8 m above floor level to the underside of the cistern to ensure a good flush. Plunger or bell type cisterns may be used for high level suites.

Low level and close coupled cisterns should be fitted according to manufacturers' instructions, and are of the plunger type. Bell type cisterns are only suitable for high level installation. Modern practice favours low level or close coupled positions because of their neater appearance and ease of accessibility for maintenance and cleaning. Inlet and overflow connections may be either side or bottom according to site circumstances.

Flushpipes

Because of the greater head available above the WC pan, high level cisterns require a flush pipe of only 32 mm in diameter, whereas low level cisterns require a flush pipe diameter of 38 mm to compensate for loss of head. Close coupled suites do not have a flush pipe. All flush pipes should be as straight as possible as bends impair the efficiency of the flush.

Flush pipes are made from various materials including:
Lead
Copper
Plastics
Galvanized mild steel
Stainless steel
Enamelled pressed steel

They are identified by the number of sections, i.e. one-, two- or three-piece, and the location where the cistern is fitted in relation to the WC pan, i.e. low level, high level, backwall fixing, side wall fixing.

The flush pipe is usually jointed to the cistern outlet in one of two ways. The first method is by a 'cap and lining' type connection, where the end of the flush pipe has a formed shoulder which is sealed on to the cistern outlet via a washer. The second method uses a compression type joint where the flush pipe passes up inside the cistern outlet and the seal is made via a neoprene 'O' ring which may be captive in the siphon outlet, or loose to be compressed via the tightening of a nut.

The flush pipe may be jointed to the WC pan by: an internal rubber or plastic cone (Figure 305); an external rubber or plastic cone (Figure 306); or a Sutton's collar (Figure 307).

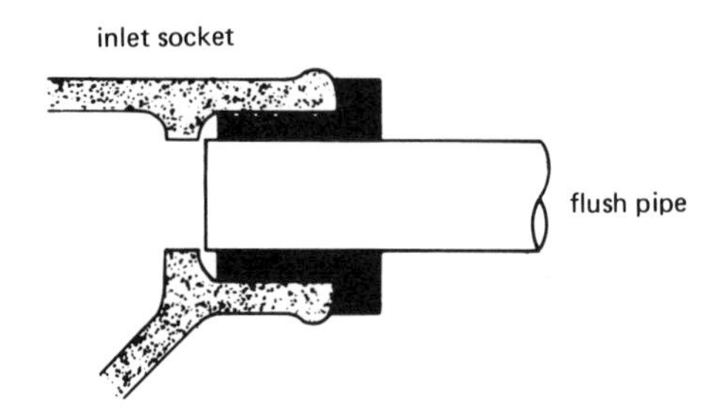

Figure 305 *Internal connector*

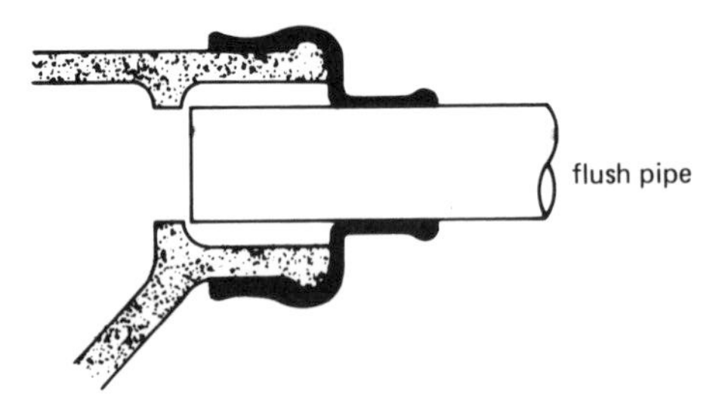

Figure 306 *External connector*

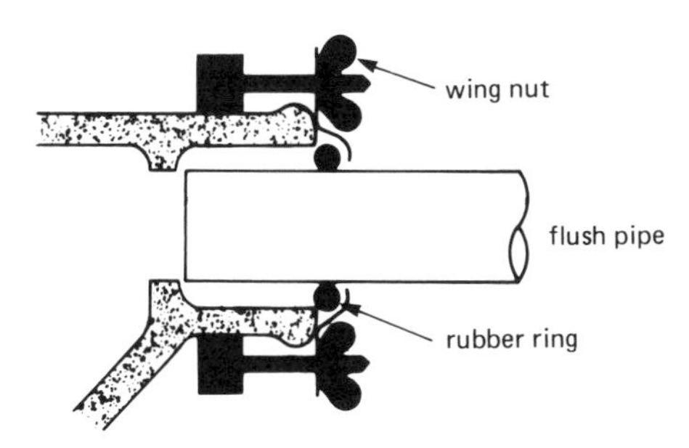

Figure 307 *Clamping ring (Suttons collar)*

With all flush pipe connections, it is essential that the pipe enters the inlet collar centrally and that no jointing compound is allowed to enter into the flushing rim. Otherwise a faulty, inefficient flush will result, and one side of the pan will receive more water than the other.

Bidets

A bidet is a sanitary fitting used for washing the lower parts of the body and in appearance resembles a pedestal WC – but is shaped differently to suit its special purpose. The bidet provides a convenient means of cleaning the excretory organs and is useful to women during periods of menstruation. A secondary but nevertheless important use of the bidet is a footbath. For identification purposes, bidets are usually classified in two distinct types:

1 submerged inlet (Figures 308 and 309).
2 over-rim-supply (Figure 310).

Most bidets are designed in such a way that the waste can be plugged and the bowl filled with water and the lower parts of the body washed much more conveniently than by use of a conventional bath. There is also not such a waste of water because the contents can be quickly run off and the ascending spray used as a swilling douche.

The tap controls are conveniently placed to enable the bather to adjust the flow rate and temperature of the water.

With through-rim-supply bidets (a type of submerged inlet bidet), the rim seat is warmed with a supply of hot water which then fills the bowl for washing. Alternatively, the water supply can be diverted through the douche, which is available in either spray or jet nozzle, whichever is preferred. Most regional water authorities have special requirements for bidets with douche attachment (submerged inlet).

Figure 308 *Vitreous china bidet with concealed mixer valve, rim supply, centre spray douche, pop-up waste and integral overflow*

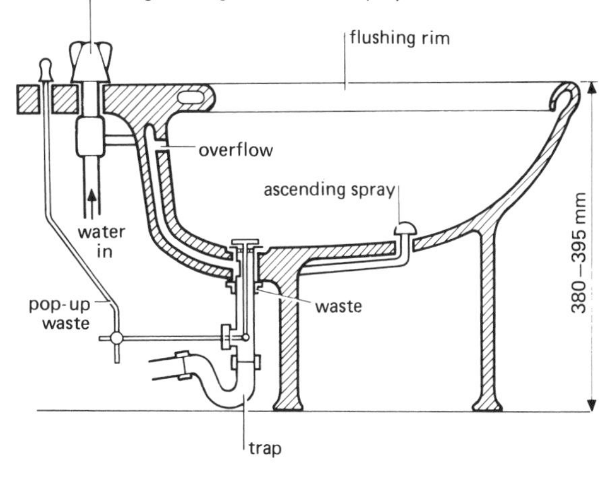

Figure 309 *Section through bidet showing flushing rim, ascending spray and pop-up waste*

Figure 310 *Vitreous china rimless bidet with over-rim-supply and integral overflow*

Bidets with over-rim-supply are simpler. They have no douche, just the bowl for washing purposes, and are supplied with water in the same manner as a wash basin.

Bidets for domestic use are made from vitreous china.

Self-assessment questions

1 The most popular pattern of ceramic sink is called:
(a) Chelsea
(b) Glasgow
(c) Belfast
(d) Croydon

2 The internal diameter of a waste pipe from a domestic sink should not be less than:
(a) 25 mm
(b) 32 mm
(c) 38 mm
(d) 50 mm

3 Certain types of sanitary appliance incorporate a built-in type of support, enabling the appliance to be fixed clear of the floor. This support is called a:
(a) lug
(b) corbel
(c) shelf
(d) bracket

4 A suitable material for the manufacture of WC pans is:
(a) vitreous china
(b) pressed steel
(c) vitrified clay
(d) galvanized mild steel

5 The depth of water seal on a washdown WC pan should not be less than:
(a) 30 mm
(b) 50 mm
(c) 80 mm
(d) 100 mm

6 An overflow from a flushing cistern should not have a bore less than:
(a) 8 mm
(b) 12 mm
(c) 16 mm
(d) 19 mm

7 An approved method of jointing a plastics flush pipe to a WC pan is to use a:
(a) sand and cement joint
(b) solvent weld joint
(c) rubber or plastics cone
(d) putty joint

8 Where movement or settlement may occur the method of jointing a WC pan to its drain or discharge pipe should be:
(a) rigid
(b) flexible
(c) permanent
(d) cement mortar joint

9 A pipe used for the conveyance of water from a flushing cistern to a WC pan is called a:
(a) feed pipe
(b) service pipe
(c) overflow
(d) flush pipe

10 The recommended capacity for a WC flushing cistern is:
(a) 4.5 litres
(b) 7 litres
(c) 9 litres
(d) 12.5 litres

8 Roofwork and sheet weatherings

After reading this chapter you should be able to:

1. Recognize types of roofing materials used by the industry, i.e. lead, copper, aluminium.
2. Appreciate thickness grade and size of sheet for differing purposes.
3. Recognize different tools and understand their use.
4. Describe methods of jointing different materials.
5. Describe methods of fixing different materials.
6. Appreciate the preparation of building surfaces to receive sheet weatherings.
7. Understand the principles of chimney weatherings.
8. Appreciate the use of sheet weathering as a damp proof course.

Introduction

The covering of complete roofs and the weathering of component parts of buildings with the use of sheet weathering material forms an important part of the work of the present-day plumber. Although the covering of large roofs and intricate weatherings has over the past few years tended to become the work of the roofing specialist, that specialist has generally progressed from the plumbing craft. In any case, the plumber must of necessity understand the principles of roofing and be able to perform to a satisfactory level in the manipulation and fixing of all the different types of material used for this purpose. There are many different types of roofing material in use today, the most common of which are:

1. Lead
2. Copper
3. Aluminium
4. Zinc
5. Non-metallic sheet (Nuralite)

In this chapter we will deal basically with the material, the tools used to work the material, joints, fixings and some of the most common weathering details on domestic properties. Lead, copper and aluminium are dealt with in detail, since these are the roofing materials encountered most often.

Lead

Sheet lead

This has for many years been used extensively for the weathering of roofs. It says volumes for the suitability of the material that it commands a large share of the market today, in spite of the many changes that have taken place and the introduction of new materials. It is still used extensively for all types of roofs with traditional methods of working and fixing, and the introduction of cladding some years ago brought about an increase in its use.

Most of the lead sheet used in building is manufactured on rolling mills and is best described as milled lead sheet. Slabs of refined lead about 125 mm thick are first rolled out to a thickness of 25 mm then cut into suitable sizes and passed backwards and forwards through the mill until reduced to the required thickness.

Milled lead sheet is manufactured to British Standard 1178 Milled Lead Sheet and Strip for Building Purposes. BS 1178 lays down requirements that control the quality of the material. These include stipulations that the material be free from defects such as inclusions and laminations, that tolerance on thickness be not more than ±5 per cent and that the chemical composition be not less than 99.9 per cent lead. The preference, as traditionally, is that the composition of lead sheet should not contain any alloying elements that would significantly affect the characteristic softness and malleability of lead.

Sizes

In the past the thickness of milled lead sheet was defined by the weight per square foot (3 lb lead, 4 lb lead, 5 lb lead, and so on). However, measurement in metric units was introduced following the publication of a metric version of BS 1178 in 1969, which provides for a range of six sizes of lead sheet defined by thickness in millimetres. For easy identification and because the range of metric sizes corresponds closely to the traditional range expressed in lb/sq ft, they have code numbers – 3, 4, 5, 6, 7 and 8. The substance of lead sheet is therefore specified by its BS code, e.g. no. 4 lead sheet (thickness 1.80 mm).

The standard range of thickness is given in Table 21.

Table 21

BS code no.	*Thickness mm*	*Weight kg/m²*
3	1.25	14.18
4	1.80	20.41
5	2.24	25.40
6	2.50	28.36
7	3.15	35.72
8	3.55	40.26

Lead sheet and strip may carry colour markings for easy recognition of the thickness in store or on site as follows:

Code no. 3 – green
Code no. 4 – blue
Code no. 5 – red
Code no. 6 – black
Code no. 7 – white
Code no. 8 – orange

Milled lead sheet is supplied by the manufacturer cut to dimensions as required or as large sheets 2.40 m wide and up to 12 m in length. Lead strip is defined as material ready cut in widths from 75 mm to 600 mm. Supplied in coils, this is a very convenient form of lead sheet for many flashing and weathering applications.

Cast lead sheet

Cast lead sheet is still made as a craft operation by the traditional method of running molten lead over a bed of prepared sand. A comparatively small amount is produced by specialist lead working firms, largely for their own use, in particular for replacing old cast lead roofs and for ornamental leadwork. There is no British Standard for this material. The available sizes of sheets made in recent years have varied from 2.75 m × 1 m to 5.5 m × 2 m. Skilled casters can cast sheets to an accuracy of 0.02 mm thickness on average in a range of thicknesses corresponding to the standard sizes for milled lead sheet code nos. 6, 7 and 8. Apart from the surface texture and lack of constancy in thickness, cast lead sheet, for all practical purposes, is not significantly different from milled lead sheet.

Machine-cast lead sheet

Thin lead sheet for building applications is made by a continuous casting process as well as by milling. A rotating water cooled drum is partly immersed in a bath of molten lead and picks up a layer of solid metal which is removed over a knife edge that scrapes the drum as it rotates. The thickness of the sheet being cast is controlled by varying the speed and temperature of the drum. Lead sheet made by this process is generally limited to thicknesses between 0.4 and 1.2 mm with a maximum width of about 1.2 m.

Properties of lead

Malleability

Lead is the softest of the common metals and in a refined form is very malleable. It is capable of being shaped with ease at ambient temperatures without the need of periodic softening or annealing, since it does not appreciably work harden. Lead sheet can, therefore, be readily manipulated with hand tools without risk of fracture and it is in taking advantage of this property that skill in working and fixing lead sheets has largely evolved. By the technique of bossing, lead sheet can be worked into the most complicated of shapes. Lead flashings can be readily dressed *in situ* to get a close fit to the structure even when the surface is deeply contoured as is the case with some forms of single-lap roof tiling.

Thermal movement

Most of the uses of lead sheet in building are those where it is fixed externally and is thus subjected to conditions of changing temperature. The coefficient of linear expansion for lead is 0.0000297 for 1 °C and it would not be rare in the summer, where the lead is continuously in the sun, for it to vary in temperature daily by 40 °C. A 2 m length of lead sheet could therefore, increase in length by 2 m × 40 × 0.0000297, i.e. about 2.3 mm. If the expansion and resultant contraction on cooling the lead cannot take place fairly freely, there will be a risk of it distorting and of a subsequent concentration of alternating (fatigue) stress which, over a long period, can cause cracking of the lead.

It is, therefore, of first importance with lead sheet fixed externally, as with other metals, to limit the size of each piece so that the amount of thermal movement is not excessive and also to ensure that there are no undue restrictions on this movement. Long experience has shown that it is quite practical to make provision for the thermal movement of lead and take full advantage of its other outstanding properties to get external leadwork that will last a long time. Recommendations for limiting the size of pieces of lead sheet are given with the descriptions in this book of the various uses. Nowadays, somewhat thinner lead sheet is used than was the common practice in the past and as a general rule the thinner the lead the smaller each piece should be.

Fatigue and creep resistance

A factor that affects the strength of lead and its dilute alloys is the size of and uniformity of the grain structure of the metal. Lead, like other metals, is crystalline in nature and grain size describes the size of the crystals. In lead, the grain is readily visible after a small magnification when the surface is treated by cleaning and etching. Basically, the purer the lead the coarser the grain and variation in its size.

The presence of very small amounts of some other metals can modify the grain structure, in particular make it smaller and more uniform so that the fatigue resistance of the metal is improved and it is then better able to cope with stresses arising from thermal movement. The cracking of the lead as a result of excessive fatigue stressing is intercrystalline.

'Creep' is the tendency of metals to stretch slowly in the course of time under sustained loading, and is a factor of significance in external leadwork. The term creep should not be applied to the slipping of lead down a pitched roof when the fixings have failed to give adequate long-term support.

Although the composition of the lead for making lead sheet for building purposes as laid down in BS 1178 of 1969 does not stipulate the inclusion of grain refining agents, the usual practice of manufacturers is to use such compositions. Usually they use a copper-bearing lead which can include up to 0.06 per cent copper in the lead of basically 99.9 per cent purity required by BS 1178. For all practical purposes this modification in composition does not affect the malleability of lead sheet.

Fatigue can occur, causing premature failure of lead sheet, because of faulty design or method of fixing. For instance, the use of oversized sheets or allowing too little freedom of movement for the sheet to expand and contract due to changing temperature, can both cause fatigue.

Where failure occurs in lead sheet on an existing building due to bad design or workmanship, it is most likely to be a fatigue failure since this can be caused by moderate oversizing of sheets or by the restriction of free movement due to incorrect fixing. Creep failure can only result from considerable oversizing of sheets. It would therefore, be preceded by a very premature fatigue failure, i.e. the sheet would fail by fatigue before the creep problem had time to become serious.

Patination of external leadwork

Lead is extremely resistant to corrosion by the atmosphere whether in town, country or coastal areas. In time, lead develops a strongly adhering and highly insoluble patina, the natural colour of which is silver-grey. Because of the insolubility of the patina, rainwater running off from weathered lead takes nothing into solution to stain or harm adjoining materials such as stonework. The patina of old leadwork often appears darker than the natural silver-grey when there is a coating of grime. Noticeably, parts of the surface of the old leadwork that are less exposed than others to the scouring action of wind and rain will appear darker and, of course, this is a condition more likely to be seen in towns than in the country or coastal areas. However, since grime forming emissions from the burning of fuel have been greatly reduced, external leadwork, like buildings in general, can be expected to present a more natural weathered appearance in the future.

While lead weathers so well it is nevertheless important to bear in mind its behaviour in the first stages of exposure. When freshly cut and exposed to the air lead forms a surface film of one of its oxides which imparts a dark grey appearance. Generally it will then slowly develop an even coloured and adherent patina by reaction with carbon dioxide and more importantly with sulphur dioxide in the atmosphere. Recent investigations have established that the permanent patina of external leadwork, even on leadwork many years old, is largely lead sulphate, irrespective of the exposure, whereas in the past it was thought to be predominantly a lead carbonate, even in town environments. However, the initial patina may begin to form somewhat patchily, particularly when the weather is showery shortly after the lead has been fixed. Such patches and streaks of light grey patina develop quite quickly. This initial patina is a lead carbonate which, while soluble in atmospheric moisture, is only loosely adherent to the lead and can wash off. Eventually the permanent patina will develop and in many forms of leadwork incidence of streaky initial patination will be of little consequence.

Compatability with other metals

The general experience is that lead can normally be used in close contact with another metal – such as copper, zinc, iron and aluminium – without corrosion by electrolysis. For example, no corrosion problems arise in the traditional use of copper nails and clips as fixings for lead and there is wide experience of the satisfactory use for lead flashings with patent glazing formed from aluminium bars. In marine and some industrial atmospheres it may be advisable to avoid direct contact between lead and aluminium because of the danger of accelerating corrosion of the aluminium and the guidance of the makers of the aluminium components should be sought.

Corrosion from timbers

Dilute solutions of organic acid leached from hardwoods can cause lead to be slowly corroded. Furthermore the corrosive effect of continuous condensation on the inner face of roofing and cladding (see below) can be exacerbated when it takes up organic acid from hardwood members of the substructure. Investigation has shown that impregnation of softwoods with preservative and fire retardant solutions does not, in itself, increase the risk of attack.

There are organic acids in cedar roofing shingles which can be taken up by light rain and dew to form a dilute acid solution which will slowly corrode lead flashings on to which it runs. In this situation it is advisable to protect the lead with a coating of bitumen paint for a few years during which natural weathering of the shingles will leach out the free acid.

Condensation

Condensation can exist in well heated buildings in which warm moist air will filter through the external walls and roof structure and, unless prevented, condense on the inner face of an impermeable cladding or roof covering. In the long term, condensation can cause significant corrosion of lead by slowly converting the metal mainly to lead carbonate. The importance of incorporating a vapour barrier in the external fabric where there is a risk of such condensation arising is well understood as is also the need to ventilate the space behind metal roof coverings and claddings where the substance is of frame construction. Past experience is that the incidence of significant corrosion of lead roof coverings and cladding through condensation is rare but the need to take these precautions against condensation should, nevertheless, be borne in mind.

Lichen on roofs

Slow corrosion of non-ferrous metals by dilute organic acids also arises with gutter linings of old buildings in country areas when lichen or similar mosses are growing on tiled or slated roofs. The attack on the metal gutter lining takes the form of narrow clean-cut grooves. What happens is that heavy dew or light rain dripping slowly off the roof picks up organic acid from the vegetable growth and where the solution falls on to the gutter lining it dissolves the normal protective patina on the metal. Repetitive dissolving and reforming of the patina results in grooves being cut into the metal, although in the case of lead it may be many years before the gutter lining is penetrated. There are modern solutions that can be applied to the roof covering to kill the lichen and prevent it growing again for a very long time. However, if the lichen has to be retained for appearance a periodic coating of bitumen paint on the affected area will arrest the attack or alternatively a sacrificial lead flashing can be fitted.

Effects of Portland cement concrete

Concretes and mortars made from Portland cement contain some free lime that can initiate a slow corrosive attack on lead in the presence of moisture. Direct contact between lead and new concrete or mortar should, therefore, be avoided in situations where drying out and carbonation of the free lime by reaction with atmospheric carbon dioxide is likely to be slow. Lead sheet built into brickwork or concrete as a damp proof course or impermeable membrane should be protected with a thick coat of bitumen paint. Flashings tucked into brick joints do not need any protection, since here carbonation of free lime is rapid and no risk of alkali attack then exists. When claddings, roof coverings and weatherings are applied to concrete surfaces, a sealing coat of a hard drying bitumen paint together with an underlay gives adequate protection during the drying out period.

Thermal insulation

Thermal conductivity of lead is 34.7 W/m °C and for practical purposes the effect of lead cladding or roofing can be ignored when calculating thermal resistance.

Table 22 *Physical properties of lead*

Atomic weight	207.2
Atomic number	82
Density	11.34g/cm^3
Coefficient of linear expansion	0.0000293 per °C
Thermal conductivity	34.76 W/m °C
Melting point	327 °C

Table 22 summarizes the physical properties of lead.

Bossing

Bossing is the term applied to the general shaping of malleable metals. In particular it is the term used to describe the shaping of lead sheet with hardwood hand tools in its application as a building material. The hardwood tools are still in current use, although quite recently similar tools manufactured from a tough plastics have become available and found to be an acceptable alternative.

Lead at ordinary ambient temperatures is only 300 °C below its melting point and for this reason it behaves in many ways at ambient temperatures similarly to harder metals at higher temperatures.

Notably, the malleability of lead without the application of heat is exceptional, and it is the easiest of the base metals to shape by bossing with hand tools.

The malleability of lead is outstanding because it has the following properties. It is the softest of the common metals; it is very ductile in that it will stretch substantially before fracture under comparatively low tensile forces; and most importantly, it does not harden significantly when worked, being relatively quickly self-annealing at ambient temperatures – the warmer the day, the more quickly it will revert to its normal softness. Thus the lead can be readily shaped by bossing without the application of heat, and yet without it cracking or becoming hard and brittle.

Malleability and, therefore, ease of bossing can vary according to the chemical composition of the metal and thus the grain structure. The purer the lead, the more malleable it will be. The leadworker should, however, find no significant difference for all practical purposes in the ease with which he or she can boss any lead sheet conforming in composition to BS 1178, or the dilute alloys specified for chemical lead sheet in BS 334, with the one exception that dilute alloys containing tellurium are work hardening to an extent that heat is needed for annealing. Hard lead, a lead/antimony alloy which is also covered by BS 334, is not sufficiently malleable for bossing in the normal way.

Leadworkers skilled in bossing can work lead sheet into the most complicated of shapes but in its practical application the basic aim is to achieve the required shape without undue thinning or thickening the substance of the lead sheet being used. In bossing there can be a surplus of lead to be bossed out and cut away or a shortage that has to be provided for when setting out the work so that the risk of thinning can be avoided.

The range of tools used for bossing lead sheet components to shape and dressing them into position are basically as follows, and they are made in different sizes to suit the thickness of lead sheet to be worked.

Bossing tools

The use of the correct hardwood (or plastics) tools for the working of sheet metals is essential. Even when using the proper tools work must be carried out using great skill and care to prevent damage or bruising to the sheet. Many different types of hardwood have been used but the most common are lignum vitae, boxwood and beech.

The wooden tools should have all the sharp edges and corners removed and to preserve them they should not be allowed to dry out excessively or become soaked with water. They should be given an initial soaking in linseed oil, followed by periodical attention. Wooden tools should not be transported together with steel tools. They should never be struck with metal tools or used in any way that may damage or score the working surface as this will be transferred to the sheet metal and detract from the finished appearance of the work.

The introduction of plastics has brought to the craft a range of tools for the working of sheet metal. These appear to be proving very satisfactory and have several advantages over the traditional wooden tools. They are not affected by atmosphere; no soaking in oil is required and they are less easily damaged.

The following are the tools most commonly used to boss lead:

Dresser	Step turner
Bossing stick	Chalk line
Bending dresser	Metric square
Bossing mallet	Straight edge rule
Tinmans mallet	Tinmans snips
Setting-in stick	Hammer
Chase wedge	Steel chisels
Drip plate	Compasses
Lead dummy	Lead welding equipment

Bossing operation (external corner)

The technique of bossing sheet lead is a skill that must be taught and learned with great patience as there is no short cut. Once it is appreciated that by proper manipulation of the tools the lead can be made to move from one point to another, a tremendous satisfaction can be achieved.

A practical demonstration by a master craftsman is by far the best way to learn this skill. The following notes should be most helpful in the setting out and the procedure to follow.

1 Using chalk, chalk line and square, set out the corner (Figure 311).
2 The two upstands form the sides of the corner. It will be evident that if the sides were turned up, the square in the corner is not required. It is therefore necessary to boss this square of lead out of the corner. To reduce the amount of work part of the corner is cut off (shaded portion in Figure 311).

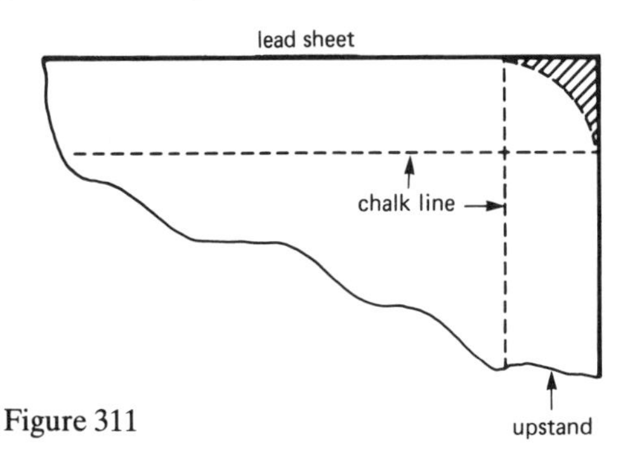

Figure 311

3 Using a piece of timber as a support place it along the fold line of the upstand and carefully raise each side (upstand) leaving the corner open. Always keep the corners rounded – no sharp angles. To assist in the lead bending at the correct line, the fold line can be set in using a setting-in stick and mallet. This must be done gently and with care.
4 A helpful step although not absolutely essential is that of raising part of the base to form what is known as a stiffener. This helps to hold the corner in place and is done by striking the lead with the dresser.
5 We are now ready to begin the bossing operation which can be carried out using any combination of the lead working tools, for example bossing mallets, bossing stick, bending dresser and lead dummy. Again as a matter of preference, lead can be placed in its normal position with the bossing operation in an upwards direction, or alternatively the lead can be turned upside down and the bossing done in a downwards direction (see Figures 312 and 313). The bossing is performed using one tool to strike the lead and the other to act as a support. The bossing must start at the base working the surplus lead upwards (note direction of arrows). The shape of the lead must be carefully checked to make sure that it is still within defined limits of that required. It is of paramount importance that creases and thickening of the lead are not allowed to take place. Should the bossing be carried out correctly this becomes obvious to the craftsman, as the corner keeps on growing as the surplus lead is worked out of the corner.

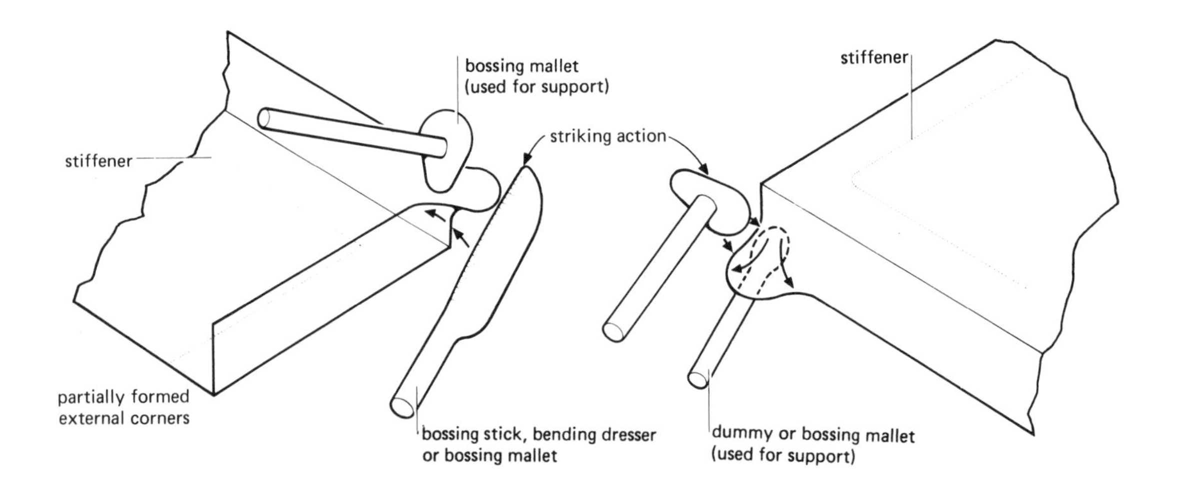

Figure 312 *Bossing an external corner*

Figure 313

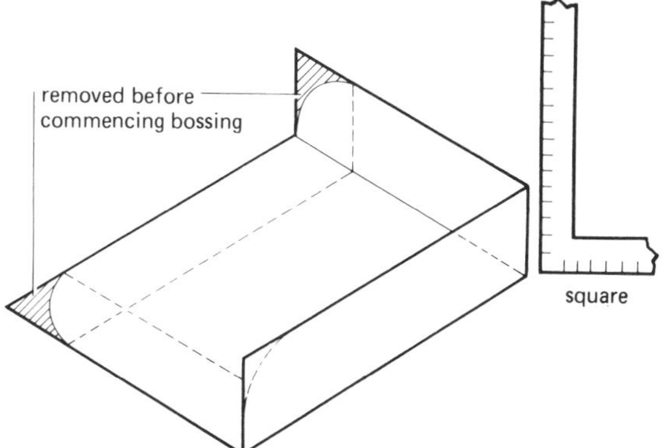

Figure 314 *One complete bossed corner. The operation is repeated until the box is complete*

6 The corner is checked for square in all directions (Figure 314). Note the corners should not have square angles but have a small radius. Finally, it should be trimmed to size.

Jointing of sheet lead

Single welt

The single welt or fold is used to weather and stiffen the edge of the sheet, for example the edge of a valley gutter, top edge of a back gutter, edge of dormer cheek etc. (see Figure 315).

Single lock welt

This is in fact two single welts locked together. It is suitable for situations where the welt will not be submerged by water because it is unlikely to remain watertight over a period of time. The method of fixing is by means of 50 mm strips of sheet copper screwed to the roof surface at 500 mm intervals. Figure 316 shows the method of making the single lock welt, while Figure 317 shows the method of fixing.

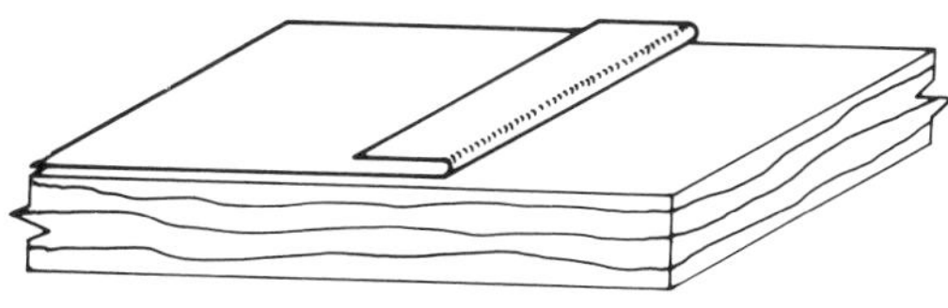

Figure 315 *Single welt*

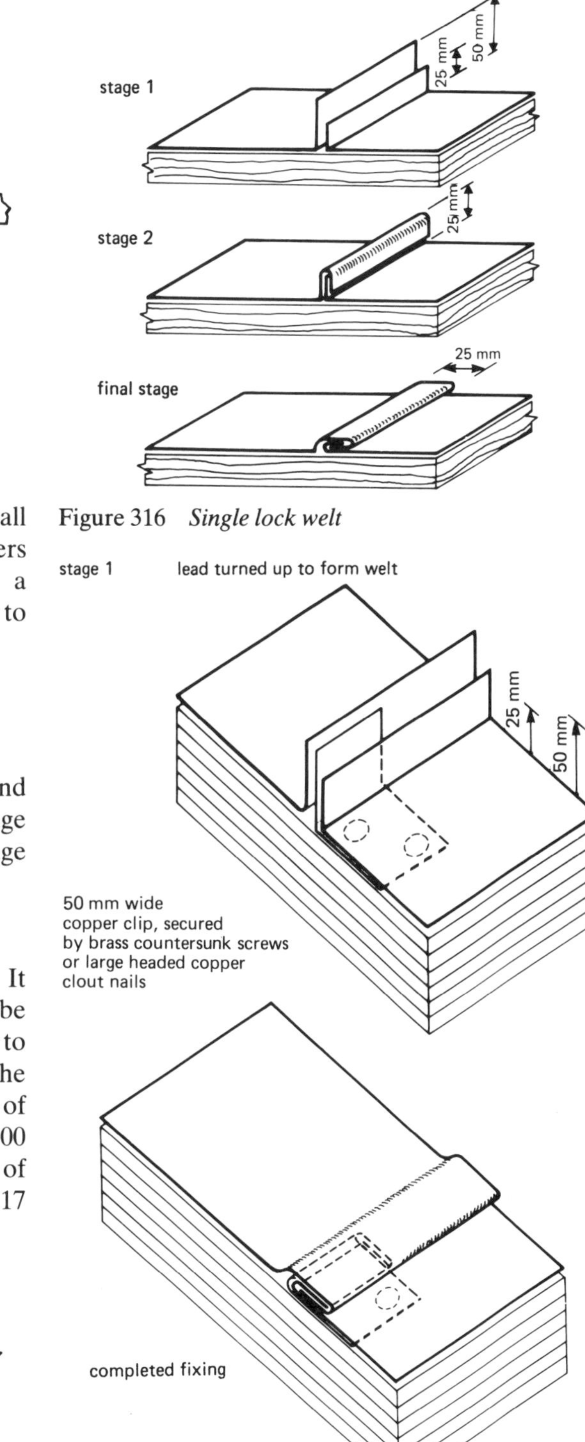

Figure 316 *Single lock welt*

Figure 317 *Method of fixing single lock welt*

Double locked welt

The double locked welt with its extra fold of metal is a water-tight joint. The method of fixing is by means of 50 mm strips of copper as detailed with the single lock welt. Figure 318 shows the method of making this welt.

Standing seam

Although this joint is not usually associated with sheet lead, it can have an application. It should however only be used in cases where it would not be walked on and never in any situation where the seam may be subjected to distortion. It is

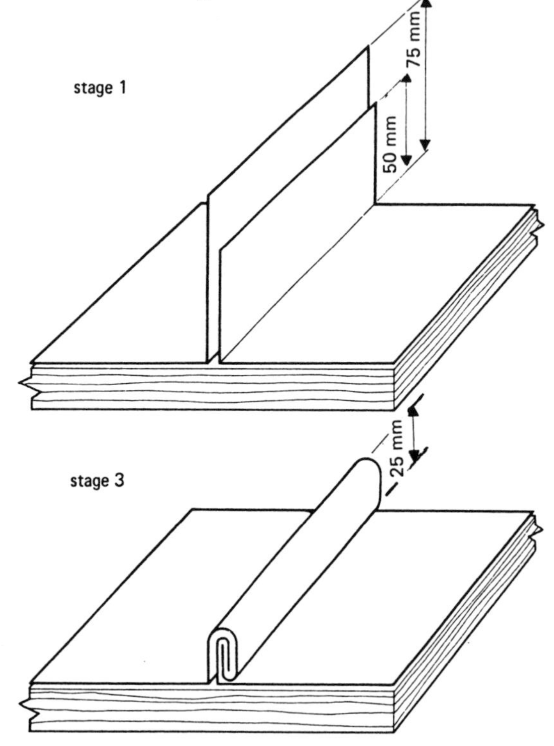

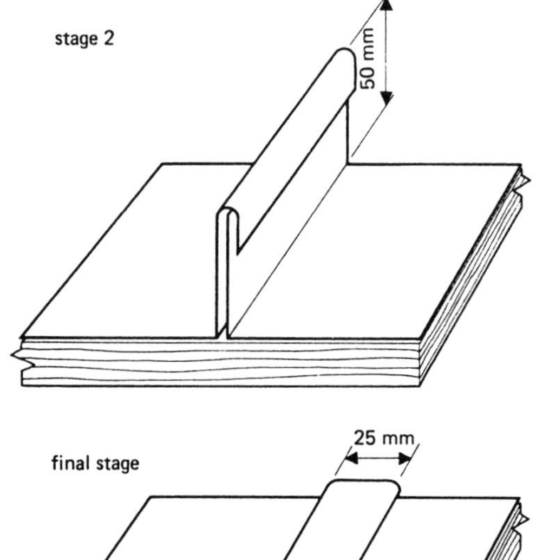

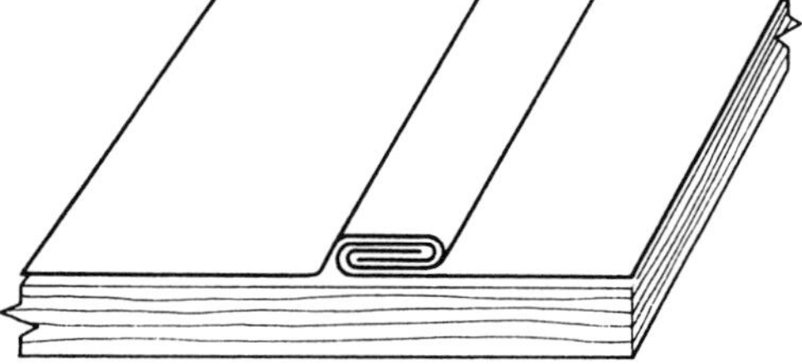

Figure 318 *Double lock welt*

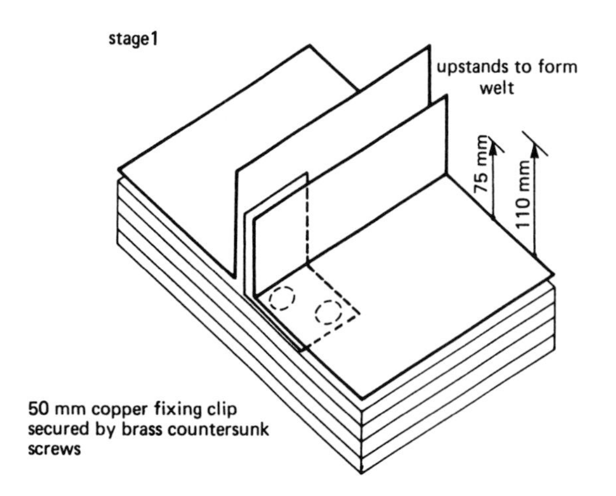

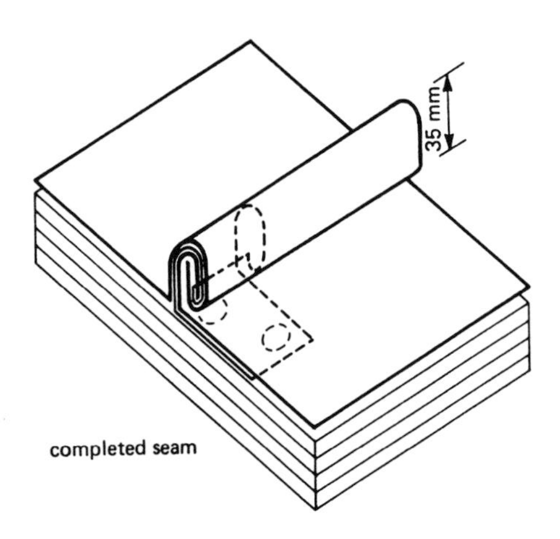

Figure 319 *Standing seam*

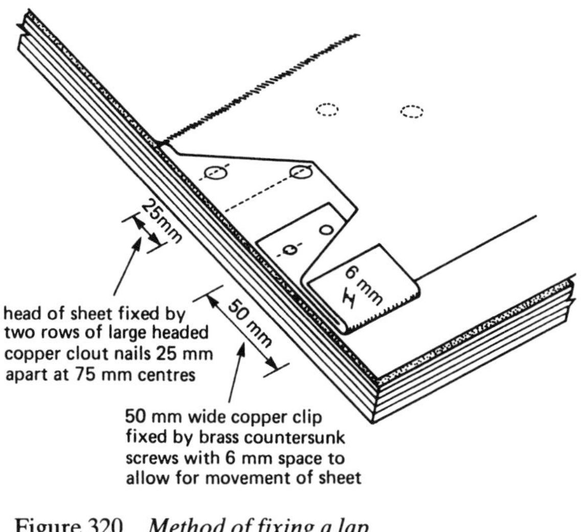

Figure 320 *Method of fixing a lap*

made and fixed similarly to the welt (see Figure 319).

Lap

This type of joint is simply the laying of the sheet to form an overlap.

The method of fixing is shown in Figure 320. Where two metals are in close contact, water can be drawn up between these two surfaces by capillary attraction. To overcome this problem the sheets must overlap sufficiently to give a vertical lap of 75 mm. It is shown in Figures 321 and 322 that the slower the pitch the greater the length of the lap necessary to fulfil the requirements.

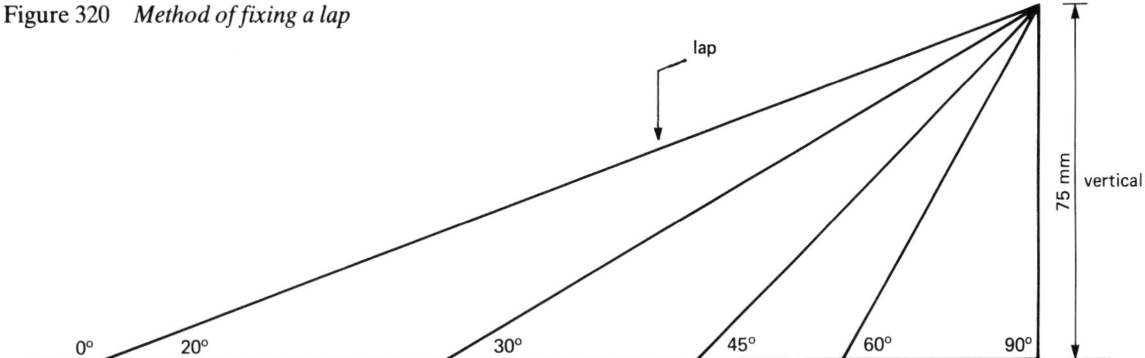

Figure 321 *The slower the pitch of the roof, the greater the length of the lap*

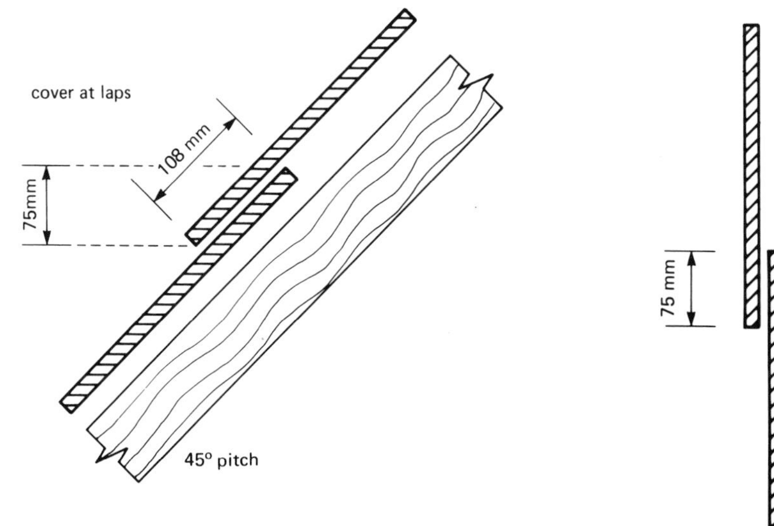

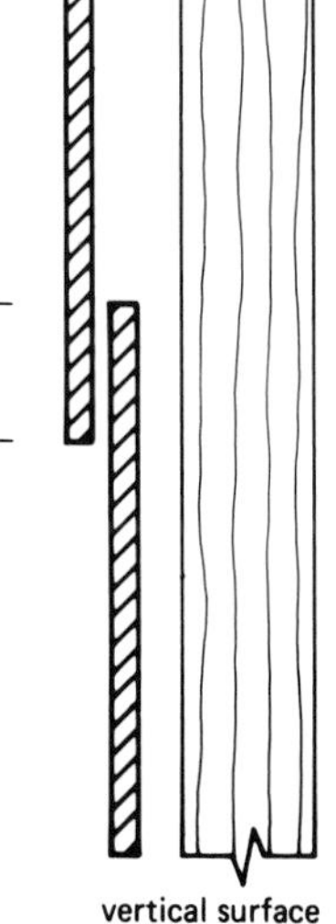

Figure 322 *Lap on 45° pitched roof*

Drips

These are used to form joints across the fall of flat and very low pitched roofs (see Figure 323). The drip should be not less than 50 mm deep or anti-capillary grooves will have to be incorporated in the joint as shown in Figure 324. By dressing the undercloak into the groove a space is formed between undercloak and overcloak. This prevents capillary attraction taking place. The top edge of the undercloak is dressed into a prepared rebate and copper nailed at 50 mm centres.

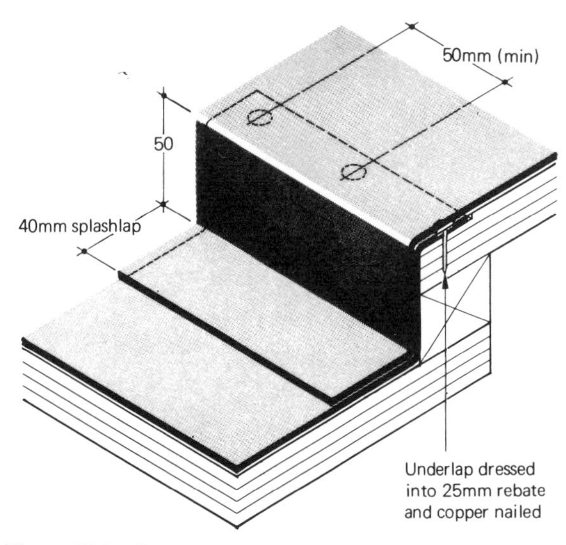

Figure 323 *Drip*

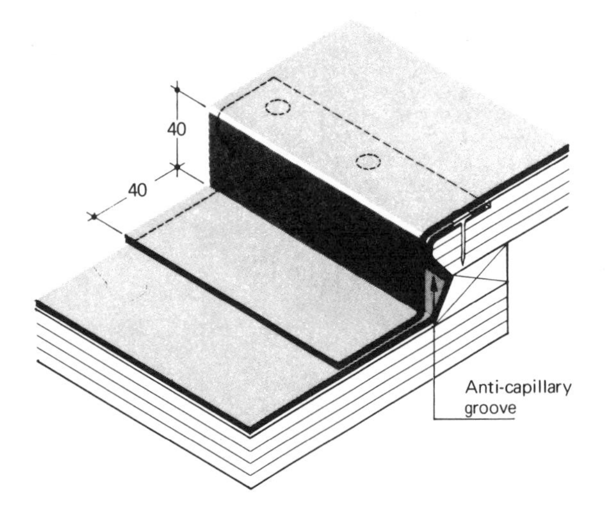

Figure 324 *Drip showing anti-capillary grooves*

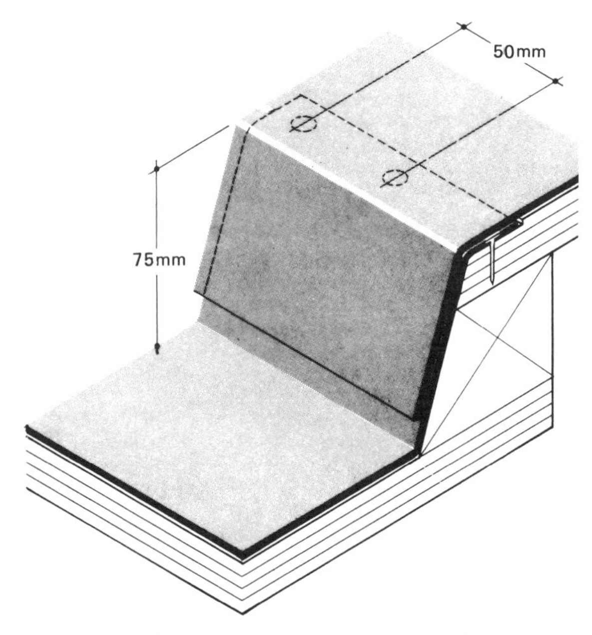

Figure 325 *Sloping drip joint with fixing and jointing method*

Figure 325 shows a sloping drip joint with fixing and jointing method.

Note: The 75 mm vertical height requirement.

Wood-cored roll

This joint is formed by dressing the edges of the adjoining panels over a shaped wooden core. The wood roll should not be too square in shape since this will prevent a good turn-in for the overcloak. For flat roofs the dimensions given in Figure 326 can be seen as standard but bigger rolls, giving a more bold appearance, are sometimes preferred for pitched roofs: the wood-cored roll is the accepted joint for flat and low pitched roofing, since it will stand up well to foot traffic. The undercloak is turned well over the roll and nailed with copper clout nails about 150 mm apart for a distance about one-third to a half the length of the panel, starting from the head. The overcloak is dressed fully over the roll, and with flat roofs it is extended as a splash lap which serves to stiffen the free edge and keep it in position. While the overcloak should be dressed to fit the shape of the roll, it should not be forced in tightly along the inside of the roll. The wood-cored roll formed in this way is most suitable for roofs up to 30° pitch,

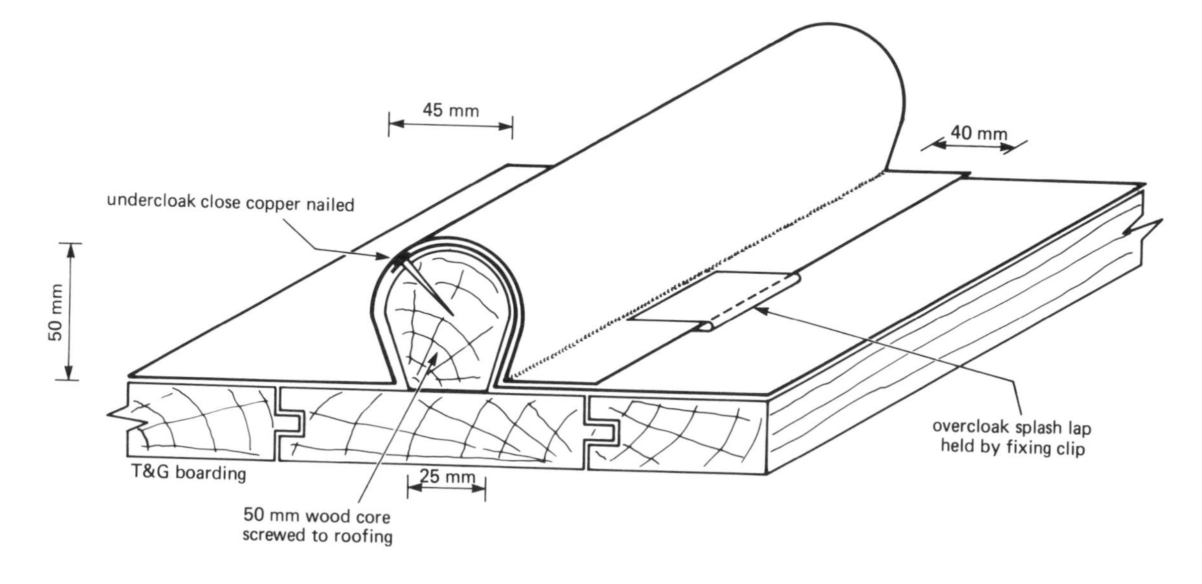

Figure 326 *Solid wood core roll*

but wood-cored rolls above this pitch are preferably formed without the splash lap and with the free edge secured with copper clips.

Hollow roll

This joint is suitable for pitched roof coverings that have few abutments or complicated joint intersections. The hollow roll is made by forming a tall welted seam and turning it into a roll (see Figure 327). The undercloak is turned up 100 mm at right angles and 50 mm wide copper clips fixed in the required positions as previously described. The overcloak is turned up 125 mm and the edge welted, not too tightly over the edge of the undercloak, then the ends of the copper clips are welted over at the same time and pinched to the lead. The prepared and clipped upstand is then turned to form a hollow roll.

Conclusion of jointing

All the above joints enable the material to move when heated. They are therefore known as expansion joints. To enable this to take place only one edge of the sheet is fixed, i.e. *undercloak*. The *overcloak* is left free to move. The overcloak may be held in place by means of clips but must never be firmly secured.

Figure 327 *Hollow roll*

Fixings

The traditional method of fixing to a timber substrate at the head of panels of lead sheet that make up a pitched roof covering (above 15° pitch) or wall cladding is with copper clout nails 75 mm apart in two rows and staggered (see Figure 328).

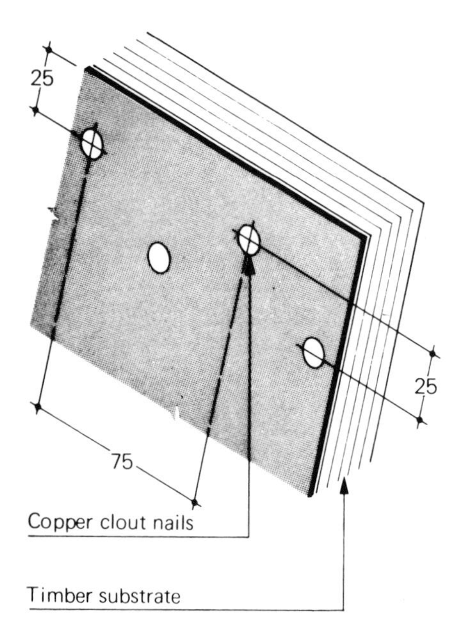

Figure 328 *Head fixing on timber substrate*

A single row will be adequate when the panel does not exceed 400 mm in height. The large heads of the nails pinch on the lead and the nailing is covered by overlapping lead sheet, and this fixing secures and supports the lead well for a very long life.

In the past, heavy lead sheet was sometimes supported at the head by dressing it over a wrought iron bar, round or rectangular in section, which was tightly secured to the substrate so that the lead was firmly pinched across the full width of the panel. However, the simpler copper nailing method is used in modern leadwork where the lead is being fixed to a substrate of timber boards, plywood or other materials into which nails can be driven and will hold firmly.

When the substrate is concrete softwood timber battens impregnated with preservative can be inset in the concrete to take the copper nailing (see Figure 329). However, in some urban areas the authority may require the lead to be fixed direct to the concrete to give maximum fire resistance. In this case brass or stainless steel screws with washers to give a good pinch on the lead are inserted into plugs in the concrete at the same spacing as for copper nails (see Figure 330).

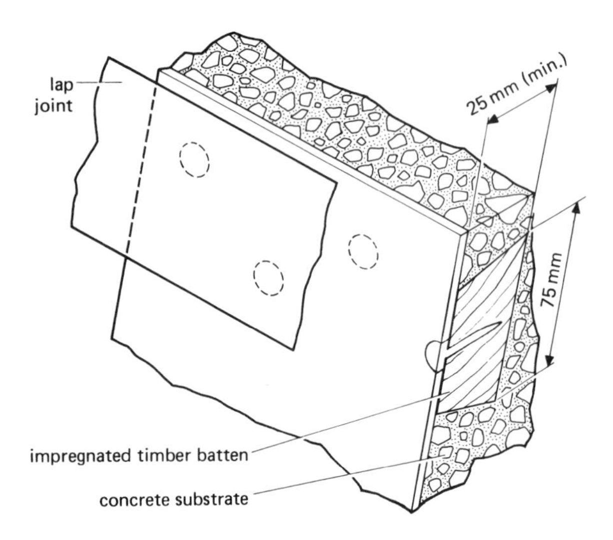

Figure 329 *Head fixing on concrete substrate*

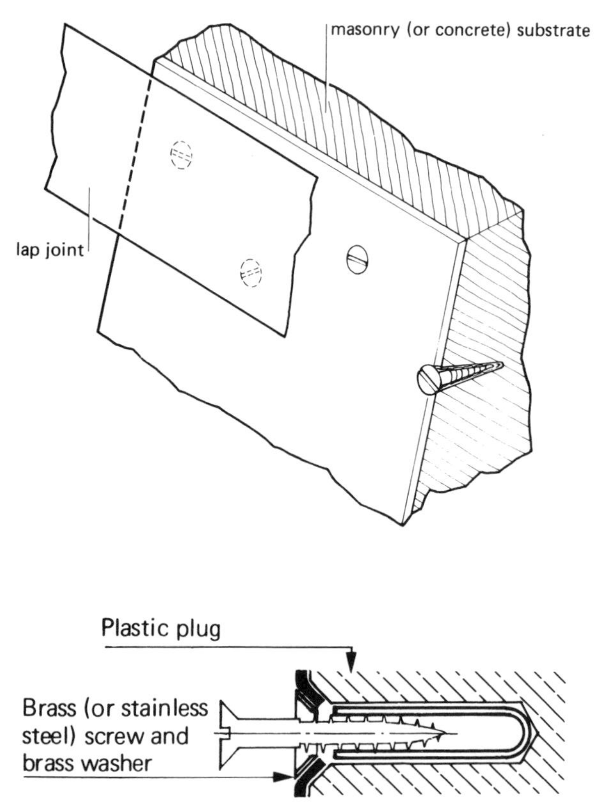

Figure 330 *Alternative method of head fixing on concrete substrate*

The plugs should be of a kind unaffected by moisture and preformed plastics plugs are commonly preferred. The heads of these screw fixings should be dished below the surface of the lead or the overlapping lead will tend to take up the shape of the projecting heads and show small bumps.

Copper clips that are incorporated in joints as fixings at the side of panels are cut from 0.6 mm thick soft temper copper sheet. They should be 50 mm wide and fixed to the substrate with three copper clout nails or two brass or stainless steel screws to each clip. In forming the joint, the lead sheet and copper clip should be lightly pinched together so that although the fold will tend to open with thermal movement of the lead, the clip will continue to give some support to the panel as well as securing it to the substrate.

Retaining clips to secure free edges against wind lift should be 50 mm wide and cut either from lead sheet or from copper sheet. Copper clips are best in situations of severe exposure, and where clips are visible (see Figure 331) they should be hot-dip coated with a high lead content solder. The retaining clip shown in Figure 331 should be fixed with some freedom between the clip and the bottom of the panel to allow for thermal movement. At the bottom of an area of wall cladding a continuous lead or copper clip can be used to provide drip edge finish.

For the lead tack method, a piece of lead 100 mm wide and about 200 mm long is lead welded to the back of the panel (see Figure 332). This tack is passed through a slot formed in the backing material, turned down on the inside and secured with brass screws and washers (see Figure 333). This type of intermediate fixing is particularly suitable when making up preformed lead faced panels for cladding.

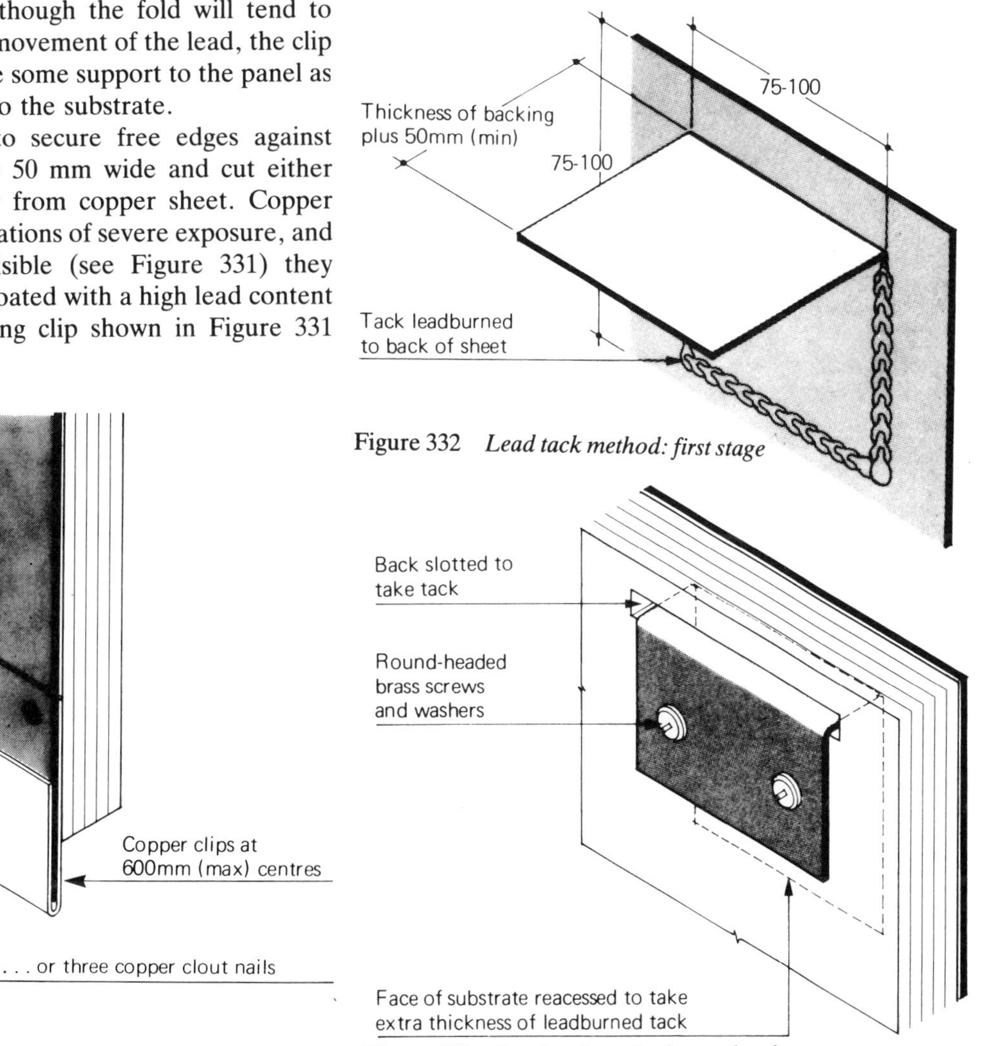

Figure 331 *Retaining clips*

Figure 332 *Lead tack method: first stage*

Figure 333 *Lead tack method completed*

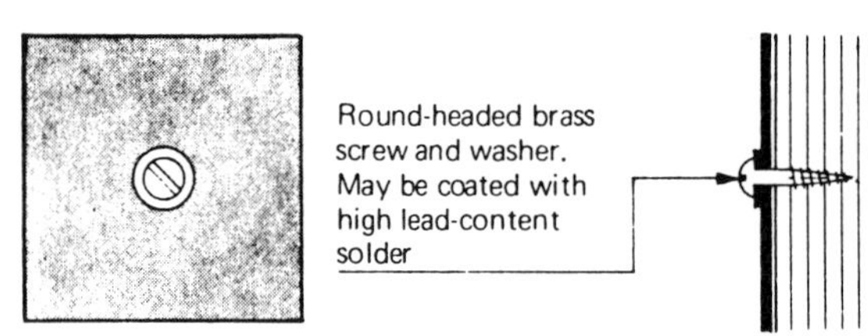

Figure 334 *Visible fixing*

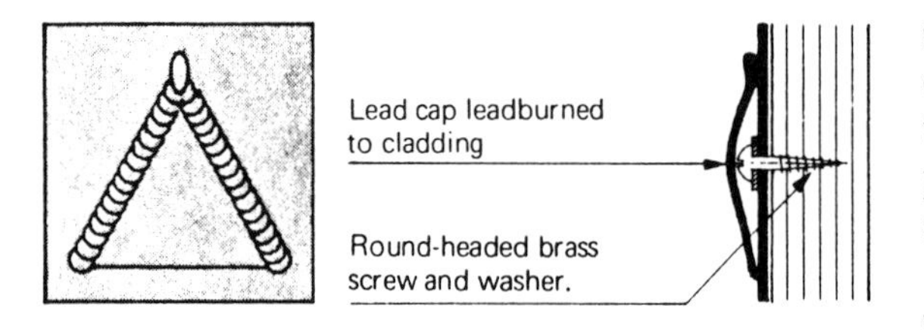

Figure 335 *Covered fixing*

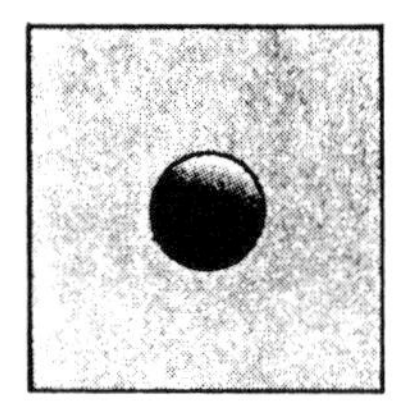
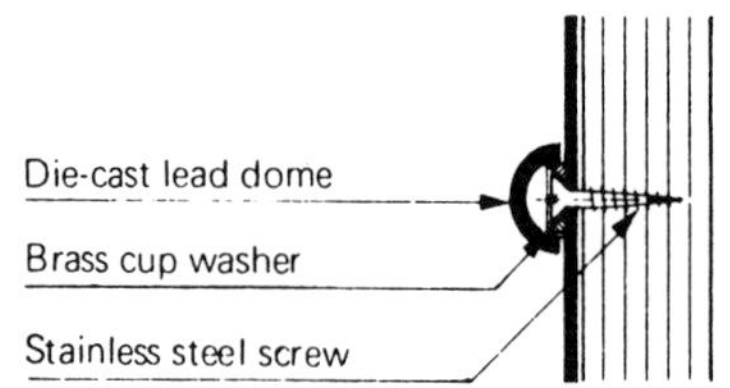

Figure 336 *Lead dome fixing*

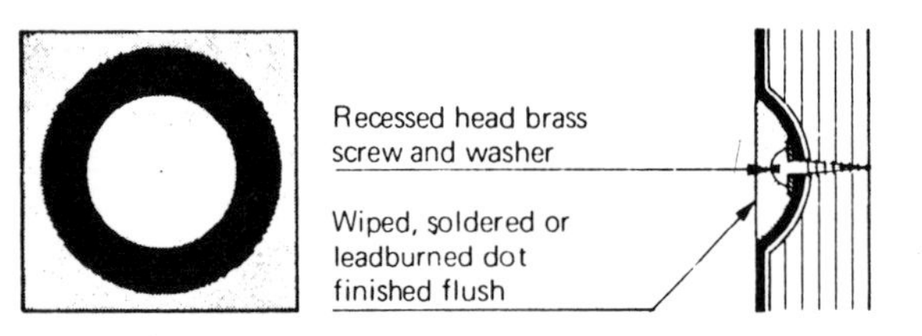

Figure 337 *Soldered dot*

Intermediate fixings
Intermediate fixings may be visible or secret. The simplest form of visible intermediate fixing is a round-head brass screw and washer (see Figure 334). It is adequately weathertight when the screw is driven in to get a good pinch on the lead and will not be visually obtrusive when the lead covering is well above ground level, particularly if the brass washers are hot-dip coated with a high lead content solder.

This fixing can be capped with a small piece of lead sheet, suitably shaped for ease of lead welding to the covering, when more positive water tightness is needed (Figure 335). Alternatively, a purpose made lead button with a special brass washer can be used if the fixing is to be ornamental (Figure 336).

Other intermediate fixings are the wiped soldered dot (Figure 337) or the lead welded dot, or a secret lead tack, lead welded to the back of the panel when the substrate is timber board, plywood or other sheet material and there is access to it from behind.

The lead welded dot is formed similarly to the wiped soldered dot, except that the dishing is slighter and the infill is lead applied by lead welding and cleaned off flush with the surface.

Preparation of roof surfaces

Substructure and backing materials
Most structural materials make suitable substrates for lead coverings applied *in situ* provided they offer a continuous smooth surface that is strong enough to support the lead (and take superimposed loadings as required) and will hold the fixings firmly and permanently. See notes on contact with other materials and similar aspects of corrosion resistance on page 167.

Timber
The traditional substrate for lead coverings fixed *in situ* is softwood boarding with an underlay. The boarding should be wrought, tongued and grooved, well-seasoned to give maximum resistance to warping and be fixed in the direction of the fall or diagonally (see Figure 338). Exterior quality plywood or blockboard is also a satisfactory timber substrate providing it is rigidly fixed, and the use of this material has tended to supersede boarding. As plywood presents a smooth level surface, an underlay may be unnecessary.

The heads of nails used for fixing boarding and other timber members should be well pushed

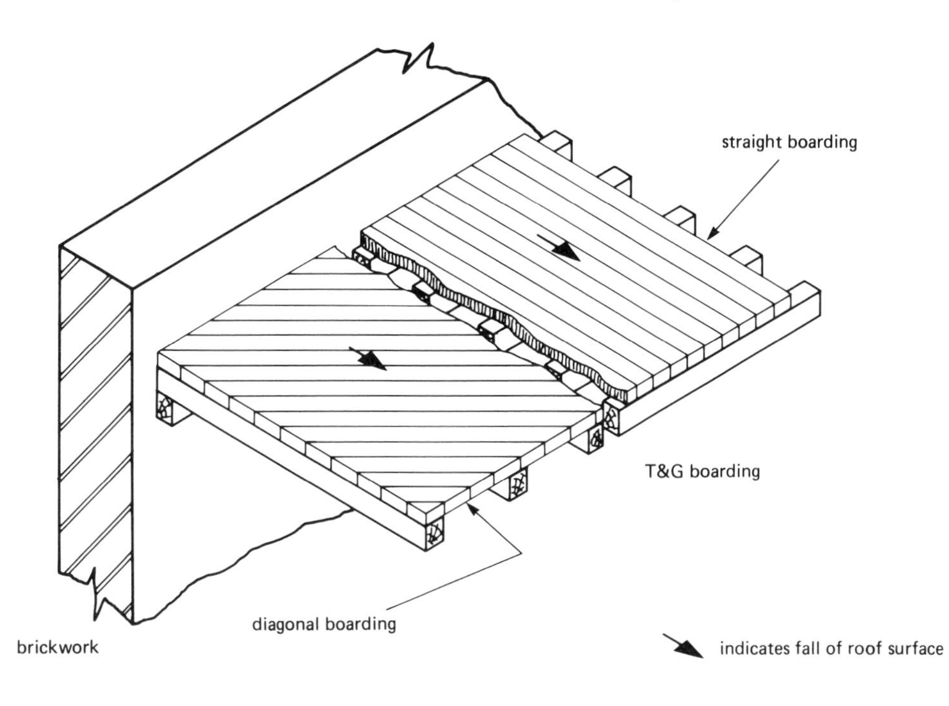

Figure 338 *Preparation of surfaces*

below the surface and similarly all screws should be counter-sunk. Sharp corners of external angles of timber substrates should be rounded off.

Concrete and masonry
The surface should have a smooth finish and an underlay should be provided both to isolate the lead from the substructure and, in the case of flat and low pitched roofs, to help the lead to move freely with temperature changes. Any materials for fixing the lead that are set in concrete or masonry should not be vulnerable to decay through the presence of retained moisture or condensation.

Other substrates
Where lead is laid on a thermal insulating material it is necessary to determine whether fixings made direct into such materials will be strong enough (equivalent at least to nail or screw fixings into timber). It is unlikely that fixings for the lead made directly into compressed cork or open texture woodwool slab will be strong enough, and timber battens will then have to be secured to the substructure in such a way that the fixings will give a long life.

With special proprietary thermal insulating substrates the question may also arise to whether the surface is firm enough to bear the dressing of the lead flat when it is being fixed.

Roof coverings

Underlays
The use of an underlay is to be recommended. This is usually a heavy type building paper or an inodorous felt. Its purpose is as follows.

1 To act as an insulator thereby preventing heat loss in winter and also preventing heat entering the building in summer.
2 To act as a sound insulator absorbing much of the noise caused by heavy storms.
3 By acting as an insulator, to also prevent water vapour (condensation) from forming on the underside of the sheet. The condensation could cause corrosion of the metal or rotting of the timber.

4 By acting as a separator between the sheet and (concrete) roof structures, to prevent the possible corrosion which may occur with new laid concrete.
5 To allow the free movement of the sheet when subjected to temperature changes.

Choice of thickness of lead sheet

Thicknesses of lead sheet normally appropriate for flat and pitched roof coverings are code nos. 5, 6, and 7. However code no. 4 may be acceptable in some cases, particularly for small roof areas. For wall claddings code nos. 4, 5, or 6 are appropriate.

Choice between these thicknesses should take into account the following factors:
Quality of the building
Necessary assurance of long life
Design of the roof or wall cladding
Size and shape of the panels required
Exposure

Sizes of lead sheet

The maximum sizes of the separate panels of lead sheet that are to form an area of roof covering, and hence the spacing of joints, will depend primarily on the thickness of lead to be used. Another factor, particularly when comparatively thin lead sheet is specified in line with modern practice, is the extent to which the covering will be exposed directly to the summer sun. The pitch of the surface is also taken into account, having in mind the need to support the lead adequately without excessive fixings.

Table 23 gives a general guide to the code of lead required for specific locations. For work in no. 4 and no. 5 lead sheet a range of maximum lengths is indicated according to the degree of exposure to the sun. The shortest length is suggested where there will be full day long exposure to the summer sun.

The most important points to bear in mind when selecting the correct code of lead are those of the size of each piece fitted and the method of fixing.

Table 23 *Recommended table of lead for various jobs*

Job	*Code*
Soakers	No. 3
Step flashings	No. 4
Aprons	No. 5
Back gutter	No. 5 or no. 6
Small flat roofs	No. 5 or no. 6
Step and cover flashings	No. 5
Valley gutter	No. 5 or no. 6
Slate pieces	No. 4
Cladding	No. 4

Weathering to chimneys

For roofs covered with slates and plain tiles, the watertight connection is made by means of both soakers and flashings. Figure 339 illustrates their use in two different types of chimney.

In example A, the chimney passes through the slope of the roof. It requires:
Back gutter (back)
Apron (front)
Flashings (sides)
Soakers (sides)

In example B, the chimney passes through the ridge of the roof. It requires:
2 aprons (front)
Flashings (sides)
Soakers (sides)

Soakers

A soaker is a thin sheet of metal, i.e. zinc, copper or lead. Part of the soaker is fitted between the slates or tiles while the other side is turned up the side of chimney or brickwork abutment (see Figure 341). These are fitted by the roofer as the tiles are fixed. The 25 mm added to the length (Figure 340) is to allow for turning or nailing at the head.

The width of the soaker must be a minimum of 175 mm. This gives an upstand of 75 mm against the brickwork and allows for 100 mm under the tiles. The length of the soaker can be obtained by calculation. First, it is necessary to find out the length of slate or tile and the lap.

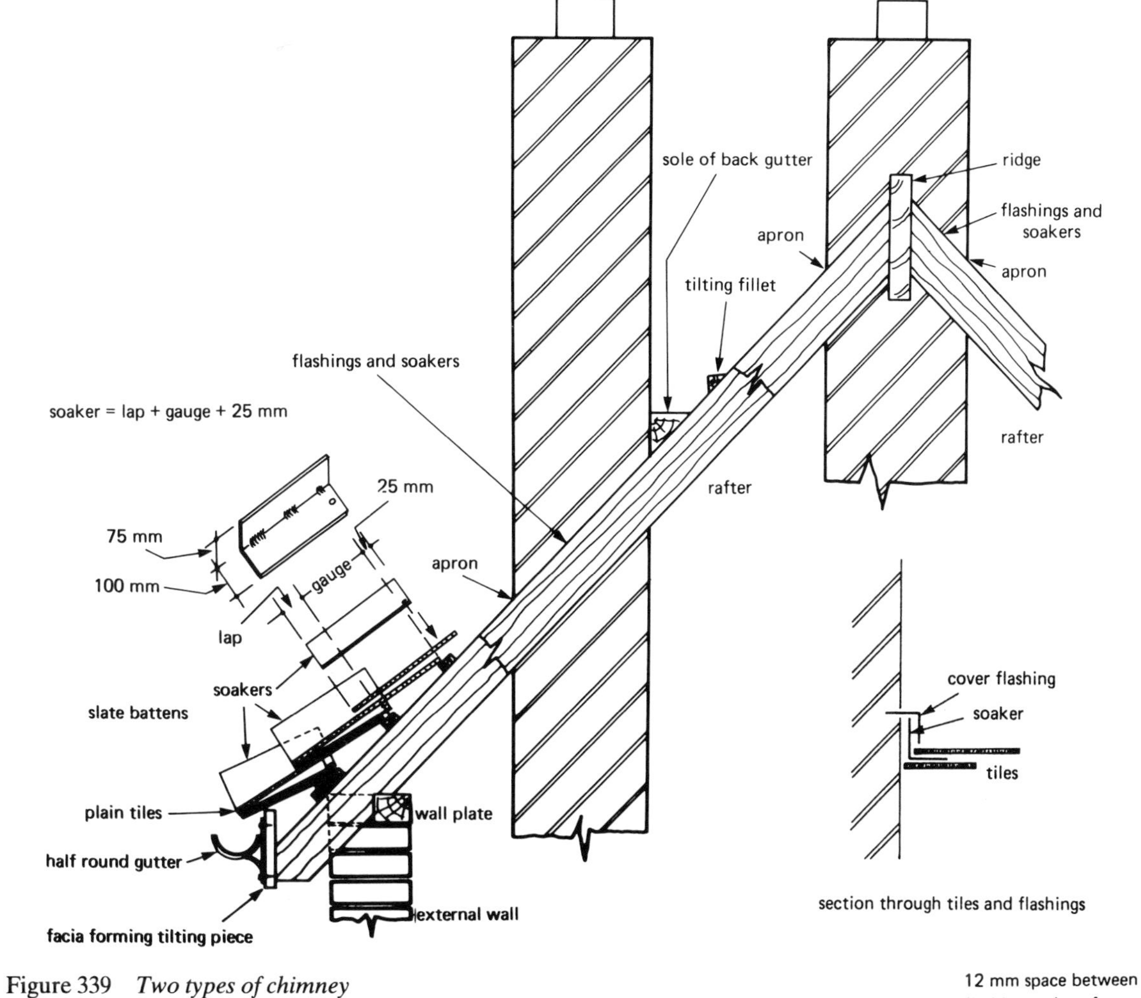

Figure 339 *Two types of chimney*

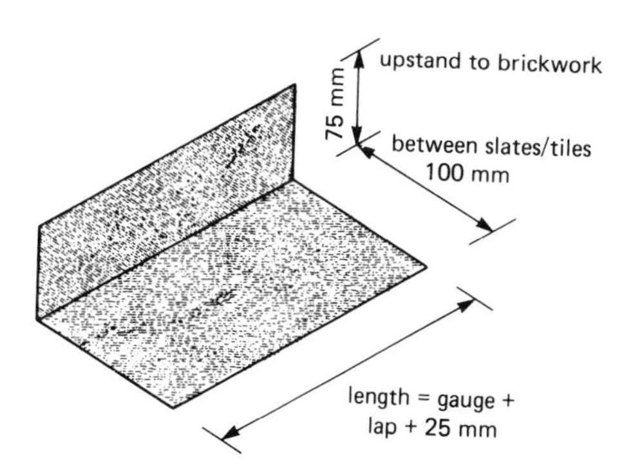

Figure 340 *Soaker*

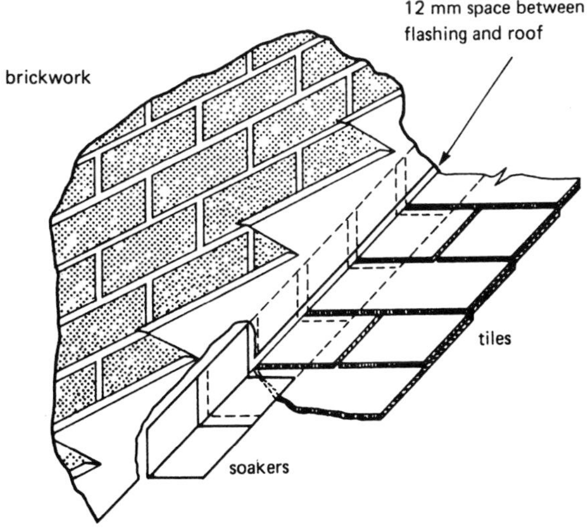

Figure 341 *Step flashing*

Example

Calculate the length of soaker required for a slated roof. The size of slates are 510 mm × 225 mm, having a lap of 76 mm.

$$\text{length of soaker} = \frac{\text{length of slate} - \text{lap}}{2} + \text{lap} + 25$$

$$= \frac{510 - 76}{2} + 76 + 25$$

$$= \frac{434}{2} + 76 + 25$$

$$= 217 + 76 + 25$$

$$= 318$$

$$\text{length of soaker} = 318 \text{ mm}$$

Slate fixing

Slate can be either head (Figure 342) or centre nailed (Figure 343). The lap allowed is 76 mm on roofs pitched up to 45°.

Slates must be laid to form a bond. Each slate must overlap another and the straight line vertical jointing should be staggered (see Figure 344).

The gauge is the centre of each batten to which the slate or tiles are fixed.

$$\text{Gauge of slate} = \frac{\text{length of slate} - \text{lap}}{2}$$

Example

Calculate the gauge of a roof to be covered with slates 510 mm × 255 mm with a lap of 76 mm.

$$\text{gauge} = \frac{510 - 76}{2}$$

$$\text{gauge} = \frac{434}{2}$$

$$\text{gauge} = 217 \text{ mm}$$

Calculate the number of soakers required for a roof of 6510 mm covered by 510 mm × 225 mm slates laid to a gauge of 217 mm.

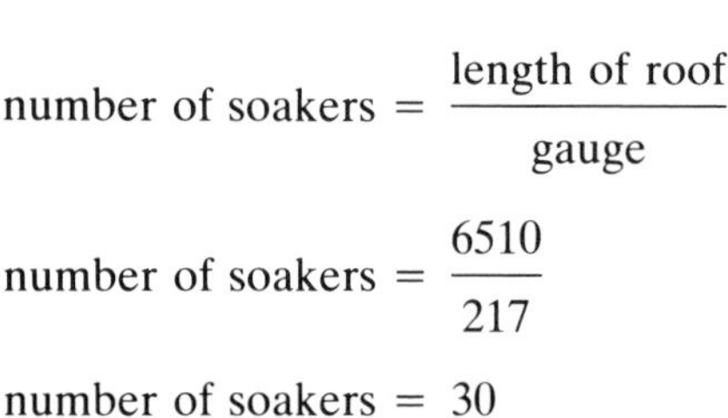

$$\text{number of soakers} = \frac{\text{length of roof}}{\text{gauge}}$$

$$\text{number of soakers} = \frac{6510}{217}$$

$$\text{number of soakers} = 30$$

It will be readily seen that the number of soakers is equal to the number of slate or tile courses.

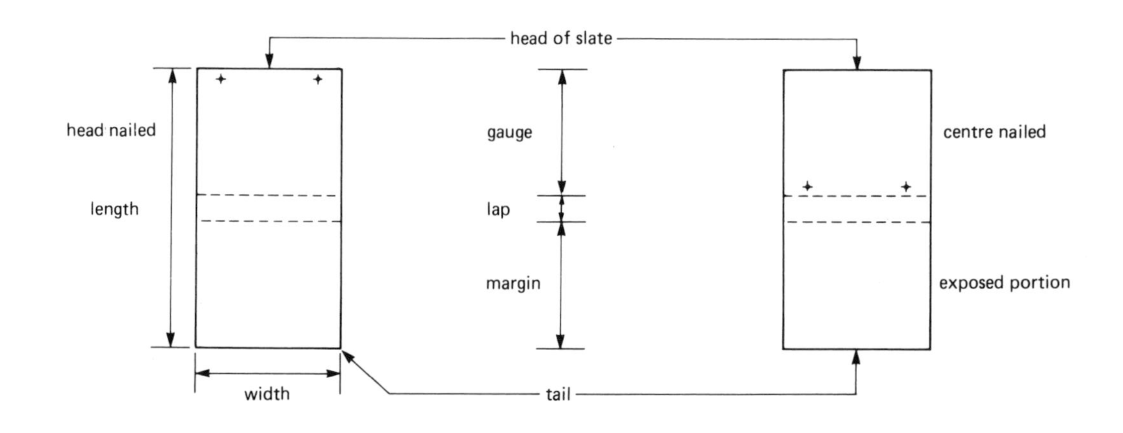

Figure 342 *Slate nailed at head*

Figure 343 *Slate nailed in centre*

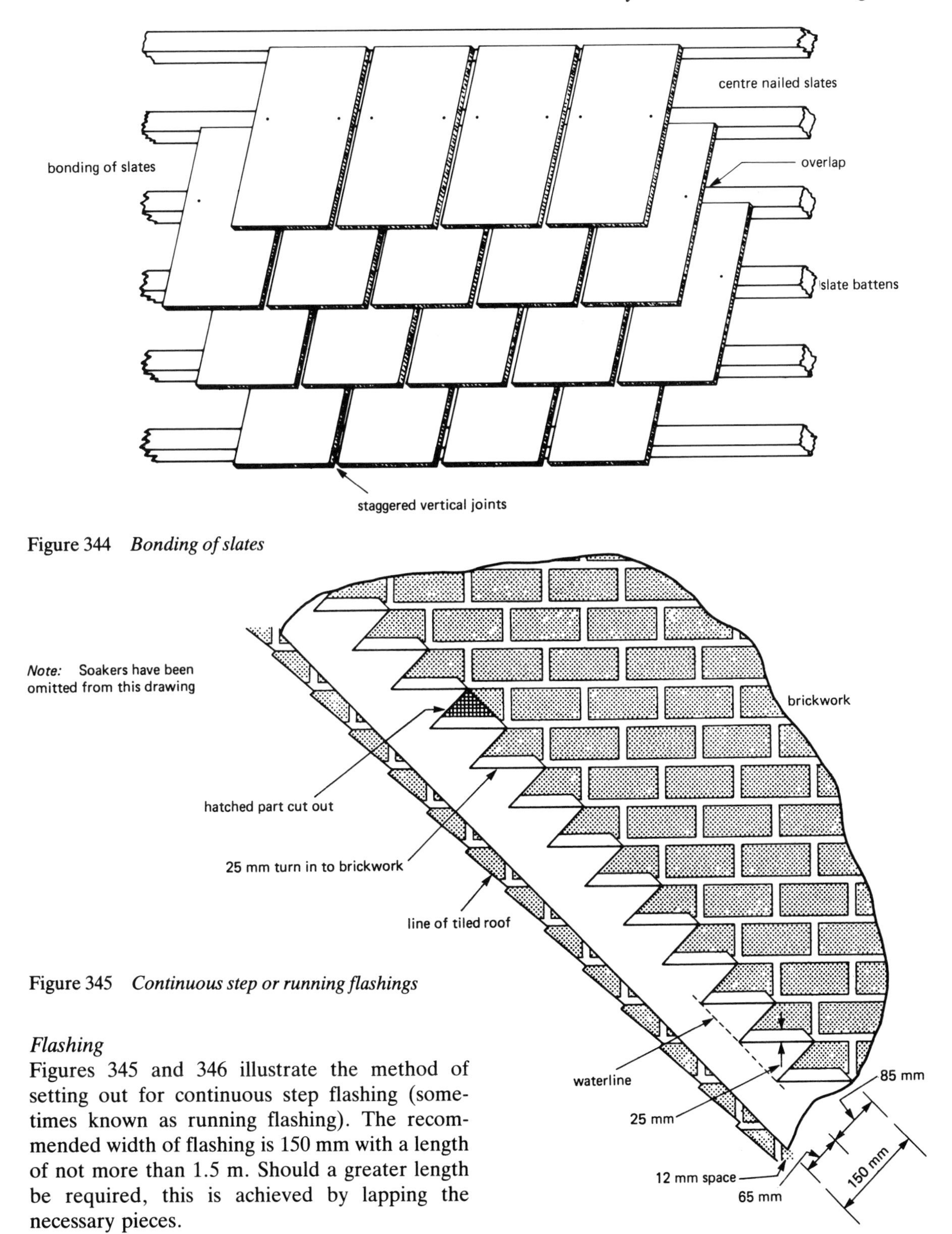

Figure 344 *Bonding of slates*

Figure 345 *Continuous step or running flashings*

Flashing

Figures 345 and 346 illustrate the method of setting out for continuous step flashing (sometimes known as running flashing). The recommended width of flashing is 150 mm with a length of not more than 1.5 m. Should a greater length be required, this is achieved by lapping the necessary pieces.

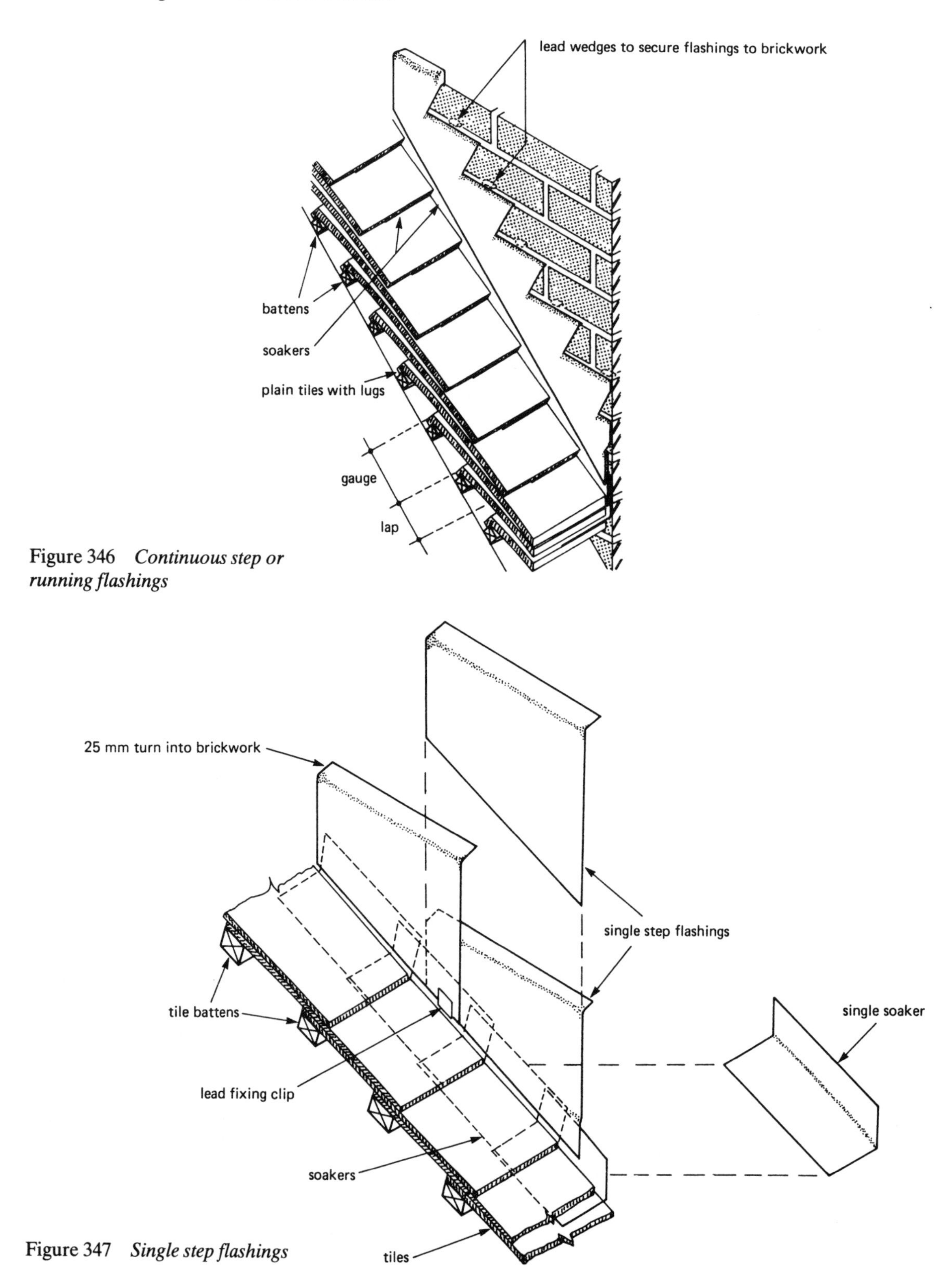

Figure 346 *Continuous step or running flashings*

Figure 347 *Single step flashings*

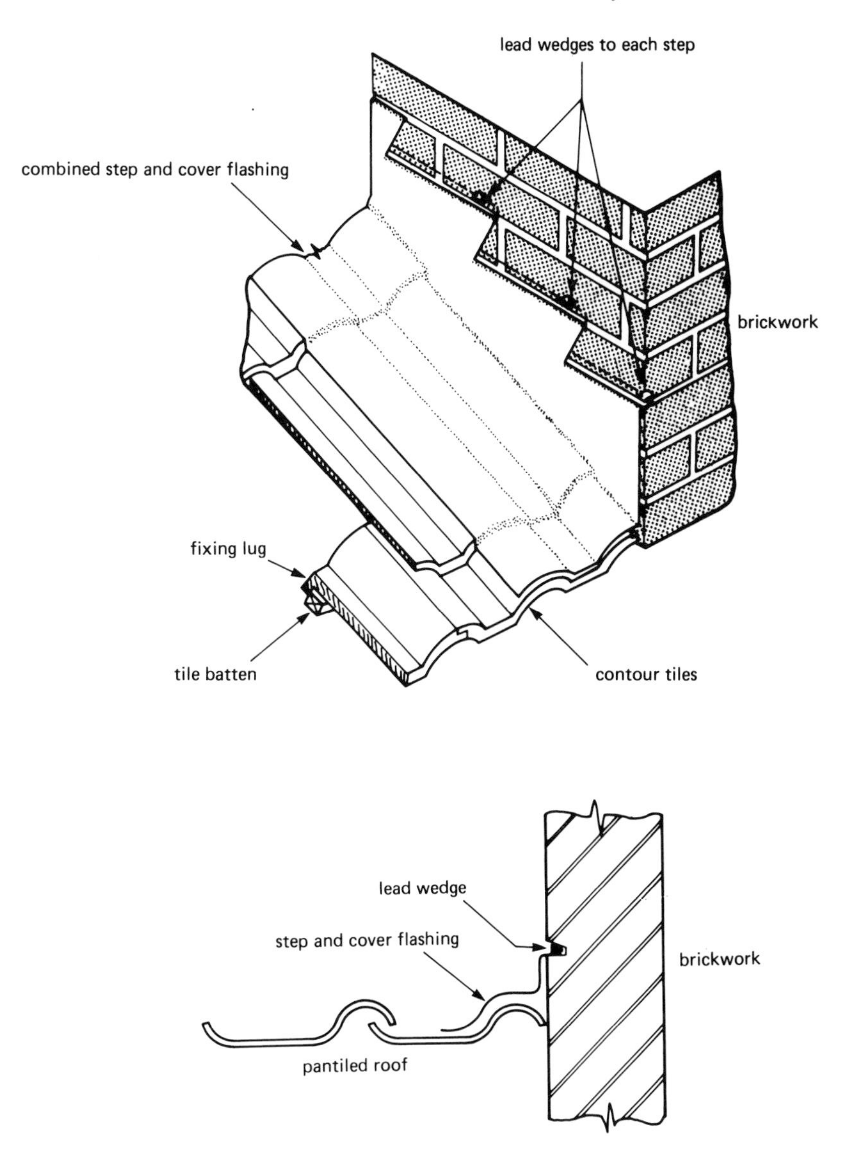

Figure 348 *Step and cover flashings*

An alternative method to continuous step flashing is that of soakers and separate step flashings (see Figure 347). This method of weathering is traditional in certain areas and of particular advantage when weathering to stone work.

Contour tiles

The weathering on roofs covered by contour tiles is achieved by a process known as step and cover flashing. A single piece of metal takes the place of both the soaker and step flashing as described for plain tiles (Figure 348). The weathering material should be dressed over a *minimum* of one contour. Two contours would be considered good practice.

Note: In all cases the weathering material must be dressed to fit snugly to the contour of the tiles.

Figure 349 *Fully weathered chimney*

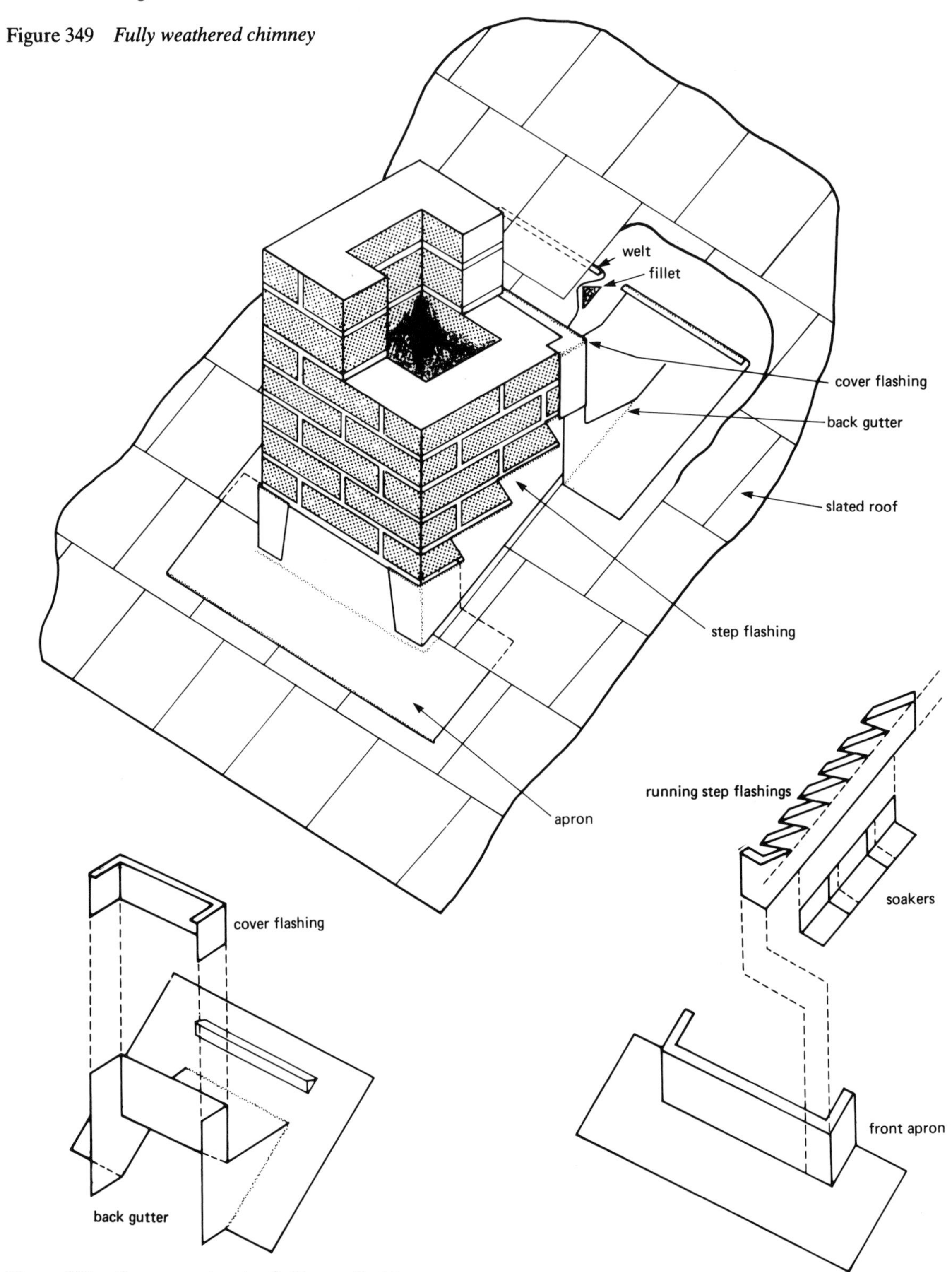

Figure 350 *Component parts of chimney flashings*

Figure 349 shows a weathered chimney, complete with all its component parts. Figure 350 shows all these parts in detail.

Copper

In designing a roof covering for a building, the architect will naturally look for a material that will give a long and reliable service. Such a material must be applied easily and quickly. It should add to the appearance of the finished job, possess a high degree of corrosion resistance, be economical and keep maintenance costs down to a minimum. Copper combines these qualities better than any other weathering material and is, therefore, an obvious choice.

There are many old buildings in existence possessing copper roofs, the age of which can be counted in hundreds of years – justifying the claim of durability for copper roofs. This long life is accounted for by the fact that copper develops, by natural processes, a surface film or patina, which forms a protection against corrosion by the effect of the sulphurous gases in the atmosphere. When exposed to the air, tarnishing takes place, resulting in a general darkening and blackening of the surface, due to the formation of copper salts. The main factor which determines the speed of this action is the quantity of sulphurous gases in the atmosphere.

When the black, tarnished layer has been formed, further changes, still caused by the sulphurous gases, gradually take place, forming an insoluble green layer on the surface. Thereafter, this film remains virtually unchanged, affording complete protection to the copper. It is not possible to forecast how long the development of the patina may take in any given district, but in London, copper laid six or seven years ago has already acquired a beautiful green film.

Properties of copper

Copper sheet and strip can be worked to conform to all the normal contours encountered in building, invariably enhancing the appearance of the building at the same time.

The physical properties of copper are such that it will remain unaffected by changes in temperature and will not creep. This allied to its resistance to corrosion ensures that a copper roof will give a long and trouble free service. Another very important advantage of copper sheet and strip is its comparatively light weight per unit area and advantage may be taken of this in designing the substructure.

Where copper is laid over timber, it has been found that boring insects, such as the death-watch beetle, do not attack the timber. This is attributed to the fact that the metallic salts formed by condensation are lethal to such insects – it is well known that copper salts are used to a large extent in timber preserving compounds.

Copper can be used on oak or other timbers without any detrimental effect on the metal. However, it is essential that a roofing sheet should never be laid direct on to an understructure but always laid with an intermediate layer of a suitable felt.

The whole of the roof covering, including clips, flashings, saddle ends, etc., but excluding continuous fixing strips, should be made from fully annealed copper sheet or strip, conforming to British Standard 2870. These recommendations apply to copper roof coverings under normal conditions. In exposed situations subject to high winds, the width and length of the bays should be reduced and/or thicker copper used. This applies especially to gables and verges. High winds may also retard the normal drainage of rainwater from the roof and this can result in the welts and seams being temporarily submerged. Under these conditions it is recommended that the welts and seams should be sealed with a non-hardening jointing compound (or mastic) before being folded.

Tools and materials

Many of the ordinary tools of the craftsman plumber, as shown and described in chapter 3 with one or two additions as detailed here, would provide a satisfactory kit of tools to enable the working of sheet copper to be carried out. In addition to the normal tools which can be readily purchased there are also a number of purpose-made ones.

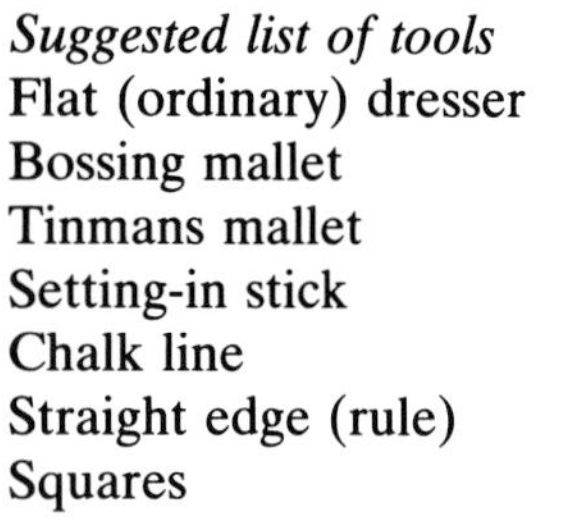

Suggested list of tools

Flat (ordinary) dresser
Bossing mallet
Tinmans mallet
Setting-in stick
Chalk line
Straight edge (rule)
Squares
Straight and curved snips
Chase wedge
Rubber mallet
Seaming pliers
Hammer
Special blocks of wood (Figure 351)
Special standing seam turning block (Figure 352)
Special dog-earing tool (Figure 353)

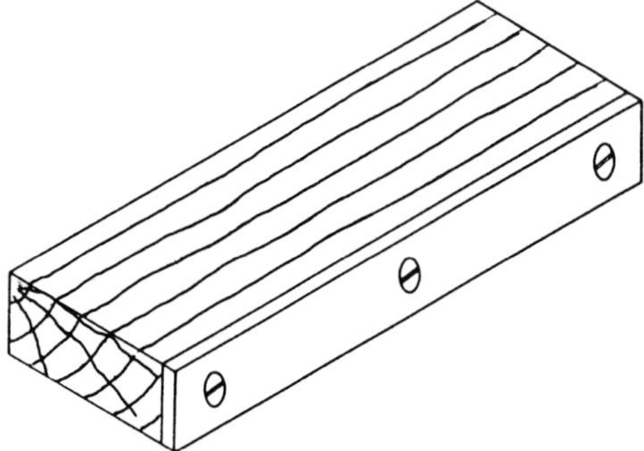

Figure 351 *Standing seam forming block*

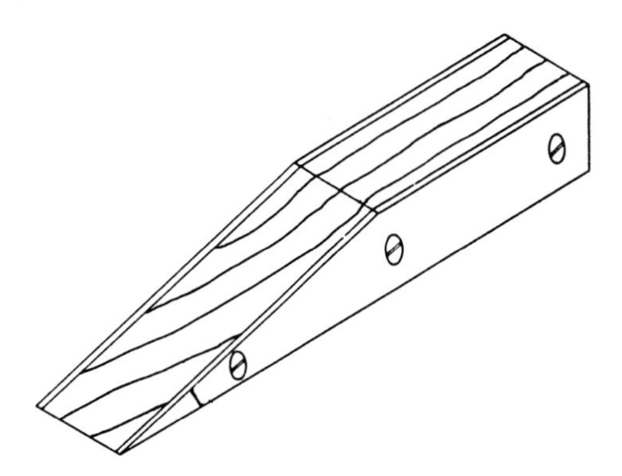

Figure 352 *Standing seam turning block*

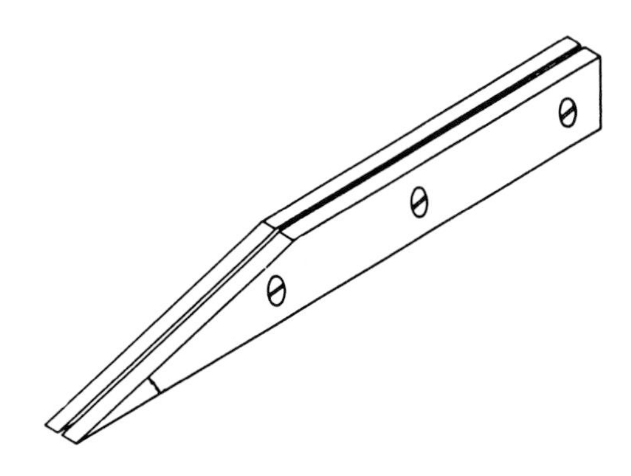

Figure 353 *Dog-earing tool*

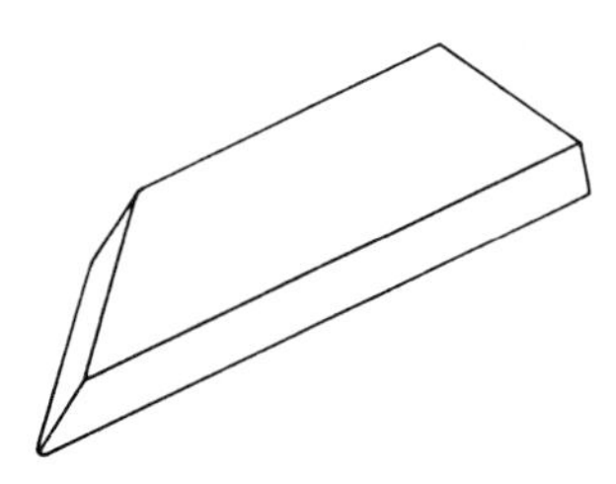

Figure 354 *Cobble tool*

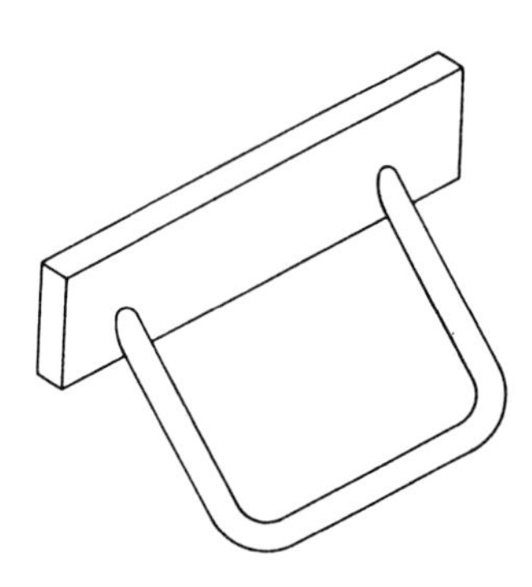

Figure 355 *Drip-edge tool*

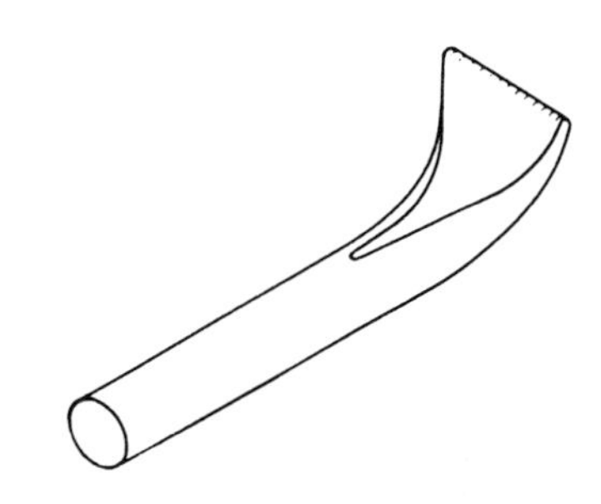

Figure 356 *Knuckle tool*

Special cobble tool (Figure 354)
Special drip edge tool (Figure 355)
Special knuckle tool (Figure 356)
Heating equipment

The above list is fairly comprehensive but is not considered complete, and will vary according to the individual craftsman and the work being carried out.

Nails

Any nails used for the fixing of the copper and felt must be made either of copper or a copper alloy such as brass. The nails should not be less than 25 mm long (measured under the head), not less than 2.6 mm thick, and weigh not less than 1.5 kg per 1000. The heads should be flat with a diameter of not less than 6 mm and the shanks barbed throughout their length.

Screws

Screws used in securing clips or other components should be made of brass. Where a batten roll joint is employed, the batten may be fastened by steel screws (or steel bolts and nut) provided that such fastenings are countersunk below the top surface of the roll and the exposed steel is suitably protected by painting with bitumen or covered with a felt ring or washer.

Clips

Copper clips of the same thickness as the roof sheeting should be used. They should be fastened to the understructure by two copper nails (or two brass screws) close to the turn up.

Clips for standing seams For standing seam systems, clips should not be less than 38 mm wide, spaced at a maximum of 380 mm centres along the length of the standing seam.

Clips along a roll Clips for rolls should not be less than 38 mm wide and spaced at not more than 460 mm centres. They should pass under the roll and be turned up on each side to retain the copper in position during fixing and service. Holes, in clips to receive screws, should be drilled or punched out with a parallel punch to the diameter of the shaft of the screw. Clips in standing seams and rolls should not be fixed closer than 75 mm from the junction with the cross welt.

Clips along ridge and hip rolls Clips should be 50 mm wide and placed two per bay. They may pass under the ridge roll and turn up on each side, or be nailed on the side of the roll.

Clips in cross welts Clips should be incorporated in all cross welts. Double lock welts require one 50 mm wide clip and single lock welts two 50 mm clips.

Clips at drips, eaves, and verges Clips should be 50 mm wide and placed two per bay in the drip edge and eaves welts, and at 300 mm centres in verge welts.

Clips against upstands Upstand clips should not be less than 38 mm wide and fixed with two copper nails (or two brass screws) at not more than 460 mm centres.

Jointing

Many of the joints used for copper are very similar to those detailed in the section on sheet lead. The only main differences are the sizes recommended for the differing materials.

There are two main systems of traditional copper roofing, the standing seam system and the batten roll system. These terms relate to the methods used to join adjacent pieces of copper in the direction of the slope with cross welts or drips for the transverse joints.

Standing seams (see Figures 357–359) running from ridge to eaves may be used on all roofs where the pitch is 6 degrees or greater, while wooden rolls can be used on all pitches. On roofs of flat or low pitch, i.e. 5 degrees or under, the sheets must be jointed by means of wooden rolls (two types of wooden roll are illustrated in Figures 360–363). The reason for this is that if standing seams are used they may be trodden flat and the joints may then allow moisture through as a result of capillarity. Also, the standing seam can be vulnerable on flat or low pitched roofs located in exposed positions. In these circumstances wind can retain the rainwater on the roof causing some sections of the seams to be submerged.

Transverse joints may be either double or single lock welts or drips depending upon the pitch of the roof.

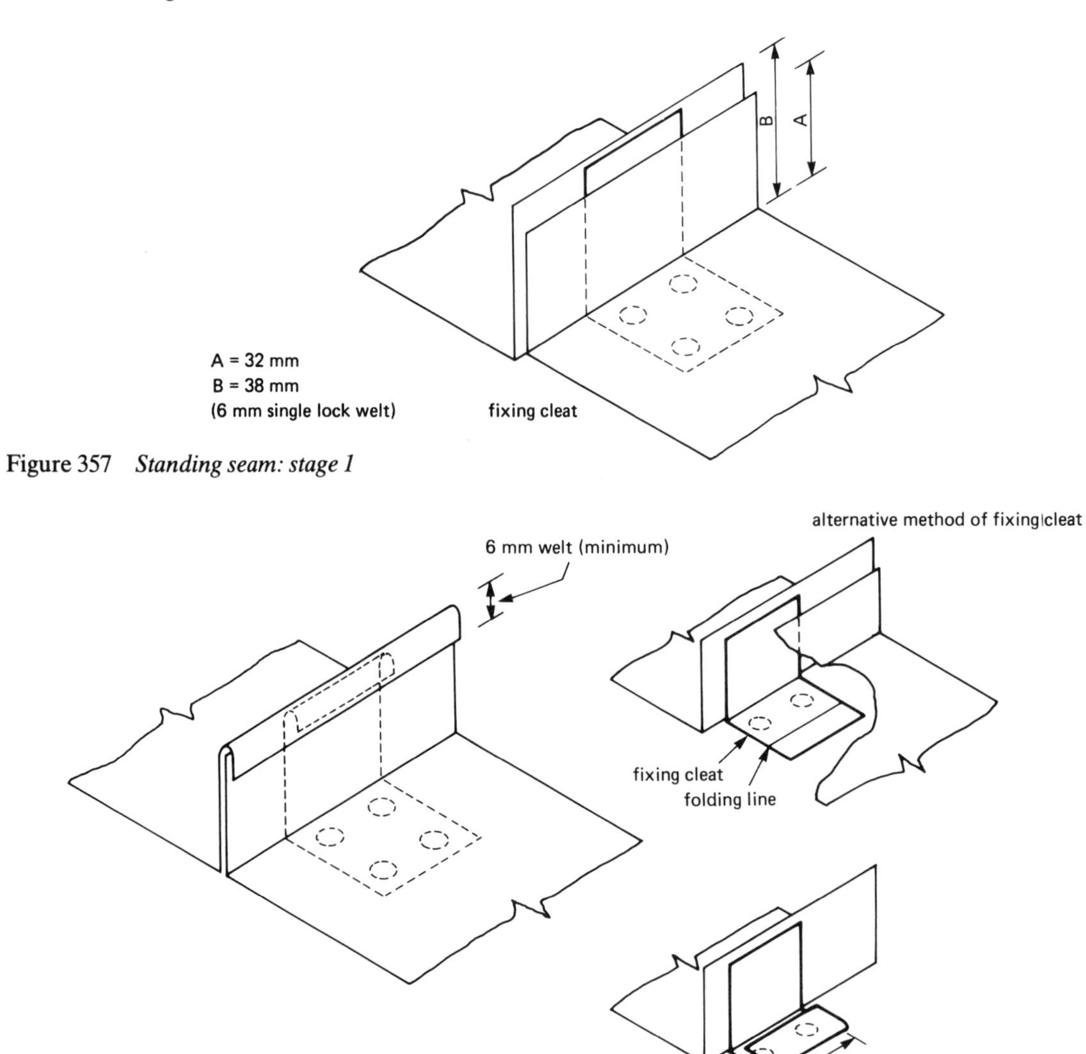

Figure 357 *Standing seam: stage 1*

Figure 358 *Standing seam: stage 2*

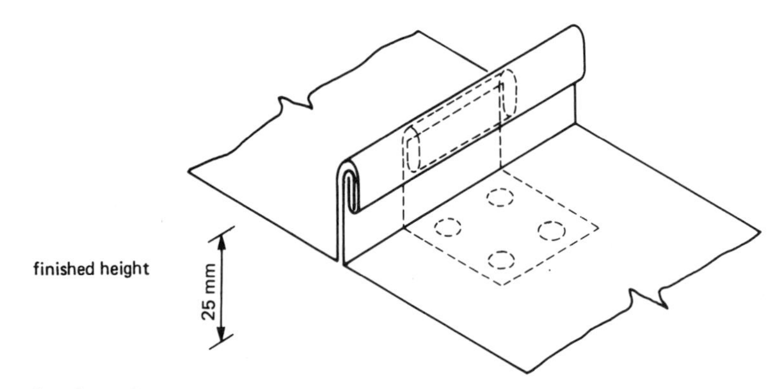

Figure 359 *Completed standing seam*

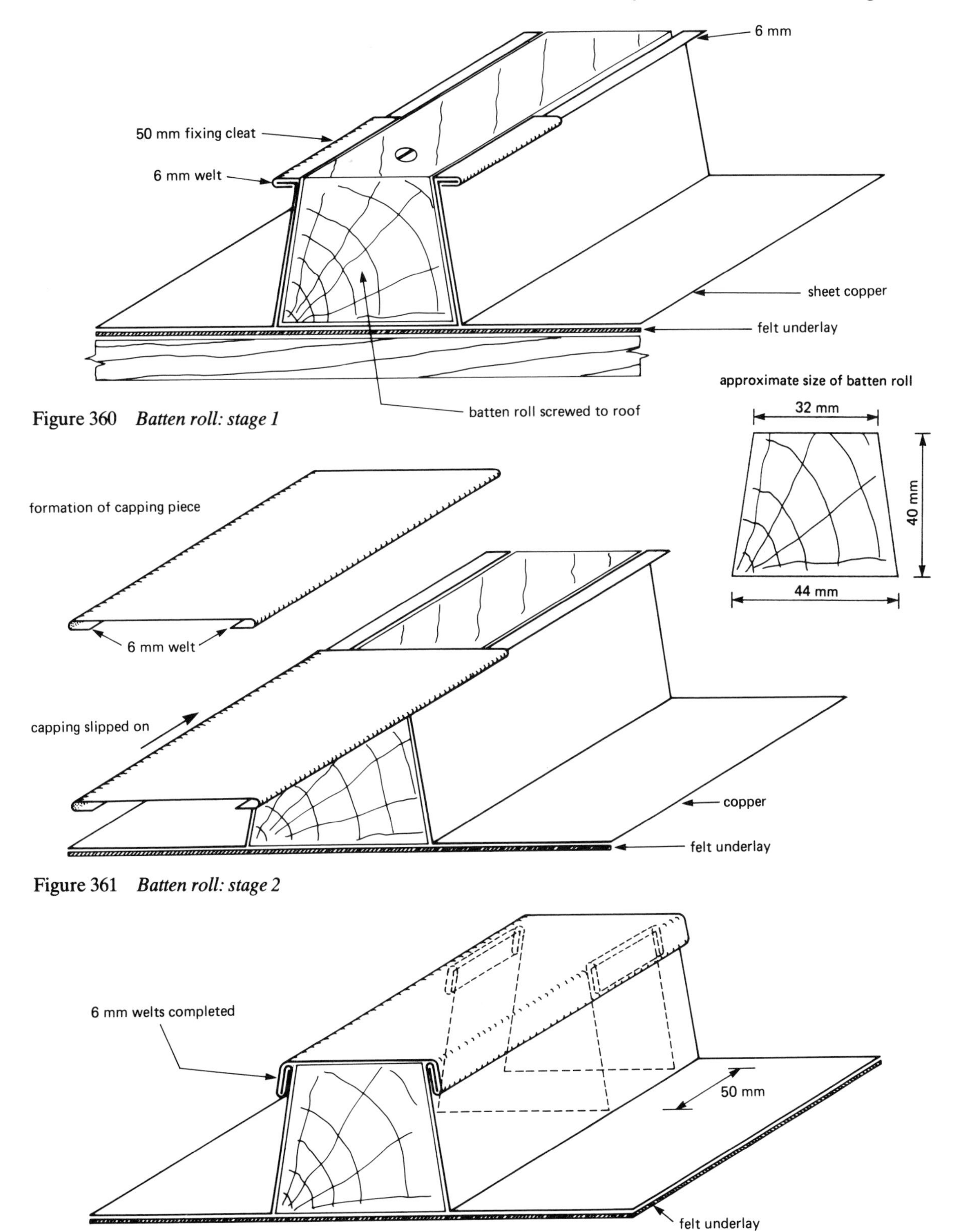

Figure 360 *Batten roll: stage 1*

Figure 361 *Batten roll: stage 2*

Figure 362 *Completed batten roll*

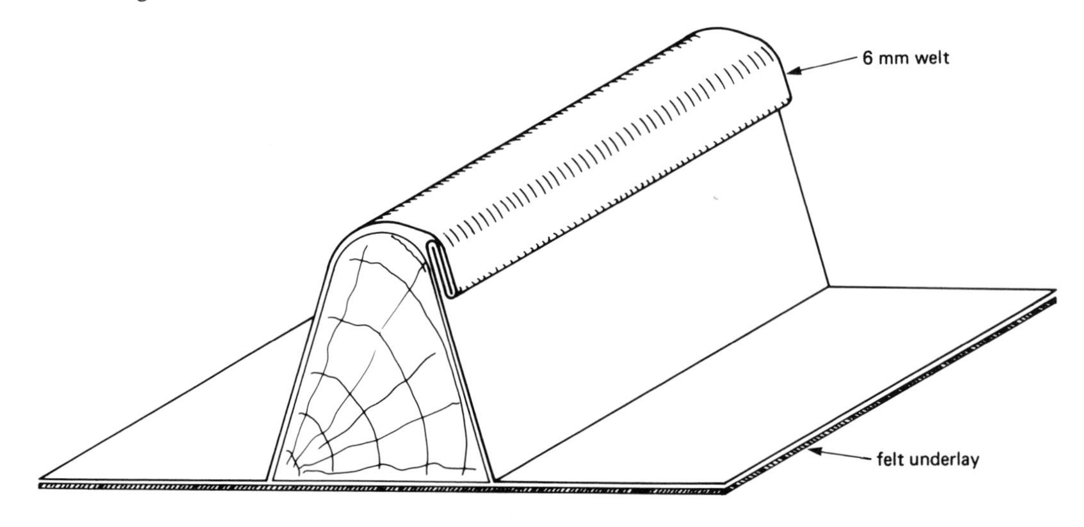

Figure 363 *Conical roll: sheet copper secured to roof by cleats as for batten rolls*

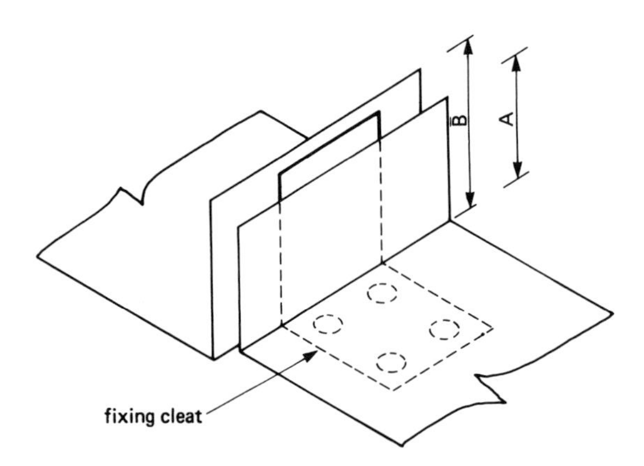

Figure 364 *Double lock welt: stage 1*

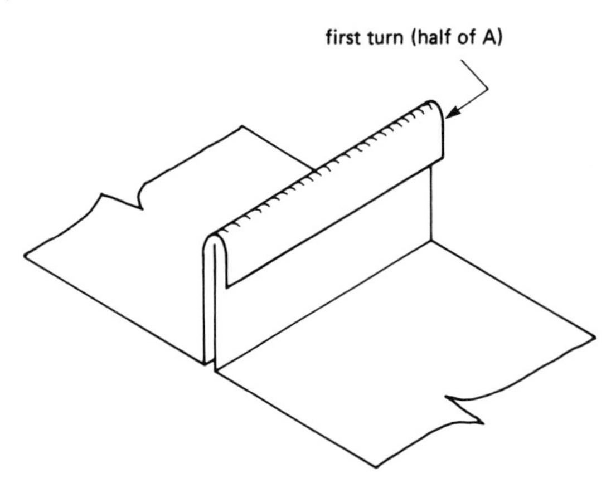

Figure 365 *Double lock welt: stage 2*

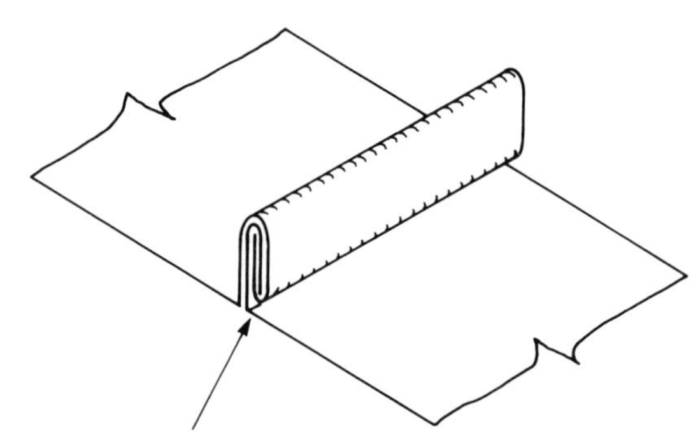

Figure 366 *Double lock welt: stage 3*

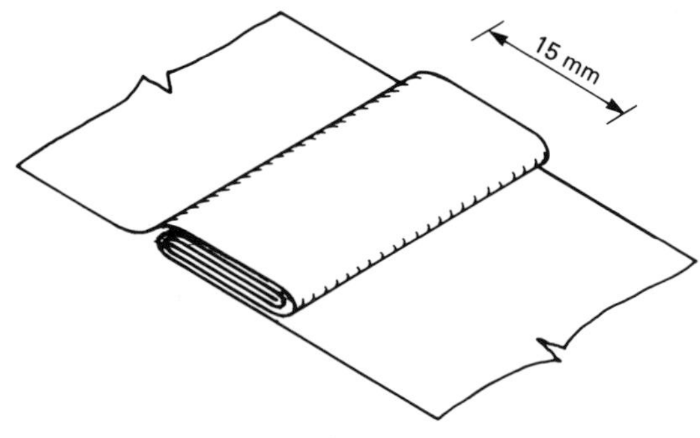

Figure 367 *Complete double lock welt*

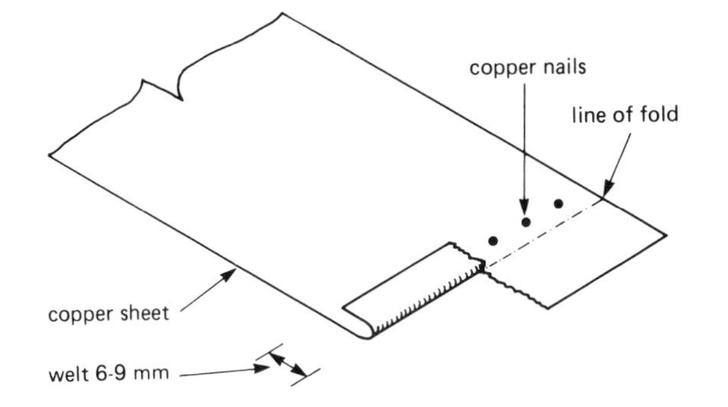

Figure 368 *Single welt: method of stiffening and fixing free edge*

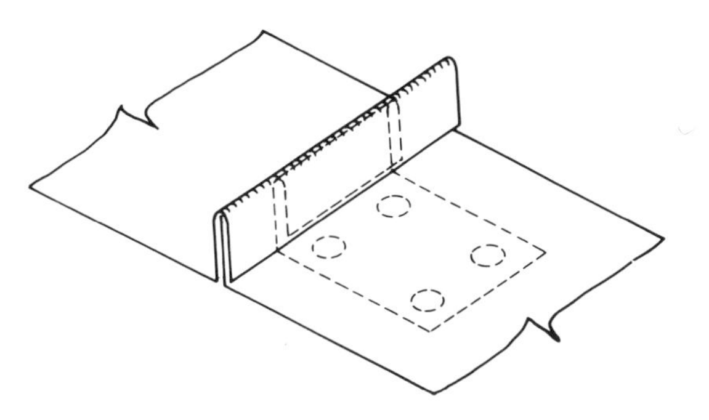

Figure 370 *Single lock welt: stage 2*

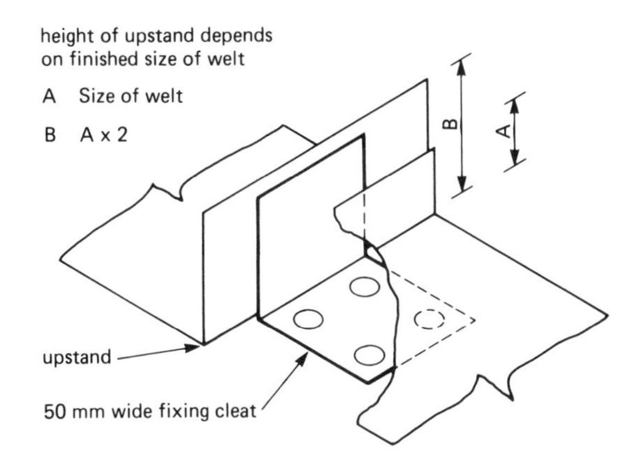

Figure 369 *Single lock welt: stage 1*

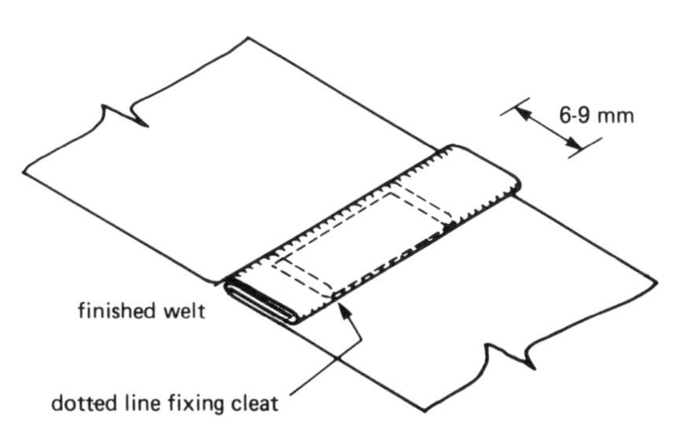

Figure 371 *Completed single lock welt*

Figures 364–367 show the method of forming a double lock welt. The finished size of the welt is variable, the suggested size being approximately 15 mm. Using this as a guide, the upstands A and B would be 30 mm and 45 mm respectively, since A is formed from two folds of the welt (15 × 2) and B from three folds of the welt (15 × 3). This is, of course, neglecting thickness of sheet.

The method of forming the single lock welt is shown in Figures 368–371.

Figures 372–375 show stage by stage the setting out, folding, and complete formation of the drip edge of a standing seam using the special forming tools shown on page 188.

Cross welts used in conjunction with standing seams must not be fixed in a line, but staggered in adjoining bays. They may, however, be used in a continuous line across a roof, where batten joints are used from ridge to eaves.

Welts used on sheet copper would normally be approximately 6 mm–9 mm.

Another method of forming joints in sheet copper work is by silver brazing. This is particularly useful in the fabrication of small weathering details such as cesspools, chute outlets, stop ends, expansion flashings, etc. It must, however, be stressed that as far as entire roof coverings are concerned, welding or brazing must not be used. The use of soft solder as a method of jointing between roofing sheets is another bad practice and should be avoided, with the possible exceptions of gutter outlets, lapped joints in eaves gutters, soil pipe flashings and small repairs on existing sheeting where there is no likelihood of stress on such a joint.

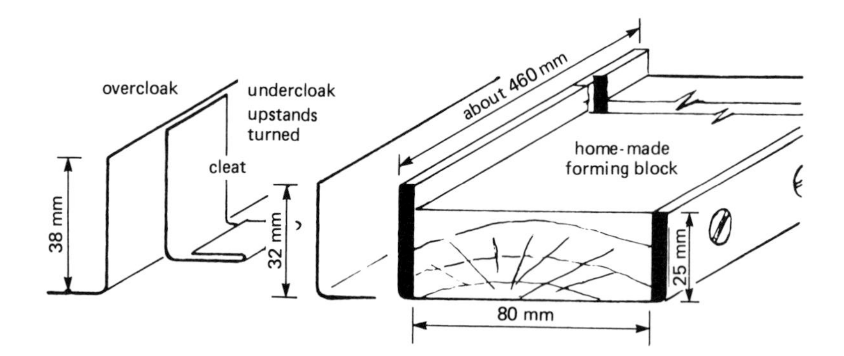

Figure 372 *Formation of drip edge on standing seam: stage 1*

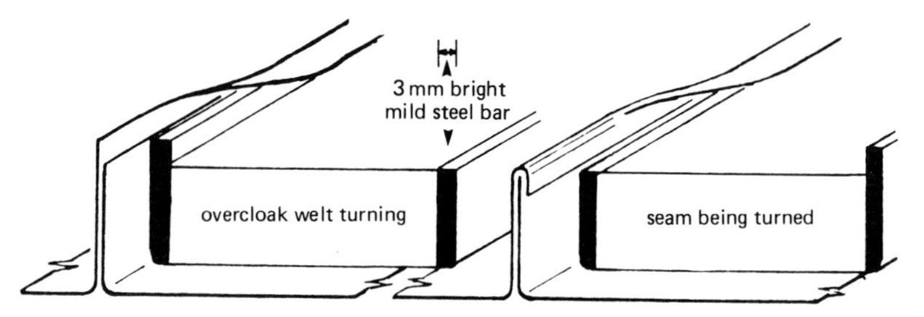

Figure 373 *Formation of drip edge on standing seam: stages 2 and 3*

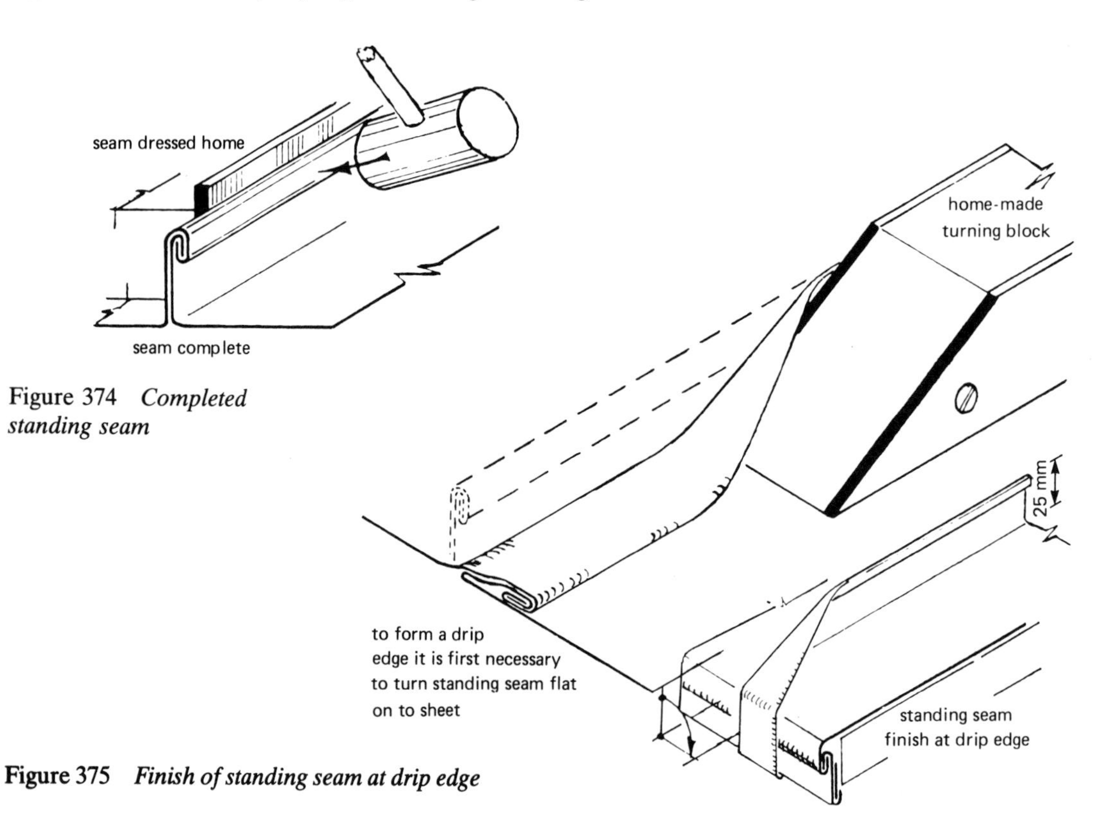

Figure 374 *Completed standing seam*

Figure 375 *Finish of standing seam at drip edge*

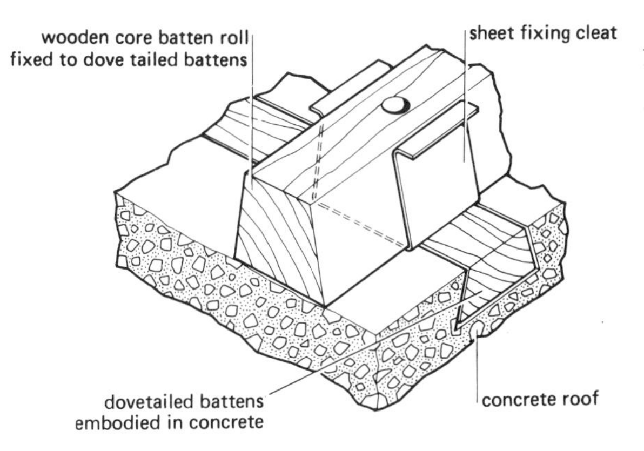

Figure 376 *Fixing copper sheet to concrete with dove-tailed battens*

Fixing

Figures 376 to 379 show methods of fixing copper sheet.

Concrete, timber and other decking materials
Reference should be made to the British Standard Code of Practice (CP 143: Part 12: 1970) for all structural aspects of fixing provisions in connection with concrete, timber and other decking materials on which sheet copper is to be laid.

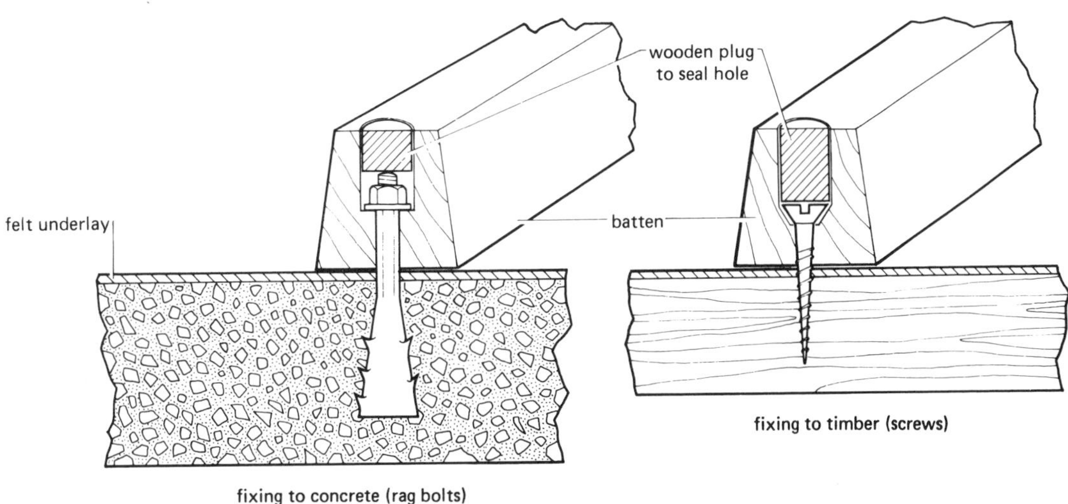

Figure 377 *Fixing copper sheet to concrete with rag bolts*

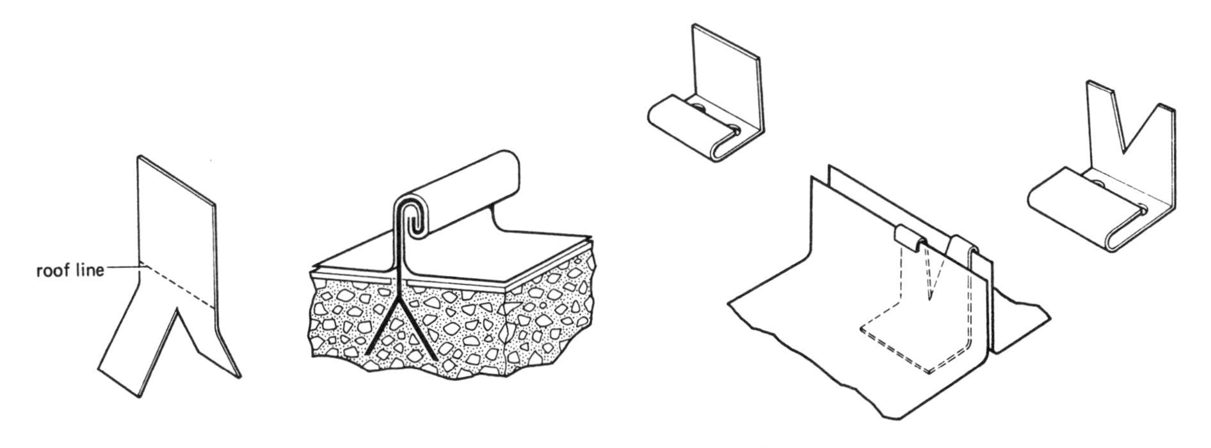

Figure 378 *Fishtailed cleats set in concrete*

Figure 379 *Alternative types of cleat*

Fixing to low density insulation boards
In the case of low density insulation boards, such as straw boards, woodwool slabs and fibre boards, a different fixing method is required from that used for securing the copper clips to timber and concrete. Two methods are recommended:
Brass bolts (with countersunk heads) of sufficient length to penetrate to the underside of the boards. The bolts should be fastened with a washer and two nuts on the underside of the board with one of the nuts acting as a lock nut;
Long brass screws (with countersunk heads) driven right through the boards and into a 25 mm thick timber block measuring 100 mm by 100 mm minimum, held on the underside.
When batten rolls are used for ridge-to-eaves joints, they should be secured either by means of bolts or, where timber rafters or frames are available beneath the decking, by long screws into the timber. In the former case, the heads of bolts should be countersunk into the batten and a large washer used between the slabbing and the locking nuts.

One point about the use of these alternative deckings calls for special comment. In general, they tend to indent when welts are being dressed tight, and additional support should be provided at these points by using a drip plate. At eaves and verge edges, fascia boards or barge boards should be fixed firmly to the understructure so that the copper can be dressed and secured. With flax board and chip board the clips may be fixed in the usual way with countersunk brass screws but it is advisable to provide a small pilot hole in these materials to ensure a satisfactory fixing.

Felt underlay
After the satisfactory preparation of the decking, a suitable felt should be laid before the copperwork is commenced. This felt should either conform to British Standard 747 (Type 4A (II) Brown no. 1 inodorous), or be a type which will not adhere to the metal or decking under temperature changes, for example Mutacel.

The felt serves four purposes:
1 It lessens the possibility of abrasion between the copper and the decking.
2 It deadens the sound of wind and rain.
3 It gives a measure of thermal insulation.
4 It prevents galvanic corrosion between the copper and any ferrous fixings used to secure the decking.

All felt laid should be covered with the copper on the same day. The felt should be laid with butt joints and secured with copper nails. On vertical surfaces extra nailing is recommended.

Flue terminals
Flues should be terminated at a sufficient height above a copper roof to ensure that the concentrated flue gases do not come in contact with the copper. Natural catchment areas adjacent to chimneys, such as gutters which collect solid products of combustion, should be protected by two coats of bituminous paint.

Copper gutters and weatherings to other roofing materials
To avoid possible corrosion problems arising from using copper with other roofing materials, it is recommended that the plumber should always seek authoritative technical advice.

Sheet aluminium

For many years metals have been used for the weathering of roof surfaces. Long life, workability, availability and cost are the basic requirements. Aluminium was first introduced as a roofing material at the end of the nineteenth century and by the middle of the twentieth century it had become a serious contender against the other metals for this type of work. The method of forming and jointing is similar in many respects to that of sheet copper using the same type of tools.

The aluminium sheets and strips most commonly recommended and used are:
1 BA super purity (99.9 per cent) aluminium (BS 1470: SI).
2 BA commercial purity (99 per cent) aluminium (BS 1470: SIC).

BA super purity offers maximum resistance to atmospheric attack, while BA commercial purity is chosen for its durability and also its lower cost.

Table 24

Thickness s.w.g.	*mm*	*Weight kg/m²*	*Width mm*	*Length m*	*Coverage m²*
20	0.91	2.48	450 600	22.3 16.8	10.2
22	0.71	1.93	450 600	28.9 21.6	13.2

Table 24 applies to BA superior purity and BA commercial purity. Roofing quality aluminium is available in 25 kg rolls. The recommended thickness is 0.9 mm.

Contact with other materials

Metals
Direct contact with copper, and copper-rich alloys such as brass, must be avoided. In no circumstances should water be allowed to drain from a copper surface on to an aluminium one. The danger of electrolytic attack is less with lead or unprotected steel, but contact faces should, nevertheless, be painted. There is also a risk of attack, particularly in industrial or marine atmospheres, by water running from lead to aluminium. If this cannot be avoided by suitable design, the lead should be painted with bituminous paint. Zinc and aluminium may be used safely together. In order to avoid contamination, tools previously used with lead or copper should be well cleaned before use with aluminium.

Cement and lime mortars
Aluminium is subject to some attack when in contact with cement, lime mortars and concrete in the presence of moisture and it is, therefore, good practice to paint the metal before embedding in joints where continuous dampness is to be expected. Where the metal is used as a coping covering, steps must be taken to avoid moisture coming in contact with the underside of the metal. This can be achieved by painting the underside with bituminous paint or by inserting an impervious roofing felt with a covering of building paper.

Damp and unseasoned hardwoods
Direct contact with certain damp or unseasoned hardwoods may cause attack on aluminium. Painting the metal where it touches the timber or boarding in such circumstances is therefore necessary. Where an inodorous felt underlay is used such painting will, of course, be unnecessary.

Jointing and fixing

Vertical joints
The two traditional systems used generally are standing seam and batten roll, both being methods of jointing the sides of sheet strip together in the ridge to eave direction to form a joint raised clear of the water flow. The choice of joint may be decided on aesthetic grounds and/or probable head of water on the roof slope. The finished height of a standing seam is approximately 22 mm and the minimum height of a batten roll is 41 mm.

Transverse joints
The pitch of the roof determines the choice of joints used to join the ends of the sheets together, or vice versa. These joints lie flat to the metal surface and the water flow passes over them. Three joints may be used: drip, double lock welt or single lock welt. Their use is recommended as follows:

From 50 mm in 3 m, which is the minimum fall for metal roofing; up to 5°, the transverse joint should be a drip.

Over 5° and up to 40°, the method of jointing is by *double lock cross welt.*

Over 40° to vertical, a *single lock cross welt* may be used.

For roofs of 20° pitch or less, all cross welts should be sealed with boiled linseed oil. The oil should be applied to the edges of the metal prior to welting.

Fixing clips
Aluminium is secured to the understructure by clips of the same gauge as the roof covering.

Clips in standing seams, batten rolls and verge welts should be spaced at 300 mm centres. In cross welts there should be one clip in every double lock cross welt spaced centrally in the bay between vertical joints. With single lock cross welts and at eaves, drips and ridge seams, two clips should be used, evenly spaced in the bay width between vertical joints.

All clips should be 50 mm wide and where used in standing seams, cross welts, eaves, verge and drips, they should be fixed to the understructure by two nails or screws set in line as close to the angle of the clip as required. Some forms of understructure will not take a secure nail or screw fixing and other methods may be required. Clips in batten rolls should be fixed underneath the rolls and not to the roll tops.

Areas

The layout and size of panels on pitched and flat roofs are governed by negative wind loading and thermal movement.

For pitched roofs in most exposed positions where standing seams or batten rolls are used the main panels are generally 1800 × 600 mm. Verge and gable end bay widths should be reduced to 300 mm. This width reduction applies for a distance of not less than 15 per cent of the length of the roof. In addition, the sheets on either side of the ridge should be half the normal length to the first cross welt. In sheltered localities lengths of metal not exceeding 300 mm may be laid between cross joints.

In flat roof construction (i.e. up to 5° pitch) in sheltered areas the length between drips may be up to 4200 mm but in general a length of 2400 mm is recommended. The smaller the drip spacing the more rapid is the run-off of water due to the 'waterfall' effect at the drips.

Soldering

Soldering of aluminium in roof work is not recommended; joints formed in this way are not permanent, and there is danger of chemical attack from entrapped corrosive fluxes.

Screws and nails

Aluminium screws or nails are recommended, although stainless steel, zinc or good quality galvanized steel may be used in certain instances. Screws or nails made from copper, or copper-rich alloys such as brass, must not be used. Specifications are given in BS CP143.

Free edges

Exposed free edges should be stiffened with a 12 mm bead.

Painting

Aluminium surfaces to be painted should be clean, and scratch-brushed or coated with a pre-treatment (etch) primer in order to provide a key. For protective painting, a bituminous paint is the most suitable. Zinc chromate, iron or zinc oxide primers may be used, but not lead based paints. Where appearance is not important a zinc chromate primer followed by a coat of aluminium paint, or alternatively an aluminium-pigmented bituminous paint, should be applied. Where essentially decorative painting is required the zinc chromate primer should be followed by compatible undercoats and top coats.

Annealing

BA super purity, BA commercial purity and BA 60 alloy harden to some extent, but they can be annealed locally with either a blow-lamp or welding and brazing torches. The correct temperature is reached when a match stalk drawn across the surface leaves a char mark; overheating should be avoided.

Marking out

Coloured crayons or grease-pencils are recommended for marking out aluminium.

Chimney weatherings

Super purity sheet aluminium can be bossed in the same manner as sheet lead to form aprons and back gutters. Aluminium is not as malleable as lead and it takes considerably more time to perform these operations. It is therefore not economical and the welting technique is universally accepted as being the best.

BA super purity (99.9 per cent) 0.9 mm thick flashing quality is recommended for chimney weathering. Figure 380 shows an exploded view

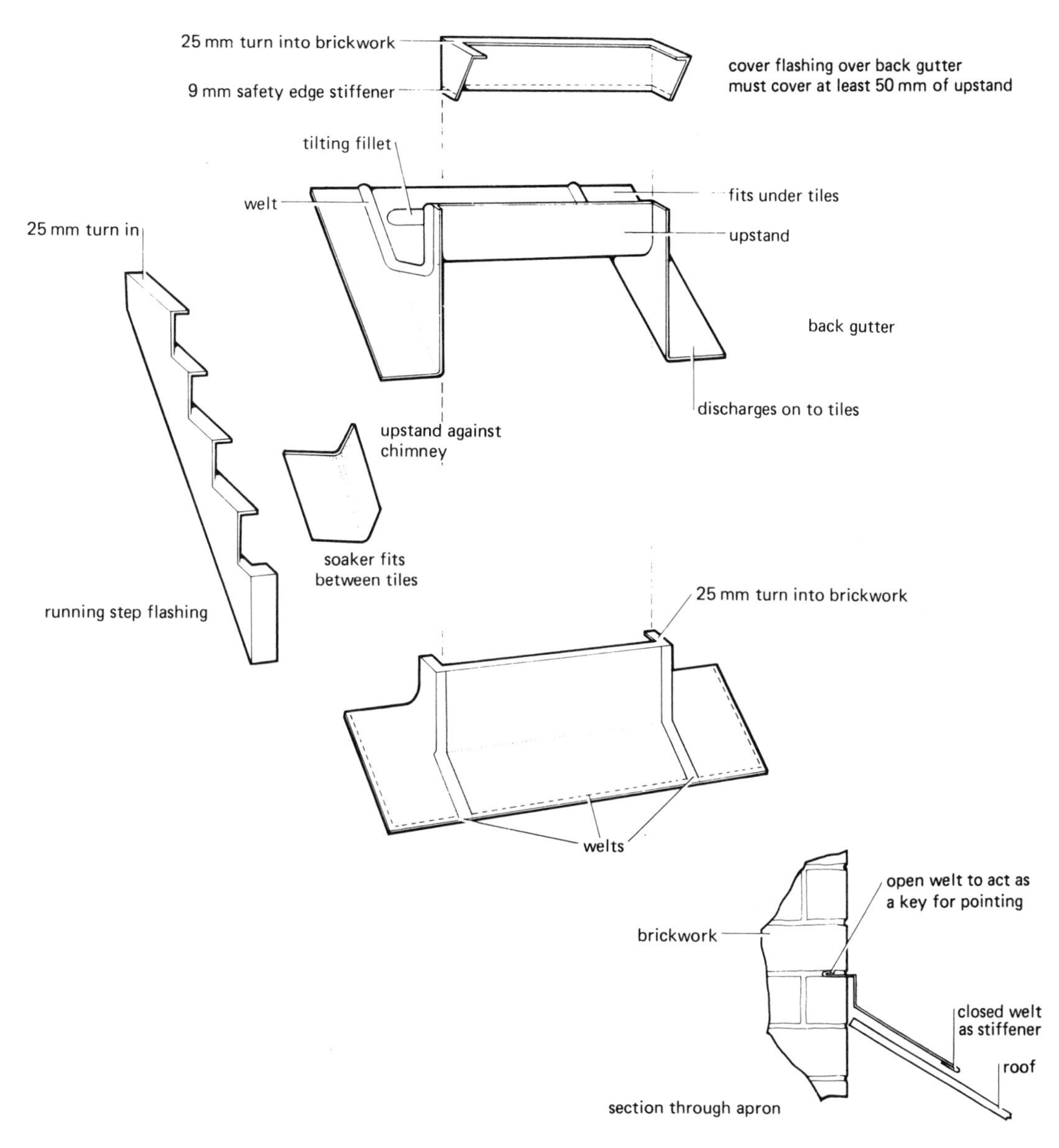

Figure 380 *Exploded view of aluminium chimney weathering*

showing the component parts (welted joints on back gutter and apron) of a chimney weathering.

Procedure
The first operation is to cut and supply for the roofer the required number of soakers. These are fitted by the roofer as he or she fixes the slates or plain tiles. The method of ascertaining the size and number of soakers is shown on page 180.

The second operation is to lay the back gutter in position behind the stack. This can be partially dressed into shape and is completed after the slating or tiling is finished.

The third operation is to dress and fit the apron to the front of the chimney, the flaps at each side fitting between the tiles while the front fits over the slates or tiles.

The fourth step is to cut and fit the running step

flashings to the sides of the chimney. These cover the upstand of the soaker, fit behind the turned end of the back gutter and over the upstand and turned ends of the apron.

The final operation is to cut, form and fit a cover flashing over the back gutter where the upstand fits against the brick chimney. All the component parts should be dressed snugly into position and held securely by wedges driven into the brick joints which are then neatly pointed with mastic or mortar.

Damp proof courses

As mentioned earlier, part of the work of the plumber is to keep the rainwater from gaining access to the building. There is another way that water may enter a building: capillarity. Capillary attraction is the phenomenon by which water can pass through a solid object, for example brick or stone, and even rise between two pieces of metal fitting close together.

Capillary attraction can also cause water to pass downwards through bricks, stone, concrete and other building fabrics and because of this it has been found necessary to install damp proof courses in chimney stacks, parapet walls and the like. Alternatively, the copings can be completely covered to overcome this downward passage of water.

Methods of overcoming capillary attraction

Cap flashing

Figure 381 shows this method, by which the coping is completely covered. Two methods, 'A' and 'B', can be used to secure the sheet to the coping.

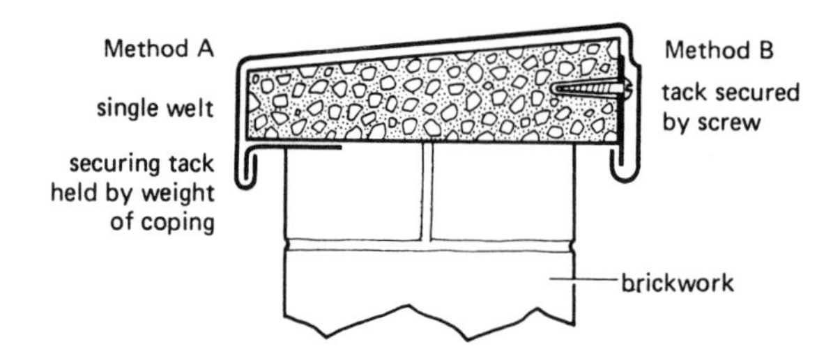

Figure 381 *Coping covered as damp proof course*

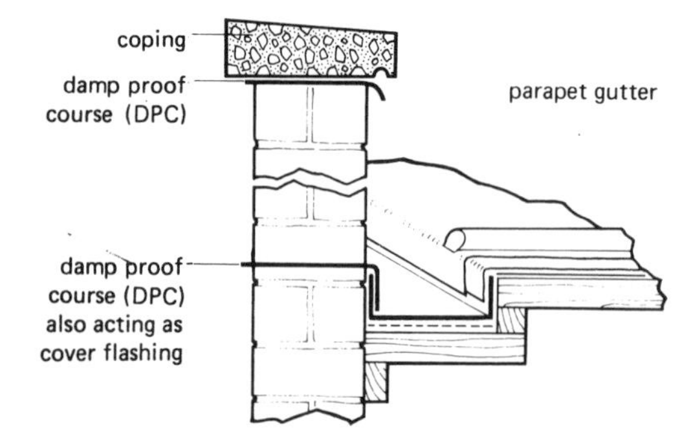

Figure 382 *Two methods of preventing the downward passage of water*

This cap flashing method of preventing capillarity is undoubtedly the best. It affords complete protection, the coping being fully covered, but at the same time maintains a bond securing it to the brick wall. It will be readily seen that where the metal weathering passes completely across the brickwork, so breaking this bond, a weakness in construction takes place even though the problem of capillarity is solved.

Figure 382 shows two other methods of preventing the downward passage of water.

Anti-capillary groove

This is a method of overcoming capillary attraction between two pieces of sheet where the vertical height is less than 50 mm. A space is created between the two pieces of material, i.e. the anti-capillary groove, so destroying the capillary action (see Figure 383).

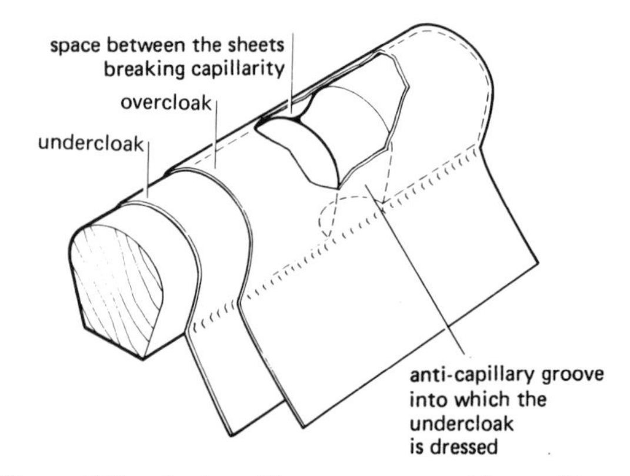

Figure 383 *Anti-capillary groove on ridge or hip*

Self-assessment questions

1 The thickness of sheet lead is recognized by a colour code. Number 5 lead is:
(a) green
(b) blue
(c) red
(d) black

2 Which of the following physical properties is possessed by sheet lead?
(a) ductility
(b) malleability
(c) tenacity
(d) elasticity

3 When laying sheet lead using lap joints, the vertical lap should be:
(a) 75 mm
(b) 50 mm
(c) 85 mm
(d) 100 mm

4 The side weathering to a chimney passing through a slated roof is known as:
(a) step and cover flashings
(b) step flashings
(c) soakers
(d) soakers and step flashings

5 The minimum width of a soaker is:
(a) 125 mm
(b) 150 mm
(c) 175 mm
(d) 200 mm

6 When fixing sheet copper to a flat roof the sheet copper is secured by means of:
(a) copper nails
(b) brass screws
(c) copper cleats
(d) wood battens

7 Before a roof is covered by sheet aluminium the roof should be covered with:
(a) mineral surface felt
(b) inodorous felt
(c) plastic sheet
(d) bitumastic sheet

8 When a metal is annealed, it becomes:
(a) softer
(b) harder
(c) brittle
(d) less malleable

9 The green coating which may form on copper roofs is an indication of:
(a) an excessive corrosive atmosphere
(b) an unusually dirty atmosphere
(c) undesirable corrosion of the metal
(d) the formation of a protective coating

10 In sheet copper and aluminium roofwork, double lock cross welts between standing seams should be staggered to:
(a) increase the roof fall
(b) give economical use of material
(c) avoid unworkable thickness
(d) permit annealing

9 Calculations

After reading this chapter you should be able to:

1. Recognize a decimal fraction.
2. Convert decimals to fractions and vice versa.
3. Add, subtract, divide and multiply using decimal numbers.
4. Understand the metric system of units.
5. Convert units within the metric system.
6. Understand the nature and function of fractions.
7. Convert fractions to percentages and vice versa.
8. Apply percentages in the solution of plumbing problems.
9. Calculate simple areas and volumes.
10. Convert quantities into given ratios.

Introduction

Taking measurements is an essential part of most plumbing jobs. Every time you take a measurement you use arithmetic, and, with practice, this becomes so automatic that you hardly notice it. Arithmetic is needed in all sorts of ways – to order the correct amounts of material without incurring waste, to confirm certain aspects of design or system layout that may arise, to obtain a comprehensive and competitive cost for a particular working process, and so on. What is important with arithmetic, as with other skills, is to understand the basic principles. If you do not find calculations easy, concentrate and practise the skill until it is understood and mastered. The following chapter contains some of the fundamental rules and methods of calculating, together with examples of how to use them.

Decimals

All metric calculations are in decimals, and the word 'decimal' means *in the order of tens*. The following sequence of numbers will illustrate the basic principles of decimals.

10,000   1000   100   10   1

0.1   0.01   0.001   0.0001

$\frac{1}{10}$ $\frac{1}{100}$ $\frac{1}{1000}$ $\frac{1}{10000}$

Each number in this series is one-tenth of the figure to its left. The decimal point separates the whole numbers from the decimal fractions. This then is the order of tens, or decimal system. The way to use this system is shown in the following examples.

Addition and subtraction of decimals
Write down the decimals in the same way as you would if you were adding or subtracting ordinary figures, that is, units under units, tens under tens, hundreds under hundreds, and so on. As a result, the decimal points are also under one another. This is most important.

Example 1
Add: 2.25, 34.7 and 30.242.

```
 2.25
34.7
30.242
------
67.192
------
```

Example 2
Subtract: 14.63 from 57.86.

```
57.86
14.63
-----
43.23
-----
```

Multiplication of decimals
It is possible to multiply by any power of 10 simply by moving the decimal point as many places to the right as there are noughts in the figure by which you are multiplying.

Example 3

(a) $4.752 \times 10 = 47.52$

(b) $4.752 \times 100 = 475.20$

(c) $4.752 \times 1000 = 4752.00$

There are two methods of multiplication. The procedure of working is the same for each, but they are set down differently. It is best to continue to use whichever method you are familiar with, but the following method is commonly used.

Example 4
Multiply: 34.16 by 3.4.
Rule A Write the figures down and multiply them as though they were whole numbers:

```
 34.16  (multiplicand)
   3.4  (multiplier)
------
 13664
102480
------
116144  (product)
```

Rule B Count the total number of decimal places (figures after the decimal points) in multiplier and multiplicand (in this example three places), and step off this number of places from the right-hand end of the product. The decimal point is then located in that position:

= 116.144

With decimal calculations, it is always a good idea to do an approximation as a check. As shown earlier, the positioning of the decimal point is important, and if the point is put one place too far to the right, then the answer would be ten times too big; if it were put one place too far to the left, it would be one-tenth less than it should be. Both errors are quite serious, but can be avoided by calculating an approximation to the answer.

Example 5
Check the answer to Example 4 by approximation. 34.16×3.4 is slightly larger than 34×3. The full, correct answer to Example 4 was 116.144. You can quickly work out that $34 \times 3 = 102$. This is only a little less than 116 and so allowing for the decimal fractions omitted in the approximation, it is obvious that the placing of the decimal point is correct. Had the answer to Example 4 been given as either 11.6144 or 1161.44, the approximation figure of 102 would have indicated incorrect location of the decimal point.

Division of decimals
Division of 10, 100, 1000, 10000 or any other power of 10 is achieved by moving the decimal point as many places to the left as there are noughts in the figure by which you are dividing as shown in Example 6. You may have to insert some noughts in the answer to fill any spaces (see Example 6(a), (c) and (d)).

Example 6

(a) 2.58 ÷ 10 = 0.258
(b) 2.58 ÷ 100 = 0.0258
(c) 7.64 ÷ 100 = 0.0764
(d) 7.64 ÷ 1000 = 0.00764
(e) 159.07 ÷ 10 = 15.907
(f) 159.07 ÷ 100 = 1.5907
(g) 159.07 ÷ 1000 = 0.15907

The method of dividing decimals when numbers involved are not powers of 10 is set out in Example 7 below.

Rule A To understand the terminology and method of working the calculation is shown as:

```
        quotient
divisor )dividend
```

Rule B If the divisor is not, as it is in Example 7, a whole number, then it must be made so by moving the decimal point to the right. The decimal point of the dividend would then also have to be adjusted by the same number of places in order to keep its value in proportion to the divisor, as the following examples indicate:

(a) 42.6 ÷ 2.3 = 426 ÷ 23 = 23)426
Both decimal places have been moved one place to the right to make the divisor a whole number and keep the dividend in correct proportion.

(b) 42.6 ÷ 2.315 = 42600 ÷ 2315 = 2315)42600
Here the decimal point has been moved three places. Noughts have been added to the dividend to fill the spaces and keep the proportion correct.

(c) 426 ÷ 2.3 = 4260 ÷ 23 = 23)4260
In this example there is no decimal point in the dividend, the point in the divisor has to be moved one place and a nought added to the dividend to keep it in proportionate value to the adjusted divisor.

Example 7

Divide: 56.37 by 122

(a) 122 into 56.37 will not go. Set out the question and put a nought above the 6, and decimal point above that in the dividend.

```
     0.
122)56.37
```

(b) 122 into 563 goes 4 times equalling 488. Put 4 after the decimal point in the quotient and subtract 488 from 56.3. This leaves 75.

```
     0.4
122)56.37
    488
    ---
     75
```

(c) 122 into 75 will not go. Bring down the 7 from the dividend. 122 goes into 757 6 times = 732. Place 6 in the quotient and subtract 732 from 757. This leaves 25. Add a nought to the 25 (this is assumed to have been brought down from the dividend). Noughts added after the last figure cannot alter the value of the dividend.

```
     0.462
122)56.37
    488
    ---
     757
     732
     ---
      250
      244
      ---
        6
```

(d) 122 into 250 goes 2 times, equalling 244. Put 2 in the quotient and subtract 244 from 250. This leaves 6. The calculation can be stopped at this point as the values involved are very small.

The answer to this problem, to three decimal places, is 0.462.

Summary of decimals

1 Decimals are fractions with denominators of 10, 100, 1000, etc. The decimal point separates the whole numbers from the fractional parts.
2 When adding or subtracting decimal numbers, the decimal points are written under one another.
3 To *multiply* by 10 move the decimal point *one* place to the *right*. To multiply by 100 move the decimal point *two* places to the *right*, etc.

4 To *divide* by 10 move the decimal point *one* place to the left, to divide by 100 move the decimal point *two* places to the *left*, etc.
5 When multiplying, first disregard the decimal points and multiply the two numbers as though they were whole numbers. To place the decimal point in the product, count up the total numbers of figures after the decimal point in both numbers, and then count off this number of figures in the product, starting from the extreme right.
6 When dividing first make the divisor into a whole number and compensate the dividend.
7 Before multiplying or dividing, always perform a rough check which will ensure the decimal point is placed correctly.

The metric system

The United Kingdom is committed to a change from the imperial to the metric system of mensuration and the SI (Système International) is the system being adopted. The change has involved a vast amount of work, much more than the relatively simple conversion of existing dimensions to their metric equivalents. This is because in many cases the existing dimensions have been altered to standardize practice and comply with commonly used sizes in countries already established in metric system usage.

All examinations of the City and Guilds of London Institute are set in metric quantities and units.

The basic SI units most frequently used in the building and construction industry are given in Table 25. The last column refers to the symbols used for each unit. These are not abbreviations and do not require a letter 's' to indicate plural. It is also unnecessary to place a full stop after them. This could be interpreted as a decimal point.

Table 25

Quantity	*Unit*	*Symbol*
Length	Metre	m
Mass	Kilogram	kg
Time	Second	s
Capacity	Litre	l

This range of units has to be extended to ensure that there is an appropriate unit suitable for either small or large quantities, and this is achieved by using a group of prefixes, the most common of which are k meaning kilo (×1000) and m meaning milli (×0.001).

Table 26 is a compilation of prefixes, including the prefix symbol and the multiplication factor of these as they vary from unity (one).

Table 27 is a compilation of metric quantities, units and symbols in general use in the building and construction industry and its associated industries.

Table 26

Prefix	*Prefix symbol*	*Multiplication factor*
tera-	T	10^{12}
giga-	G	10^{9}
mega-	M	10^{6}
kilo-	k	10^{3}
hecto-	h	10
	unity (one)	
deci-	d	10^{-1}
centi-	c	10^{-2}
milli-	m	10^{-3}
micro-	u	10^{-6}
nano-	n	10^{-9}
pico-	p	10^{-12}

Table 27

Quantity	*SI unit*	*Unit abbreviation*
Linear	Millimetre	mm
Linear	Centimetre	cm
Linear	Metre	m
Linear	Kilometre	km
Area	Millimetre squared	mm^2
Area	Centimetre squared	cm^2
Area	Metre squared	m^2
Area	Kilometre squared	km^2
Volume	Millimetre cubed	mm^3
Volume	Centimetre cubed	cm^3
Volume	Metre cubed	m^3

Volume	Kilometre cubed	km^3
Capacity	Litre	l (better written in full to avoid confusion with 1)
Time	Second	s
Rate of flow	Litre per second	l/s
Volume rate of flow	Cubic metre per second	m^3/s
Pressure	Newton per square metre	N/m^2
Pressure	Newton per square millimetre	N/mm^2
Pressure	Kilonewton per square metre	kN/m^2
Pressure	Kilonewton per square millimetre	kN/mm^2
Pressure	Meganewton per square metre	MN/m^2
Pressure	Bar (100 kN/m^2)	bar
Pressure	Millibar (100 N/m^2)	mb
Pressure	Pascal (1 N/m^2)	p
Force	Newton (derived from kgm/2)	N
Energy/work	Joule (derived from kgm/2 or Nm)	J
Quantity of heat	Joule	J
Power	Watt (derived from Nm/s or J/s	W
Electrical current	Amperes	A
Electrical potential	Volt (derived from W/A)	V
Electrical resistance	Ohm (derived from V/A)	Ω
Luminosity	Candela	cd
Temperature	Degree Celcius (normal temperature use	°C
Temperature	Degree Kelvin	K

Fractions

A fraction represents a part of a whole. The circle in Figure 384 has been divided into eight equal parts. Each part is called one-eighth of the circle and is written as ⅛.

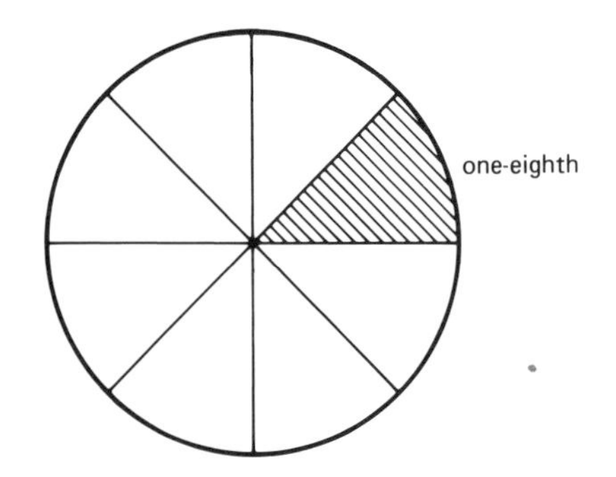

Figure 384

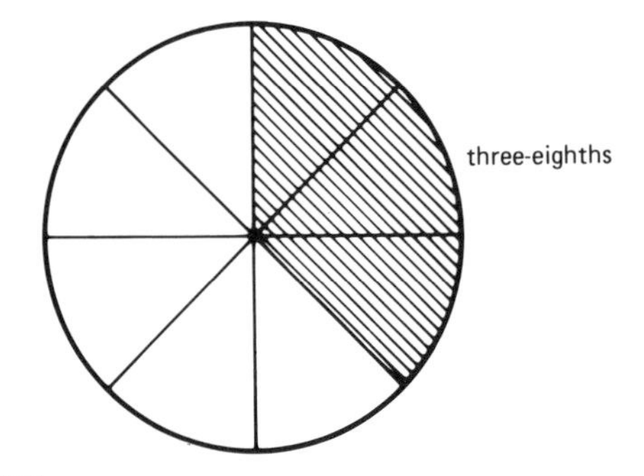

Figure 385

The number 8 below the line shows how many equal parts there are and is called the *denominator*. The number above the line shows how many of the equal parts are taken into consideration and is called the *numerator*. If three of the eight equal parts are taken then we have taken ⅜ of the circle (Figure 385). The two fractions ⅛ and ⅜, which have the same number as their denominators, are said to have a *common denominator*.

From the introduction above we can see that a fraction is always part of something. The number below the line (the denominator) gives the fraction its name and tells us the number of equal parts into which the whole has been divided. The top number above the line (the numerator) tells us the number of equal parts that are to be taken. For example, the fraction 7/10 means that the whole has been divided into ten equal parts and that seven of these parts are to be taken into consideration.

The value of a fraction is unchanged if we multiply or divide both its numerator and denominator by the same amount.

$\frac{3}{5} = \frac{12}{20}$ (by multiplying the numerator and denominator by 4)

$\frac{12}{42} = \frac{2}{7}$ (by dividing the numerator and denominator by 6)

Example 8
Write down the fraction $\frac{3}{5}$ with a denominator of 25.

In order to make the denominator 25, we must multiply the original denominator of 5 by 5 because 5 × 5 = 25. Also remembering that to leave the value of the fraction unchanged, we must multiply both numerator and denominator by the same amount:

$$\frac{3}{5} = \frac{3 \times 5}{5 \times 5} = \frac{15}{25}$$

Exercise 1
Write down the following fractions with the denominator stated:

(a) $\frac{3}{5}$ with denominator 20

(b) $\frac{1}{9}$ with denominator 63

(c) $\frac{2}{3}$ with denominator 12

(d) $\frac{5}{7}$ with denominator 35

(e) $\frac{3}{8}$ with denominator 64

(f) $\frac{4}{5}$ with denominator 15

(g) $\frac{6}{18}$ with denominator 54

(h) $\frac{1}{6}$ with denominator 78

Reducing a fraction to its lowest terms
Fractions like $\frac{5}{8}$, $\frac{7}{16}$ and $\frac{9}{20}$ are said to be in their *lowest terms* because it is not possible to find a number which will divide exactly into both the numerator and denominator. However, other fractions like $\frac{8}{12}$, $\frac{9}{18}$ and $\frac{21}{24}$ are not in their lowest terms because they can be reduced further by dividing both numerator and denominator by some number which divides exactly into both of them. Thus:

$\frac{8}{12} = \frac{2}{3}$ (by dividing both numerator and denominator by 4)

$\frac{9}{18} = \frac{1}{2}$ (by dividing both numerator and denominator by 9)

$\frac{21}{24} = \frac{7}{8}$ (by dividing both numerator and denominator by 3)

Sometimes it is possible to divide the numerator and denominator by more than one number to reduce the fraction to its lowest terms, for example:

$\frac{105}{168} = \frac{35}{56}$ (by dividing with 3)

$\frac{35}{56} = \frac{5}{8}$ (by dividing with 7)

therefore it is possible to reduce $\frac{105}{168}$ to $\frac{5}{8}$ (its lowest terms).

Exercise 2
Reduce the following fractions to their lowest terms:

(a) $\frac{12}{16}$

(b) $\frac{5}{25}$

(c) $\frac{9}{15}$

(d) $\frac{8}{64}$

(e) $\frac{20}{60}$

(f) $\frac{140}{210}$

(g) $\frac{180}{240}$

(h) $\frac{132}{198}$

Types of fractions
If the numerator of a fraction is smaller than its denominator, the fraction is called a *proper fraction*. Therefore, $\frac{1}{3}$ $\frac{3}{4}$, and $\frac{7}{8}$ are all proper fractions. Note that a proper fraction has a value which is less than 1.

If the numerator of a fraction is greater than its denominator, then the fraction is called an *improper fraction*. Thus $\frac{4}{3}$, $\frac{7}{4}$ and $\frac{11}{8}$ are all improper fractions. Note that all improper fractions have a value which is more than 1.

Every improper fraction can be expressed as a whole number and a proper fraction. These are sometimes identified as mixed number fractions. Thus $1\frac{1}{3}$, $1\frac{3}{4}$ and $1\frac{3}{8}$ are all mixed number fractions.

To convert an improper fraction to a mixed number fraction, it is necessary to divide the numerator by the denominator.

Lowest common multiple (LCM)

The LCM of a set of numbers is the *smallest* number into which each of the given numbers will divide. Thus the LCM of 4, 5 and 10 is 20 because 20 is the smallest number into which these numbers will divide exactly.

Lowest common denominator (LCD)

When it is necessary to compare the values of two or more fractions the easiest way is to express the fractions with the same denominator. This common denominator should be the LCM of the denominators of the fractions to be compared and it is called the LCD.

Addition of fractions

The procedure to be used when adding fractions is as follows:

1 Find the lowest common denominator of the fractions to be added.
2 Express each of the fractions with this common denominator.
3 Add the numerators of the new fractions to give the numerator of the answer. The denominator of the answer is the lowest common denominator found in step 1.

Example 9

Find the sum of: $\frac{3}{4}$ and $\frac{2}{7}$

First find the lowest common denominator (this is the LCM of 4 and 7). It is 28. Now express $\frac{3}{4}$ and $\frac{2}{7}$ with a denominator of 28.

$$\frac{3}{4} = \frac{3 \times 7}{4 \times 7} = \frac{21}{28}$$

$$\frac{2}{7} = \frac{2 \times 4}{7 \times 4} = \frac{8}{28}$$

Adding the numerators of the new fractions:

$$\frac{21}{28} + \frac{8}{28} = 1\frac{1}{28}$$

Subtraction of fractions

The method is similar to that used in addition. Find the common denominator of the fractions and after expressing each fraction with this common denominator, subtract.

Example 10

Simplify: $\frac{7}{8} - \frac{2}{5}$

The LCM of the denominator is 40.

$$\frac{7}{8} - \frac{2}{5} = \frac{5 \times 7 - 8 \times 2}{40} = \frac{35 - 16}{40} = \frac{19}{40}$$

Multiplication

When multiplying together two or more fractions it is necessary to first multiply all the numerators together and then multiply the denominators together. Mixed number fractions must always be converted into improper fractions.

Example 11

Simplify: $\frac{3}{7} \times \frac{5}{8}$

$$\frac{3}{7} \times \frac{5}{8} = \frac{3 \times 5}{7 \times 8} = \frac{15}{56}$$

Example 12

Simplify: $3\frac{2}{3} \times \frac{2}{5}$

$$3\frac{2}{3} \times \frac{2}{5} = \frac{11 \times 2}{3 \times 5} = \frac{22}{15} = 1\frac{7}{15}$$

Division

To divide by a fraction, all we have to do is to invert it and multiply.

Example 13

Divide: $\frac{3}{5}$ by $\frac{2}{7}$

$$\frac{3}{5} \div \frac{2}{7} = \frac{3}{5} \times \frac{7}{2} = \frac{3 \times 7}{5 \times 2} = \frac{21}{10} = 2\frac{1}{10}$$

Example 14

Divide: $1\frac{4}{5}$ by $2\frac{1}{3}$

$$1\frac{4}{5} \div 2\frac{1}{3} = \frac{9}{5} \div \frac{7}{3} = \frac{9}{5} \times \frac{3}{7} = \frac{27}{35}$$

Summary of fractions

1 The denominator (bottom number) gives the fraction its name and gives the number of equal parts into which the whole has been divided. The numerator (top number) gives the number of equal parts that are to be considered.
2 The value of a fraction remains unaltered if both the numerator and the denominator are multiplied or divided by the same number.
3 The LCM of a set of numbers is the smallest number into which each of the numbers of the set will divide exactly.
4 To compare the values of fractions which have different denominators express all the fractions with the LCD and then compare the numerators of the new fractions.
5 To add fractions together, express each of them with their LCD and then add together the resulting numerators.
6 To multiply fractions, multiply the numerators together and then multiply the denominators together.
7 To divide fractions, invert the divisor and then proceed as in multiplication.

Percentages

A percentage is used to compare a given quantity to 100 parts. So eight per cent, or 8%, means 8 parts out of 100 parts. Percentage means 'per hundred' and indicates how many $\frac{1}{100}$ parts are present.

The arithmetical sign for percentage is %.

$$1\% = 1 \text{ in every } 100 = \frac{1}{100}$$

$$26\% = 26 \text{ in every } 100 = \frac{26}{100}$$

$$50\% = 50 \text{ in every } 100 = \frac{50}{100} = \frac{5}{10} = \frac{1}{2}$$

$$80\% = 80 \text{ in every } 100 = \frac{80}{100} = \frac{8}{10} = \frac{4}{5}$$

It can be seen from the above that a percentage can be expressed as a fraction simply by placing the percentage as a numerator above a denominator of 100. Note that in some cases it is possible to cancel the fraction down to express it in its lowest terms.

When converting a fraction into a percentage, we multiply it by 100.

$$\frac{1}{4} = \frac{1}{4} \times 100 = 25\%$$

$$\frac{2}{5} = \frac{2}{5} \times 100 = 40\%$$

$$\frac{17}{20} = \frac{17}{20} \times 100 = 85\%$$

Another mathematical exercise involves converting a percentage to a decimal. This is done by moving the decimal point two places to the left, for example:

1% = 0.01
26% = 0.26
50% = 0.50
80% = 0.80

It follows that to convert a decimal to a percentage we must move the decimal point two places to the right (i.e. multiply by 100), for example:

0.34 = 34%

0.652 = 65.2%

0.045 = 4.5%

0.207 = 20.7%

Percentage of a quantity

It is easy to find the percentage of a quantity if we first express the percentage as a fraction.

Examples

(a) What is 10% of 80?
Expressing 10% as a fraction gives $\frac{10}{100}$ and the problem then becomes:

What is $\frac{10}{100}$ of 80?

$$10\% \text{ of } 80 = \frac{10}{100} \times 80 = 8$$

(b) What is 25% of £50?

$$25\% \text{ of } £50 = \frac{25}{100} \times £50 = £12.50$$

(c) What is 54% of 300 mm?

$$54\% \text{ of } 300 \text{ mm} = \frac{54}{100} \times 300 \text{ mm} = 162 \text{ mm}$$

(d) What is 7% of 42 kg?

$$7\% \text{ of } 42 \text{ kg} = \frac{7}{100} \times 42 \text{ kg} = 2.94 \text{ kg}$$

Summary of percentages

1 Percentages are fractions with a denominator of 100.
2 To convert a fraction into a percentage, multiply it by 100.
3 To convert a percentage into a decimal, move the decimal point two places to the left.
4 To convert a decimal to a percentage, move the decimal point two places to the right.
5 To find the percentage of a quantity, first convert the percentage into a fraction and then multiply the quantity by the fraction.

Area and volume

Area

Area is a measure of the amount of surface and is measured in square units. A square which has sides of 1 metre length has an area of 1 square metre as is shown by Figure 386.

The notation for 1 square metre is 1 m^2. The small number 2 is called the index of m and it indicates that two units of metre have been multiplied together.

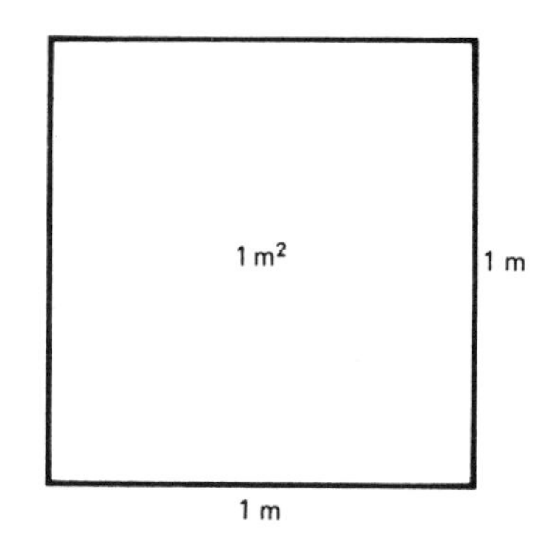

Figure 386

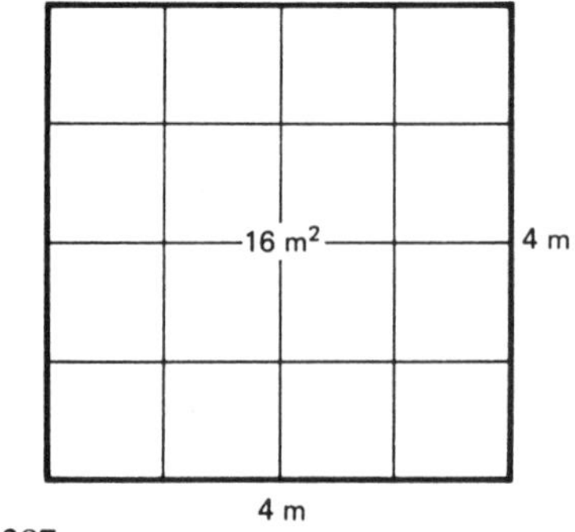

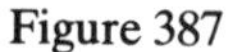

Figure 387

Similarly, a square with sides of 4 m length would have an area of 16 m^2 (Figure 387). Thus the area of a plane figure is measured by seeing how many square units it contains, and is arrived at by multiplying the lengths of any two sides together, for example, in Figure 387:

$4 \text{ m} \times 4 \text{ m} = 16 \text{ m}^2$

Area of a rectangle

A rectangle is a four sided figure with one pair of opposing sides longer than the other pair of opposing sides. The long sides are identified as *length*, the shorter sides as *breadth*.

The rule for finding the area of a rectangle is length × breadth. This rule applies to any size of rectangle.

Thus the area of a rectangle = length × breadth (see Figure 388). The area of the rectangle in Figure 388 is length × breadth, i.e.:

$5 \text{ m} \times 3 \text{ m} = 15 \text{ m}^2$

In using this formula the units of length and breadth must be the same, that is they both must be in metres, centimetres or millimetres.

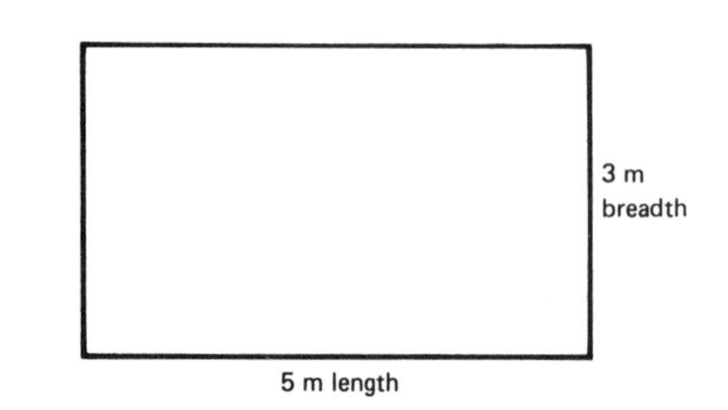

Figure 388

Example 15
Find the area of a piece of sheet metal measuring 1320 mm × 432 mm. Express the answer in square metres.

In questions of this type it is best to express each of the dimensions in metres before attempting to find the area.

$$\text{area} = 1320 \text{ mm} \times 432 \text{ mm} = \frac{1320}{1000} \times \frac{432}{1000}$$

$$\text{area} = 1.32 \text{ m} \times 0.432 \text{ m}$$

```
 1.32
 0.432
 -----
 52800
  3960
   264
 -------
0.56924
 -------
```

$$\text{area} = 0.56924 \text{ m}^2$$

Volume
The volume of a body is a measure of the space it occupies. This measurement of space is done in three dimensions. Volume is measured in cubic units, for example m^3, cm^3, mm^3.

A cube has sides of 1 metre (e.g. Figure 389) is said to occupy one cubic metre. The notation for 1 cubic metre is 1 m^3. The small number 3 is called the index of m and it indicates that three units of metre have been multiplied together.

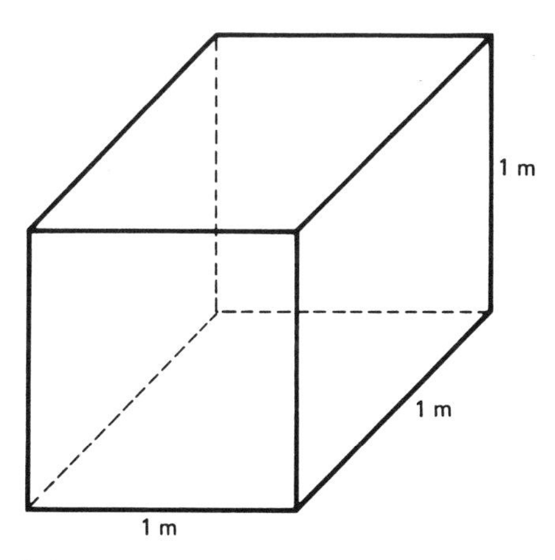

Figure 389

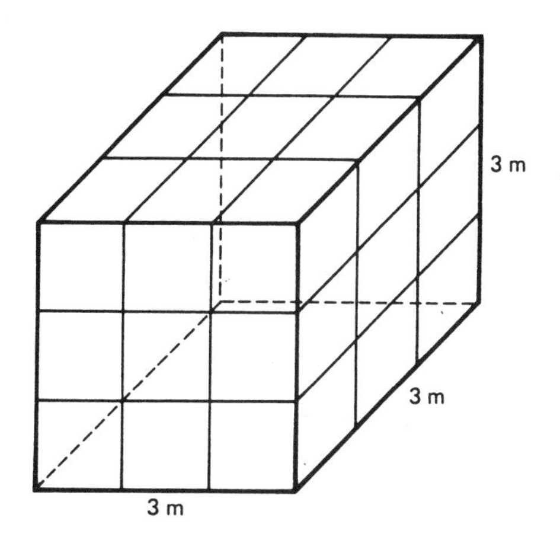

Figure 390

Similarly a cube with sides of 3 m length (e.g. Figure 390) would have a volume of 27 m^3.

Figures shaped like those in Figures 389 and 390 are called *cuboid*. The volume of a cuboid = length × breadth × height. All three dimensions must be in the same units of length, that is metres, centimetres, or millimetres.

Example 16
Find the volume of a cistern with is 1.5 m long, 1.2 m wide and 1 m deep.

$$\text{volume} = 1.5 \times 1.2 \times 1$$

$$\text{volume} = 1.8 \text{ m}^3$$

You will notice that different terminology is sometimes used to describe various situations: e.g. long instead of length, wide instead of breadth, and deep instead of depth or height. These variations do not alter the formula process or the method of calculating the answer.

Example 17
Calculate the volume of a hot water tank which measures 750 mm × 680 mm × 620 mm. In questions like these it is easier to express the dimensions as a decimal of a metre. This removes the problems of multiplying large figures.

$$\text{volume} = 750 \text{ mm} \times 680 \text{ mm} \times 620 \text{ mm}$$

divide each dimension by 1000 to convert to m

$$\text{volume} = 0.75 \times 0.68 \times 0.62$$

```
  0.75
  0.68
------
  4500
   600
------
0.5100
------
  0.51
  0.62
------
  3060
   102
------
0.3162
------
```

volume = $0.3162\ m^3$

If the area of the base of a cuboid is known it is possible to determine its volume by multiplying this base area with the height of the cuboid.

volume of cuboid = area of base × height.

Example 18
A cold water cistern is 1.25 m high and has a base area of $2.6\ m^2$. What is its volume in m^3?

volume of cistern = area of base × height

volume = $2.6\ m^2 \times 1.25\ m$

```
  2.6
  1.25
-----
 2600
  520
  130
-----
3.250
-----
```

volume = $3.25\ m^3$

Summary of area and volume

1 The area of a figure is measured by seeing how many square units it contains. 1 square metre is the area contained in a square of 1 m side. 1 square centimetre is the area inside a square whose side is 1 cm. 1 square millimetre is the area inside a square whose side is 1 mm.
2 Area of a rectangle = length × breadth.
3 Area of a square = side length ^{2}.
4 The volume of a body is measured by seeing how many cubic units it contains.
5 Volume of a cuboid = length × breadth × height, or
= area of base × height
6 The units used to calculate area and volume must be expressed in the same terms, e.g. millimetres, centimetres, or metres.

Ratio

A ratio is a comparison of two numbers. A fraction is an example of a ratio, i.e. ⅓ = 1:3 ratio.

The colon (:) in a ratio separates the two numbers. The above example indicates that a fraction is changed to a ratio by writing the numerator and denominator (in that order) and separating the numbers by a colon.

A decimal or percentage can be expressed as a fraction and both of these can then be presented as ratios, e.g.:

0.5 = ½ = 1:2

$4.5 = 4\tfrac{1}{2} = \frac{9}{2} = 9:2$

$10\% = \frac{1}{10} = 1:10$

75% = ¾ = 3:4

Since a ratio can be regarded as a fraction, multiplying or dividing both terms of a ratio by the same number does not change the value of the ratio, e.g.:

2:3 = 12:18 (multiplying both sides by 6)

14:7 = 2:1 (dividing both sides by 7)

The two examples below show how to separate a quantity according to a given ratio.

Example 19
£50 is to be divided into two portions in the ratio 3:7. How much is in each portion?

Total number of parts 3 + 7 = 10.

One portion is $\frac{3}{10}$ of the money = $\frac{3}{10} \times$ £50 = £15.

The other portion is $\frac{7}{10}$ of the money = $\frac{7}{10} \times$ £50 = £35.

Answer = £15 and £35.

A similar method is used if more than two portions are involved.

Example 20
£46 has to be split between three apprentices in the ratios 3:9:11. How much does each receive?

Total number of parts 3 + 9 + 11 = 23.

apprentice 1 receives $\frac{3}{23}$ of the money $\frac{3}{23} \times$ £46 = £6

apprentice 2 receives $\frac{9}{23}$ of the money $\frac{9}{23} \times$ £46 = £18

apprentice 3 receives $\frac{11}{23}$ of the money $\frac{11}{23} \times$ £46 = £22

answer = £6, £18 and £22.

Self-assessment questions

1 Add 3.165 and 5.207.

2 Subtract 46.72 from 73.91.

3 Multiply 16.04 by 81.33.

4 Divide 24.12 by 3.

5 Reduce $\frac{15}{35}$.

6 Simplify $\frac{5}{8} - \frac{1}{5}$.

7 Calculate 25% of £42.64.

8 Calculate the area of a surface which measures 3.5 m × 2.4 m.

9 Calculate the volume of a space measuring 1.6 m × 700 mm × 650 mm.

10 Divide £108.90 into the ratio 2:7.

10 Science

After reading this chapter you should be able to:

1 Define heat and temperature.

2 Understand the measurement of heat and temperature.

3 Define 'specific heat'.

4 Understand the term 'transmission of heat'.

5 Explain conduction, convection and radiation.

6 Understand the expansion characteristics of various substances.

7 Understand the nature of pressure in liquids and gases.

8 Calculate basic pressure calculations related to water.

9 Explain the terms 'density' and 'relative density'.

10 Understand the use of heat-sensitive crayons and paints.

Heat

Heat is a form of energy, and energy means a capacity for doing work.

All substances contain heat. Even ice, which we associate with cold, is quite warm compared to liquid air, which exists at 166 °C below the freezing point of water.

The molecules which make up substances are always vibrating to and fro. They need energy for this work, and heat provides it. The more energy the molecules possess, the more vigorously and further apart they will be able to vibrate. As the heat input into a substance increases the molecules step up their vibratory rate, and weaken their cohesive bonds, often bringing about a change of state, i.e. solid to liquid or liquid to gas.

For example, ice is water in a solid state. It only has a small heat content so its molecules hardly vibrate at all, and since they are close together the cohesion between them is strong and ice is therefore rigid. If heat is applied to ice, the molecules gain energy, vibrate, and weaken their cohesive bonds. The (solid) ice changes into (liquid) water. Further heating of the water increases the heat energy until the molecules vibrate so strongly that they actually jump out of the water to form a gas (steam).

The measurement of heat

As explained previously, heat is the name given to energy which is in the process of moving from one place to another as the result of a temperature difference between them. Since heat is a method of transferring energy, it is measured in *joules*, the same as any other kind of energy.

Heat capacity.

The heat of a body is defined as the heat required to raise its temperature by 1 °C. Therefore, the unit of heat capacity is the joule per degree C (J/°C).

Specific heat capacity

If we take equal quantities of water and oil and warm them in separate containers, but by the same flame, we may find that the oil temperature may rise by 15 °C in five minutes but the water may only rise by 8 °C in the same period of time. Since the supply of heat is the same in both cases, it is clear that oil has a lower heat capacity than water (see Figure 391).

When comparing the heat capacities of different substances we talk of their *specific heat capacities*.

Definition

The specific heat capacity of a substance is defined as the heat required to raise unit mass of it through 1 °C, and the unit of specific heat capacity is the joule per kilogramme degree C (J/kg°C).

Table 28 shows specific heat capacities for various substances. It will be seen that water has the unusually high specific heat capacity of 4.186 J/kg°C. Very few substances have a higher value than this.

Table 28 *Specific heat capacities in kJ/kg°C*

Water	4.186	Iron	0.460
Methylated spirit	2.400	Zinc	0.397
Ice	2.100	Copper	0.385
Air	1.046	Brass	0.380
Aluminium	0.887	Tin	0.234
Cast iron	0.544	Lead	0.125
Mild steel	0.502	Mercury	0.125

The specific heat capacities of all substances vary slightly as their temperature changes but Table 28 is sufficiently accurate for present needs.

From Table 28 it can be seen that only a small amount of heat is required to raise 1 kg of mercury 1 °C. This means that it is very sensitive to temperature change, which is one reason why it is used in thermometers.

Lead too has a low specific heat capacity, together with tin, although tin requires more heat to raise its temperature than does lead. Consequently tin takes longer to cool down, and this explains the problem of runs on the bottom of a wiped soldered joint. The cooling lead solidifies in the solder joint while the tin is still liquid and can run over the surface of the joint.

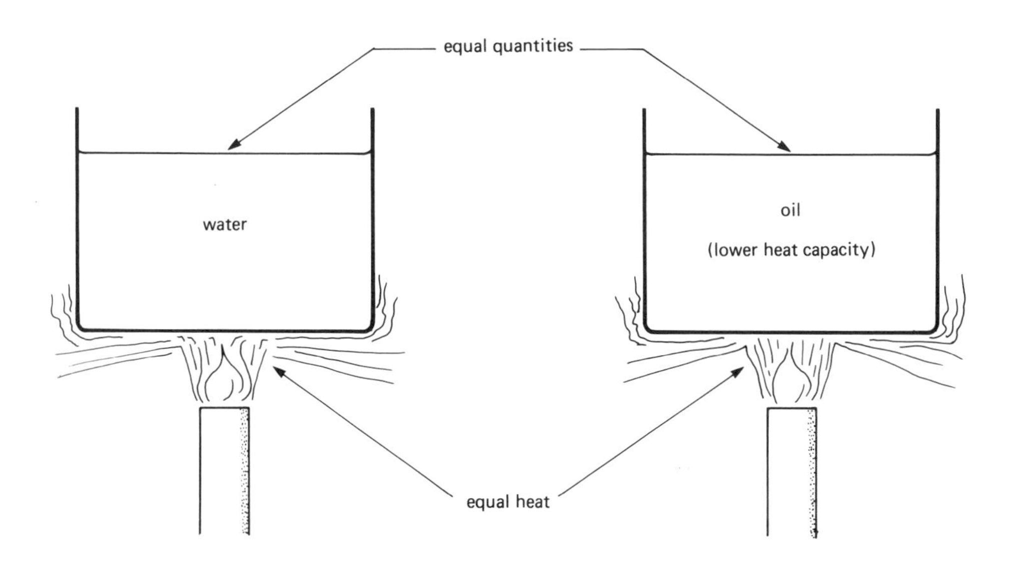

Figure 391

The knowledge of heat capacity and specific heat is necessary to complete calculations related to domestic hot water and central heating systems. The formula generally used is as follows:

quantity of heat (kJ) =
mass (kg) × specific heat (kJ/kg) ×
temperature change (°C)

Example 1
Calculate the quantity of heat required to raise 80 litres of water from 10 °C to 55 °C.

quantity of heat = mass × specific heat × temperature change

= 80 × 4.186 × 45

= 15069 kJ

Example 2
18 kg of cast iron is heated through 75 °C. How many joules of heat energy are absorbed by the cast iron?

quantity of heat = mass × specific heat × temperature change

= 18 × 0.544 × 75

= 734 kJ

Example 3
25 litres of water cool from 62 °C to 15 °C. What quantity of heat is given off?

quantity of heat = mass × specific heat × temperature change

= 25 × 4.186 × 47

= 4918 kJ

In the above examples the answers are shown as whole. For the purpose of simplicity decimal portions have been omitted. No allowance was made for thermal efficiency of the water heating apparatus or for the loss of heat to the surrounding air while heating was taking place. These are, of course important points, the consideration of which affect the design and installation of hot water and central heating systems.

Temperature
The quantity of heat a substance contains and its temperature are two quite different things. Consider a bucketful of hot water and a red hot steel wire. The water contains more heat energy than the wire, but the wire is at a higher temperature (see Figure 392).

The temperature of a substance is its degree of hotness, and this is measured by means of a thermometer.

Thermometers
There are several different types of thermometer available. The most common depend on the expansion of a liquid when heated, others on the expansion of a compound strip of two metals.

Most students will be familiar with mercury or alcohol thermometers. These usually have spherical bulbs and are mounted on metal, boxwood or plastic scales. Thermometers used in laboratories

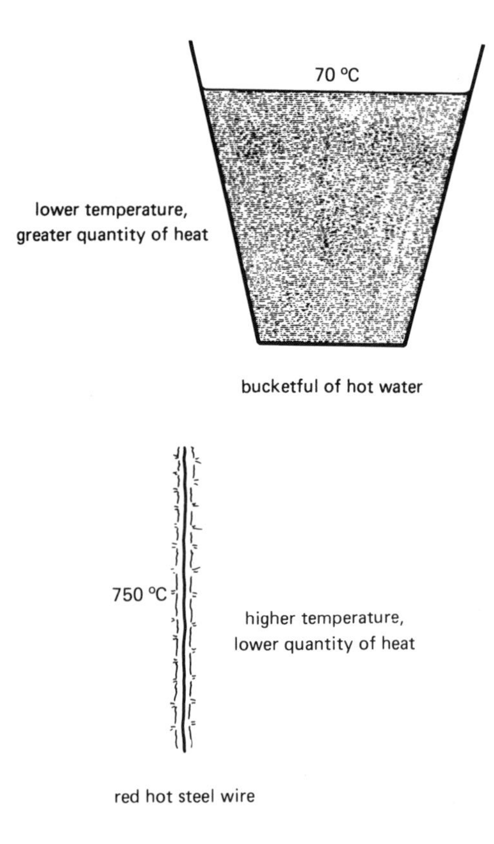

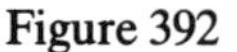

Figure 392

have cylindrical bulbs for easy insertion through holes in corks and have their scales engraved directly on the stem.

The principle of the graduation on all types of thermometers is to choose two easily obtainable fixed temperatures and use these upper and lower fixed points and to divide the interval between them into a number of equal parts or degrees. The upper fixed point is the temperature of steam from water boiling under standard atmospheric pressure. The temperature of the boiling water itself is not used for two reasons. First, any impurities which may be present will raise the boiling point. Secondly, local overheating may occur accompanied by bumping as the water boils. The temperature of the steam just above the water will be constant.

The lower fixed point is the temperature of melting ice. The ice must be pure, since the impurities will lower the melting point.

Temperature scales

The difference between the two fixed points is divided into 100 equal parts, each called a degree. The ice point is 0 °C, and the steam point 100 °C. This method of subdividing was suggested by a Swedish astronomer named Celsius, and is now called the Celsius scale (see Figure 393).

Another method of dividing the difference between the two fixed points is to use the absolute or Kelvin scale (Figure 394). Each single degree on these scales equals the same temperature interval as each single degree on the Celsius scale but freezing point is 273.15 K with boiling point 100 degrees higher at 373.15 K.

Absolute zero (0 K) is theoretically the lowest possible temperature that can ever be reached. The conversion of temperature from °C to K is completed by adding 273.15. Conversion from K to °C is the reverse, i.e. subtract 273.15.

No temperature on the Kelvin scale is negative

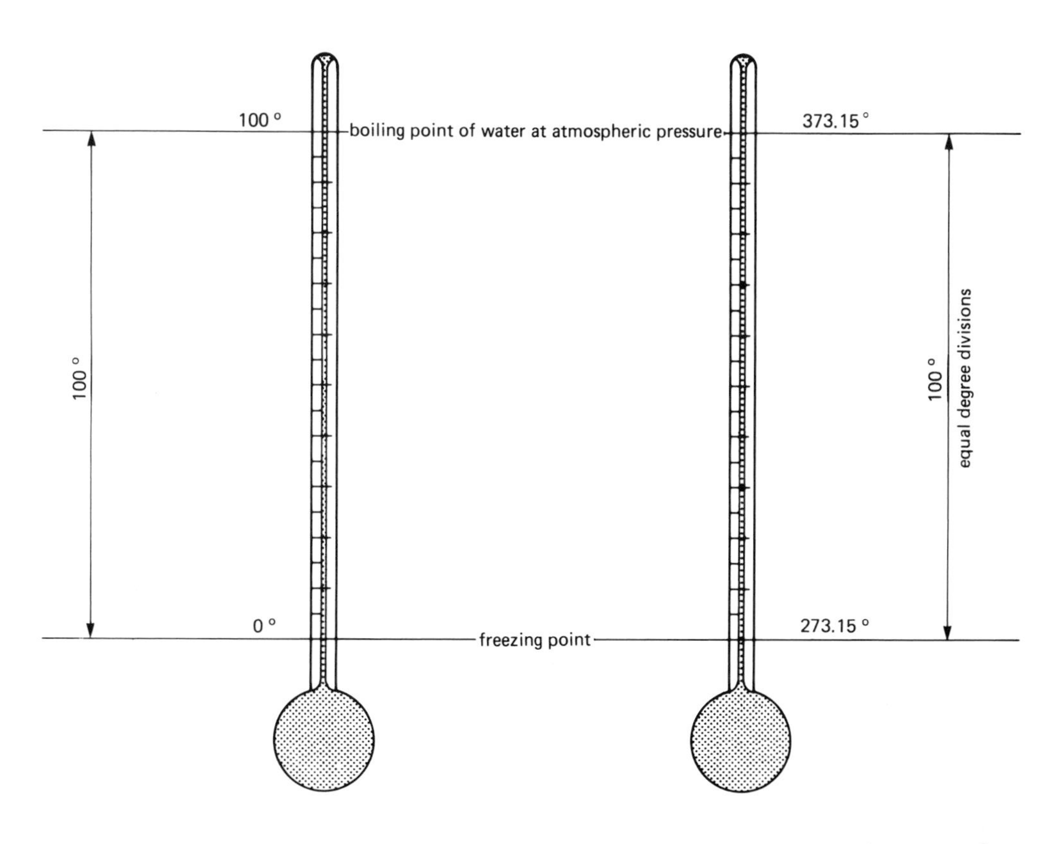

Figure 393 *Thermometer graduations: Celcius*

Figure 394 *Thermometer graduations: Kelvin*

but temperatures on the Celsius scale become negative once they drop below 0 °C.

Thermometers enable us to determine the temperature of substances with very great accuracy. The liquid most commonly used in a thermometer is mercury, because:

1 A small increase in its temperature causes a comparatively large expansion of the mercury.
2 Equal increases in its temperature result in equal amounts of expansion.
3 It remains liquid over a wide range of temperature.
4 The mercury is easily visible in the thermometer.

Measurement of temperatures in a workshop

Many of the processes carried out by a plumber require materials or components to be heated to certain temperatures. The methods used to measure these temperatures may have to be precise or of an approximate value dependent upon the circumstances.

Some metals, such as steel, have noticeable colour changes as their temperature alters, each colour indicating an approximate temperature range, as shown in Table 29.

In the case of copper the only noticeable effect is the change in the degree of redness, i.e. from dull red to a light orange. Other metals such as aluminium have no noticeable colour change and this method cannot be applied to them.

Heat-sensitive materials in the form of paints or crayons are available to cover temperature ranges from 40 °C to 1370 °C. When metal objects suitably marked with one of the above materials attain the appropriate temperature, the marking material changes its colour. For example, a brown marking will change to green at 350 °C. Once this change in colour has taken place the heat-sensitive material will not change again regardless of any temperature change.

Table 29

Colour	*Approximate temperature (°C)*
Brilliant white	1500
Bright white	1400
White	1300
Bright orange	1200
Orange	1100
Bright cherry red	1000
Cherry red	900
Brilliant red	800
Dull red	700
	600
Faint red	500

Transmission of heat

If a steel rod is pushed into a flame and left there for a time the section of rod not in the flame becomes warm. Heat travels through metal by a process called *conduction*. This process is complex. It differs between metals and non-metals, and only a brief explanation can be given here.

When a metal is heated the free electrons which it contains begin to move faster: the hot electrons move towards the cooler parts of the metal and at the same time there is a slower movement of cooler electrons in the reverse direction.

To a much smaller extent, heat is transmitted through a metal by vibrations of the atoms themselves, passing on energy from one to the other in the form of waves. These waves are in tiny energy packets and are called phonons. In non-metals which have no free electrons, heat energy is conducted entirely by phonons.

Most metals are good conductors of heat. Copper is exceptionally good. Other substances such as wood, cork and air are bad conductors.

Good and bad conductors have their uses. The bit of a plumber's soldering iron is made of copper, so that as it is used and its tip cools through contact with the work, heat is rapidly conducted along the body of the bit to restore the temperature of the tip and maintain it above the melting point of the solder being used.

Bad conductors have a very wide application. In plumbing work these materials with low thermal conductivity are generally used to prevent heat loss from pipes or to insulate systems to prevent freezing during very cold weather.

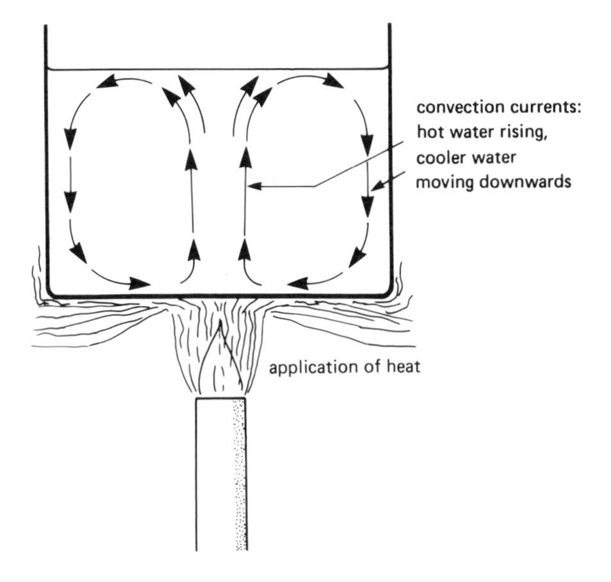

Figure 395 *Convection currents*

Conduction of heat through liquids and gases

All common liquids, with the exception of mercury, are poor conductors. Nevertheless, heat can be moved very quickly through liquids such as water by a different process called *convection*. When a vessel containing water is heated at the bottom (Figure 395) a current of hot liquid moves upwards and is replaced by a cold current moving downwards. Unlike conduction, where heat is passed on from one section of the substance to another, the heat here is actually carried from one place to another in the liquid by the movement of the liquid itself.

The same process occurs when a gas is heated, although gases are far poorer conductors of heat than liquids (see Figure 396).

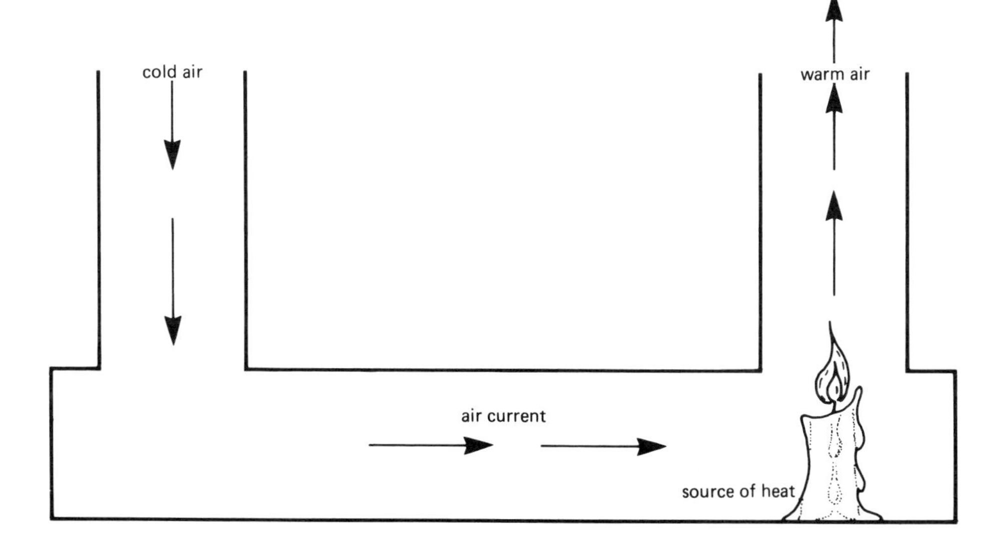

Figure 396

Convection

When a quantity of water near the bottom of a vessel is heated it expands. Since its mass remains the same, it becomes less dense, and therefore it rises. Thus a warm convection current moves upwards. On the other hand, if some water in a vessel is heated at the top, the liquid there expands and stays floating on the denser water beneath. Convection currents are not set up, and the only way heat can travel downwards under these conditions is by conduction.

This explains why heated water circulates in hot water supply, and gravity heating systems. The movement of warmed air in a room follows the same principles, and will do so as long as there is a difference of temperature between the rising and falling gas streams.

When the temperatures become equal the process ceases. It therefore follows, that the greater the temperature variation, the quicker will be the circulation (see Figure 397) within the liquid or gas. Both conduction and convection are ways of conveying heat from one place to another and require the presence of a material substance either liquid, solid or gas.

There is a third method of heat transmission which does not require a material medium. *Radiation* heat consists of invisible waves or lines of heat energy which are able to pass through a vacuum. They will also pass through air without appreciably warming it. These waves are partly reflected and partly absorbed by materials and objects upon which they fall. The part which is absorbed becomes converted into heat, this process is called *radiation*.

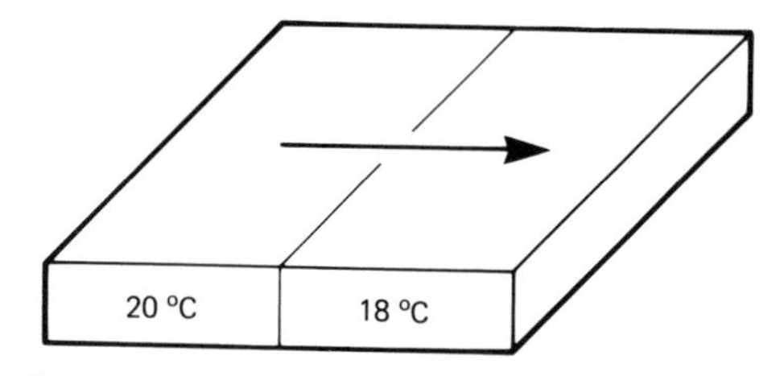

small temperature difference = low rate of heat movement

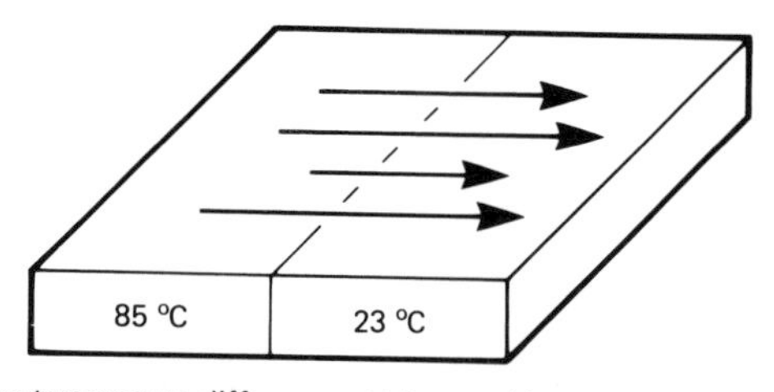

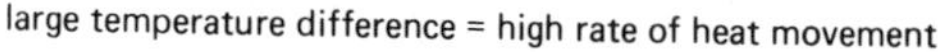
large temperature difference = high rate of heat movement

Figure 397

The rate at which a body radiates heat depends on its temperature and the nature and area of its surface. A body absorbs most heat when its surface is dull black and least when its surface is highly polished (see Figure 398). The polished surface is therefore a good *reflector* of heat.

The radiators of a hot water central heating system, despite their name, in fact emit their heat mainly by conduction and a smaller amount by radiation and convection.

The effect of heat on solids, liquids and gases

With a few exceptions, substances expand when they are heated, and very large forces may be set up if there is an obstruction to the free movement of the expanding or contracting material.

Heat may also bring about a change of state; for example, solder, which is normally a solid metal, becomes a liquid when sufficiently heated, and changes its physical state.

Heat can also accelerate or bring about a chemical change as is produced when hydrogen and oxygen are burned in the correct proportions to produce water.

Expansion of various substances

When rods of different substances but of the same length are heated evenly, experiment shows that their expansion is not equal (see Figure 399). Aluminium expands about twice as much as steel. Brass expands about one and a half times as much as steel. An alloy of nickel and steel known as invar has a very small expansion when its temperature rises and for this reason is used in thermostats and watches.

Plumber's solder expands less than lead. There is very little difference between them, but it is enough to set up damaging stresses between sheets of lead and the solder used to join them. This occurs especially where such solder joints are

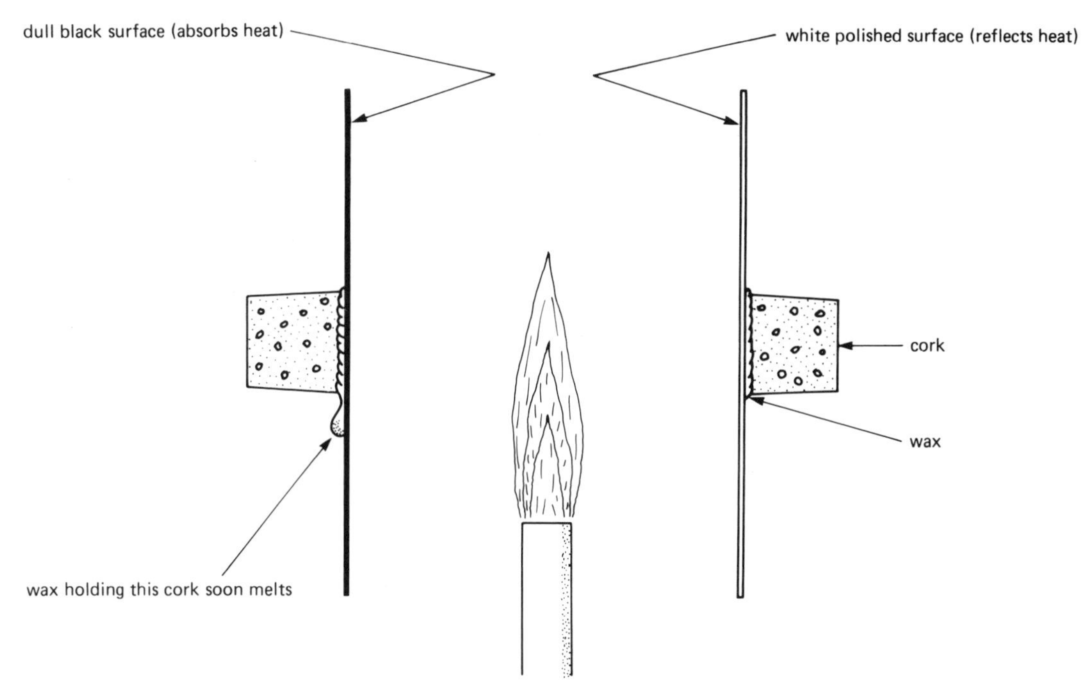

Figure 398 *Heat reflection*

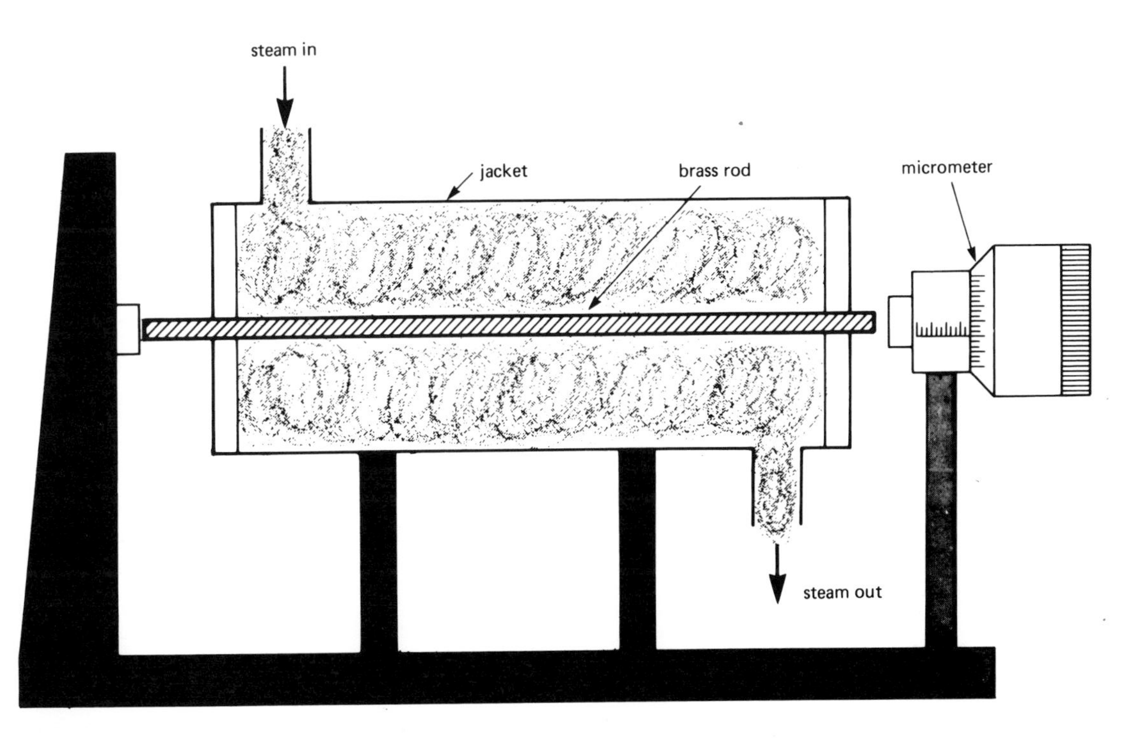

Figure 399 *Laboratory apparatus to measure thermal expansion*

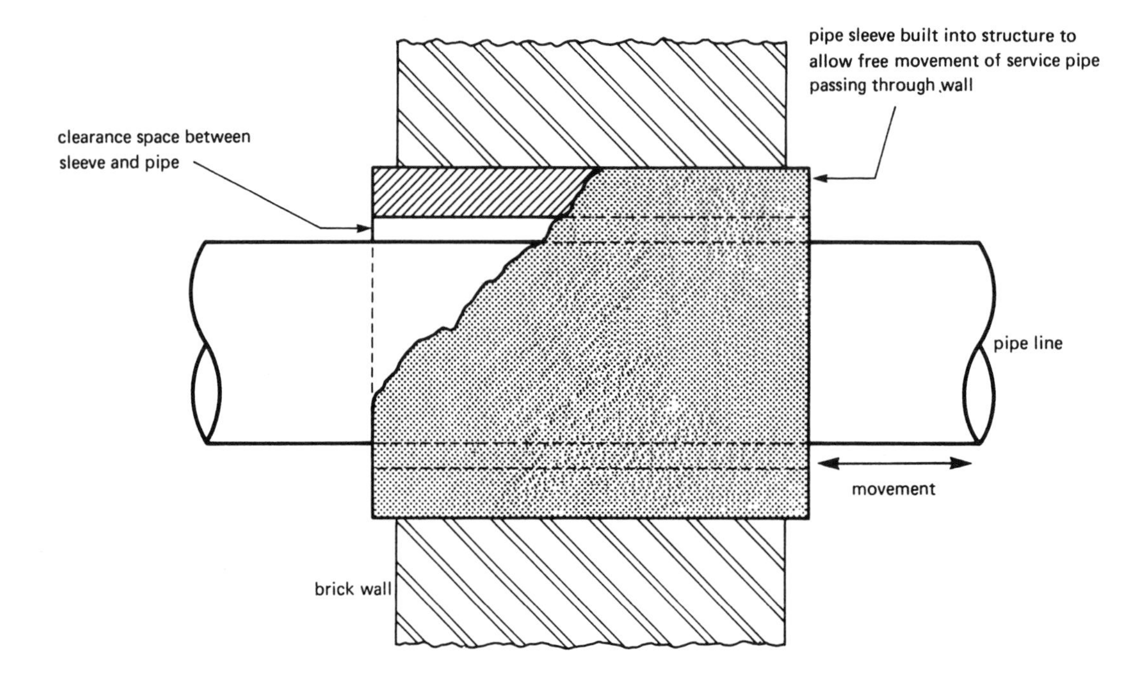

Figure 400 *Sleeving of pipelines through building structure*

exposed to wide variations in temperature such as those between day and night temperatures on a roof. For this reason it is not considered good practice to use solder to join sheets of lead used for roof weathering.

Plastics materials such as polyvinyl chloride and polythene are now used extensively in plumbing systems, and it is worth noting the comparatively high rates of expansion of these and similar materials. Small as the changes in material size appear, the fact remains that they happen and can exert a considerable pull or push on other substances that try to restrain this movement. Unless suitable allowances (as shown in Figure 400) are made and precautions taken to accommodate thermal movement, damage and inconvenience will occur.

Applications of thermal expansion

Although expansion can be troublesome in plumbers' work it often proves very useful. For example, the bimetallic strip has many useful applications including flame failure devices, electric and gas thermostats.

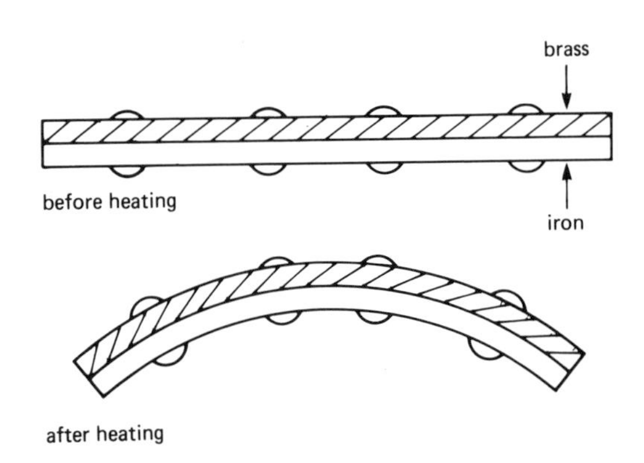

Figure 401 *Bimetallic strip*

Figure 401 shows how the different rates of expansion of iron and brass can be made use of to form a bimetal strip. Equal lengths are riveted together. On heating, the brass expands more than the iron and the strip forms a curve with the brass on the outside.

Figure 402 shows the principle of a thermostat which makes use of the different rates of thermal expansion in metals. This type of thermostat could be used for controlling the temperature of a room, or the water in a hot water storage vessel.

Coefficient of thermal expansion
The amount a substance expands when it is heated depends upon the properties of the substance: whether it is solid, liquid or gaseous, and upon the amount of heat it absorbs.

The amount that solids expand for each °C rise of temperature is fairly constant and therefore easily measured. Liquids and gases do not respond quite so conveniently, and water behaves in a manner which is termed anomolous (Figure 403).

Thermal expansion affects all the dimensions of a material. Length, width and thickness all increase as the temperature of the material rises. But since most of the materials we deal with are much longer than they are wide or thick, at this stage only length or linear expansion will be dealt with.

Definition
The coefficient of linear expansion of a solid substance is the fraction of its original length by which the substance expands per degree rise in temperature, or will contract when its temperature decreases by 1 °C.

The length can be measured by using any convenient unit, although generally metres are preferred. Whatever unit length is used, the fraction that the material expands for 1 °C change

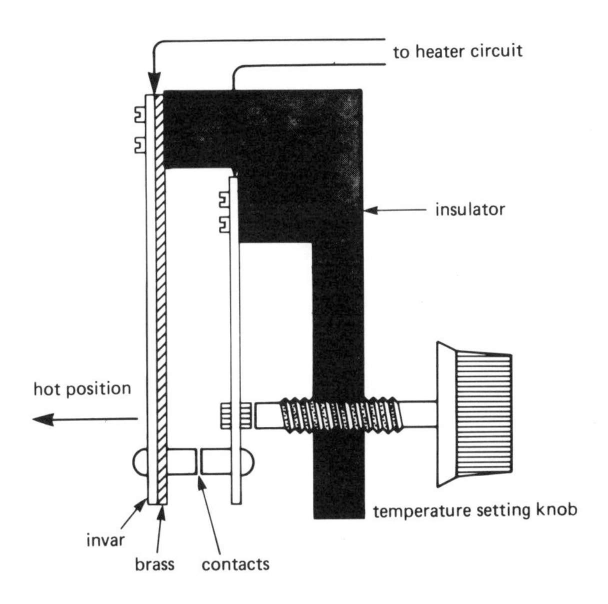

Figure 402 *Thermostat*

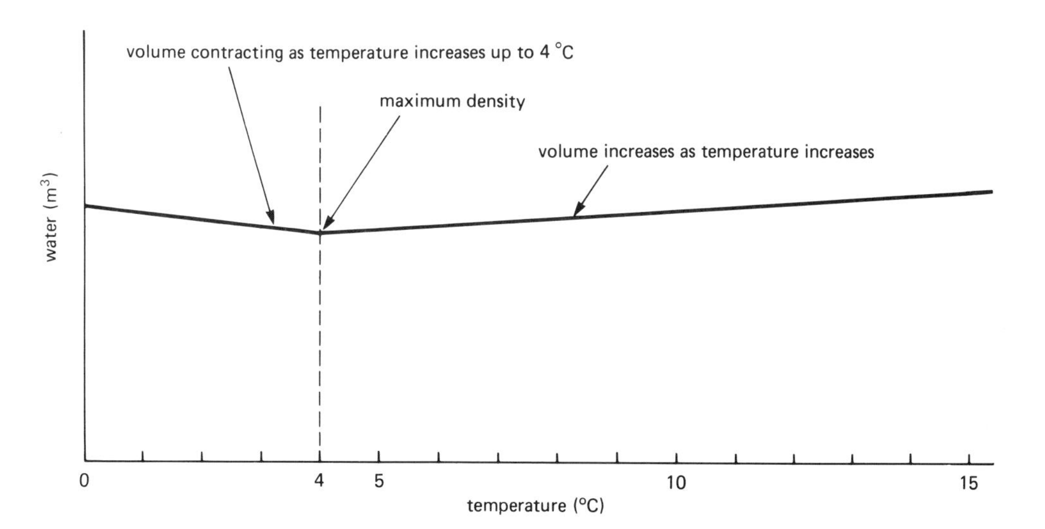

Figure 403 *Variation of water volume with temperature*

Table 30 *Linear coefficients of thermal expansion*

Material	*Variation per unit of length for one degree Celsius temperature change*
Lead	0.000029
Zinc	0.000029
Aluminium	0.000026
Plumbers' solder	0.000025
Tin	0.000021
Polythene	0.000018
Copper	0.000016
Iron	0.000011
Cast iron	0.000011
Mild steel	0.000011
Mercury	0.000005

of temperature must also be measured as a fraction of that unit. Table 30 shows the coefficients of linear expansion of various materials.

To calculate the amount of expansion or contraction which occurs in materials as their temperature changes, the following formula may be used:

change in length = length of material × temperature change in material × coefficient of linear expansion

The following examples will show the application in relation to plumbers' work.

Example 4

A copper hot water pipe 15 m long is filled with water at 10 °C. By how much will the length of this pipe increase when it carries hot water at 70 °C?

change in length = 15000 mm × 60 °C × 0.000016

= 900000 × 0.000016

= 14.4 mm increase in pipe length

Example 5

A polythene waste pipe has received a discharge of hot water at 55 °C, and is allowed to cool to 15 °C. When the pipe was at its hottest it measured 4 m in length. How much will it shorten or contract?

change in length = 4000 mm × 40 °C × 0.00018

= 160000 × 0.00018

= 28.8 mm decrease in length

Expansion of liquids

Different liquids have different thermal expansions (see Figure 404) and unlike solids have no fixed length or surface area, but always take up the shape of the containing vessel. Therefore in the case of liquids we are only concerned with volume changes when they are heated.

Definition

The coefficient of expansion of a liquid is the fraction of its volume by which it expands per degree rise in temperature.

Any attempt at very accurate measurement of the expansion of a liquid is complicated by the fact that the vessel which contains the liquid also expands.

However, since all liquids must always be kept in some kind of vessel or container it is just as useful to know the apparent expansion of a liquid. This is the difference between its real expansion and the expansion of the vessel, and is accurate enough for plumbers' work.

The unusual expansion of water

Not all substances expand when they are heated. Over certain temperature ranges they contract. Water is an outstanding example. If we take a quantity of water at 0 °C range and begin to apply heat the water contracts over the temperature range 0 °C – 4 °C (Figure 403).

At about 4 °C the water reaches its smallest volume which means it is at maximum density. If we continue to apply heat to raise the temperature the water expands.

The peculiar expansion of water has an important bearing on aquatic life during very cold

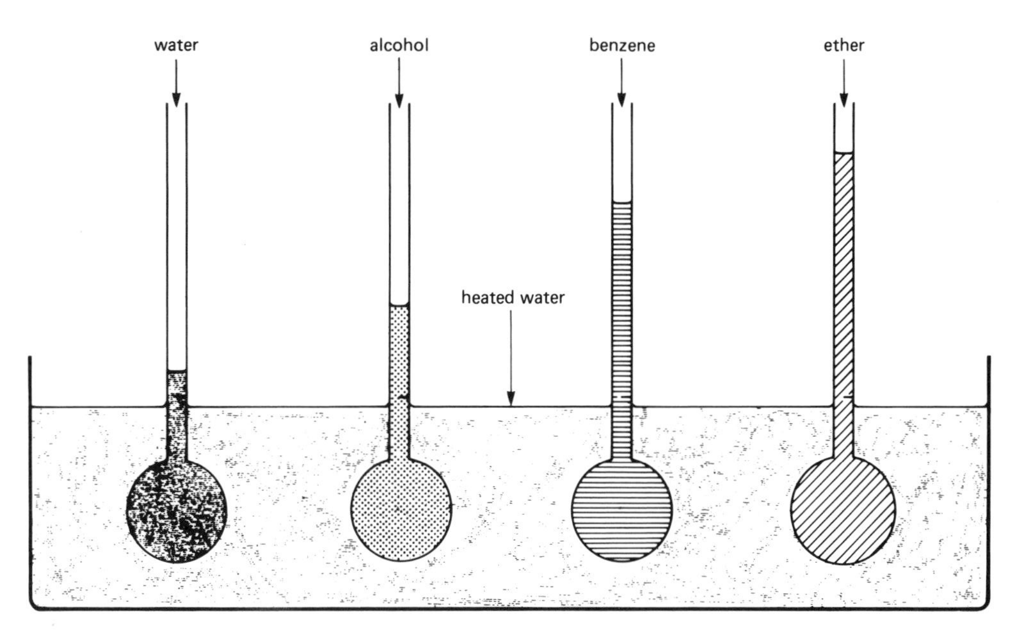

Figure 404 *Comparison of expansion for different liquids*

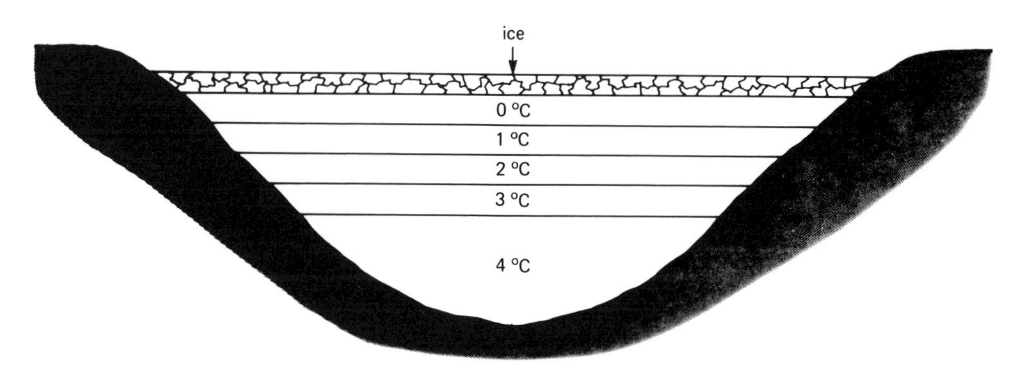

Figure 405 *Temperatures in an ice covered pond*

weather (see Figure 405). As the temperature of a pond falls, the water contracts, becomes denser and sinks. A circulation is set up until all the water reaches its maximum density at 4 °C. If the temperature continues to drop, any water below 4 °C will stay at the top due to its lower density. In due course ice forms on the top of the water, and after this the water beneath can only loose heat by conduction. This explains why only very shallow water is likely to freeze solid.

As mentioned previously a rise in temperature causes objects to expand and a fall of temperature causes contraction. This rule applies to gases, most liquids and solids, but the effect is much more marked in the case of gases than in the case of the other two.

When the coefficient of linear expansion of a solid was defined previously, it was not specified that the original length of the substance should be measured at 0 °C. This is due to the smallness of the linear expansion whether we start with the substance at 0 °C or at normal temperature.

Expansion of gases

In the case of gases, however, the situation is different. The expansion which occurs in gases is very much larger than that of solids, and consequently the value of the coefficient obtained will depend on the starting temperature. Therefore, in order to make an accurate comparison between different gases, the coefficient is always calculated in terms of an original volume at 0 °C.

The volume coefficient of expansion of a gas is defined as follows.

Definition

The coefficient of expansion of a gas at constant pressure is the fraction of its volume at 0 °C by which the volume of a fixed mass of gas expands per °C rise in temperature.

The original experimental work on this subject was carried out towards the end of the eighteenth century by the French scientist Jacques Charles. The results obtained are generally known as Charles's Law:

The volume of a fixed mass of gas at constant pressure expands by $\frac{1}{273}$ of its volume at 0 °C per °C rise in temperature.

Pressure in gases and liquids

The nature of gas pressure

Any gas consists of a collection of molecules of a particular kind which are in a state of rapid motion. The fact that the molecules are in motion is evident from the fact that if a small quantity of an odorous gas, such as acetylene or propane, is liberated at any point in a workroom or laboratory the smell of the gas soon pervades the whole room.

If the gas is confined in a closed vessel, some of the moving molecules strike the sides of the vessel and each impact exerts a small force upon the side. The number of molecules of gas inside the vessel will normally be very large and, on the average, the same number of molecules will strike a given area on the sides of the vessel each second, so producing a steady pressure.

If the gas contained is compressed by some mechanical means, its molecules will be pressed closer and closer together and the cohesive force between them becomes stronger. Eventually, if sufficient pressure is applied so that the intermolecular cohesive force is strong enough to hold the molecules close together, the gas will change to a liquid. If the compression force is released, the liquid will return to its gaseous state.

This fact is used to separate the gases which make up the atmosphere. If air is sufficiently compressed under controlled conditions of pressure and temperature, it will change from a gaseous to a liquid state. When the compression forces are released, the liquid reverts to a gas. As the elements which make up air become gases at different temperature, it is possible to collect them separately, and store them in cylinders for use in welding, and other industrial processes. In this way oxygen, argon and nitrogen gases are obtained.

Pressure can also affect the process of a physical change. For example, water boils and changes from a liquid to a gas at 100 °C at standard atmospheric pressure (100 kN/m^2). If a pressure greater than this is acting on the water, then it will not boil until it has reached a much higher temperature and will not boil until it has gained sufficient heat energy to do so. Likewise reduced pressures (below standard atmospheric pressure) reduce its boiling point temperature.

Definition

Pressure is defined as the force acting normally per unit area. (Here the word 'normally' means vertically.)

The SI unit of pressure is *newton per metre*^{2} (N/m^2).

Relation between pressure and volume

The pressure of a gas decreases in the same proportion as its volume increases. From this statement it is clear that if we multiply the varying volumes of a given mass of gas by the corresponding pressures, any decrease in the value of one of them will be exactly counterbalanced by the increase in the value of the other, and the result will always be the same.

Expressed mathematically, this may be stated as:

$$P_1V_1 = P_2V_2 \text{ (temperature constant)}$$

where P_1 and P_2 are two pressures and V_1 and V_2 the corresponding volumes of a given mass of gas.

This result is known by the name of its discoverer as Boyle's Law.

Boyle's Law

The volume of a given mass of gas is inversely proportional to its pressure, if the temperature remains constant.

'Inversely proportional' is the mathematical expression of the fact that as the pressure *increases* the volume *decreases* in the same proportion.

Water pressure

Water pressure is caused naturally by the weight of water which, under the influence of the earth's gravitational force, exerts pressure on all surfaces on which it bears (atmospheric pressure).

The molecules of which water is composed are held together by cohesion. This cohesive force is stronger in liquids than in gases but not as strong as in solids. This fact means that water molecules can move with relative freedom, and the force of gravity tends to pull them down in horizontal layers so that the surface of a liquid subject to the pressure of the atmosphere is horizontal. Therefore water flows to find its own level in irregular shaped vessels, pipes and cisterns (see Figure 406).

The fact that the liquid stands at the same vertical height in all the tubes whatever their shape confirms that, for a given liquid, the pressure at a point within it varies only with the vertical depth of the point below the surface of the liquid.

Liquids exert a pressure in the same way that air does.

The pressure in a liquid increases with depth. This may be shown by means of a tall vessel full of water with side tubes fitted at different heights (Figure 407). The speed with which water spurts out is greatest for the lowest tube, showing that pressure increases with depth.

Intensity of pressure

This is defined as that force created by the weight of a given mass of water acting on 1 unit of area (m^2). Or:

$$\text{intensity of pressure} = \frac{\text{force}}{\text{area}}$$

and, since force is measured by newtons:

$$\text{intensity of pressure} = \frac{\text{newtons}}{\text{area}}$$

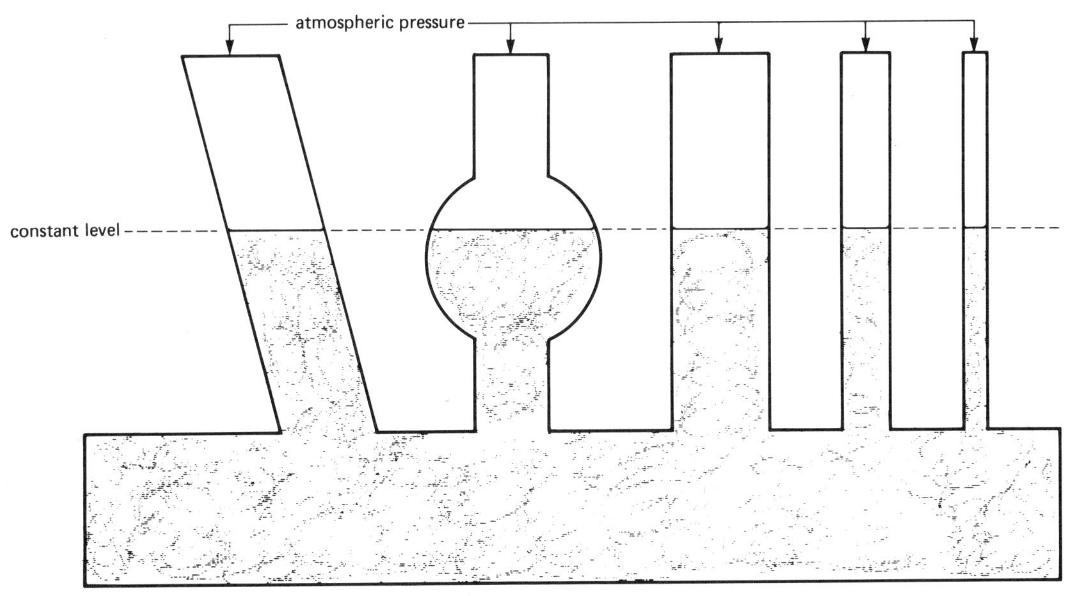

Figure 406 *Apparatus used to show that water flows to find its own level*

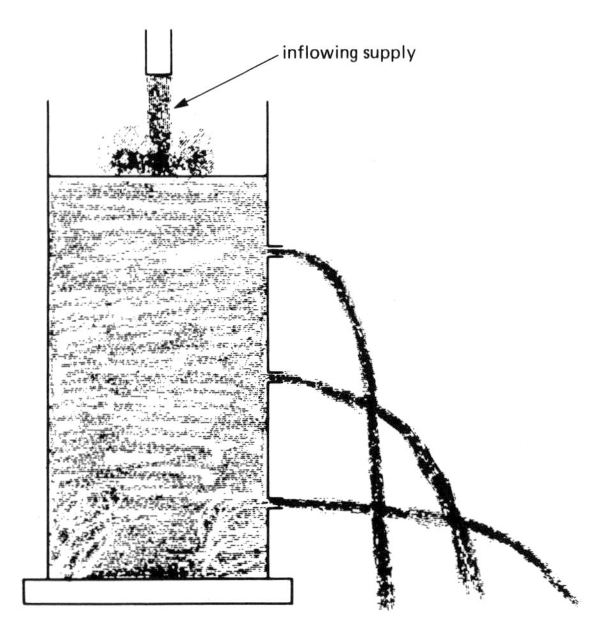

Figure 407 *Spout can, indicating that pressure in a liquid increases with the depth of the liquid*

The newton is of very small value in numerical terms, and it is generally more practical to use the kilonewton (kN = 1000 newtons) when dealing with pressures.

1 litre of water weighs 1 kg and the force produced by this mass of water equals 9.8 newtons. If we consider a cube of water measuring 1 m × 1 m × 1 m (1 m^3), this volume of water contains 1000 litres and weighs 1000 kg. Therefore, the intensity of pressure on the base of this cube which has an area of 1 m^2 will be 1000 × 9.8 newtons or 9.8 kN/m^2.

So, if we visualize a column of water 1 m in height or depth and call this distance head we can derive the following basic formula for calculating pressure:

intensity of pressure = head (m) × 9.8 kN/m^2

Example 6

Calculate the intensity of pressure on the base of a hot water tank subjected to a head of 4 m.

intensity of pressure = head (m) × 9.8 kN/m^2

= 4 × 9.8

= 39.2 kN/m^2

The example shows that a 4 m head of water will exert an intensity of pressure of 39.2 kN on a base area of 1 m^2. What must be considered now are areas larger or smaller than 1 m^2 which leads us to the total pressure acting upon an area.

If the pressure shown in the example had been acting upon a base of 2 m^2 then the total pressure acting upon the whole area would be 2 × 39.2 kN or 78.4 kN. At this point it should be noted that the area symbol (m^2) is left out of the answer because with total pressure calculations we are not relating the pressure to 1 m^2 but to an area larger. The same rule applies to areas smaller than 1 m^2, which leads to the formula for calculating total pressure:

total pressure = intensity of pressure × area acted upon

Example 7

Calculate the total pressure on the base of the hot tank in Example 6, if the base has an area of 0.3 m^2

total pressure = intensity of pressure × area acted upon

= 39.2 × 0.3

= 11.76 kN

Density and relative density

Definition

The density of a substance is defined as its mass per unit volume.

Equal volumes of different substances vary considerably in mass. For example, aluminium alloys are as strong as steel, but volume for volume weigh less than half as much. This lightness or heaviness of different materials is known as *density*.

One method of finding the density of a substance is to take a sample and measure its mass and volume. The density is then calculated by dividing the mass by the volume.

The densities of all common substances, solids, liquids and gases, and all chemical elements have been determined and are to be found in books of

Table 31 *Densities in kg/m³*

Aluminium	2.7×10^3	Mercury	13.6×10^3
Brass (varies)	8.5×10^3	Steel (varies)	7.8×10^3
Copper	8.9×10^3	Water (at 4 °C)	1.0×10^3
Ice (at 0 °C)	0.92×10^3	White spirit	0.85×10^3
Lead	11.3×10^3	Zinc	7.1×10^3

chemical and physical data, although some of particular interest to the plumber are listed in Table 31.

Water has a density of 1 g/cm³ or 1000 kg/m³ at 4 °C. Mercury is a metal which is liquid at normal temperatures and this has a very high density of 13.6 g/cm³. It is a very useful substance and is widely used in thermometers, switches and laboratories.

It is important for construction workers to appreciate density, particularly when related to various building materials and claddings. The known volume of any part of the structure can be multiplied with the density of the material to give the mass, and hence the weight. Such information is essential for calculating foundations, tank and cistern supports, etc.

Relative density

Definition

The relative density of a substance is the ratio of the mass of any volume of it to the mass of an equal volume of water. Or:

$$\text{Relative density} = \frac{\text{mass of any volume of the substance}}{\text{mass of an equal volume of water}}$$

In normal weighing operations the mass of a body is proportional to its weight, so it is also true to say:

$$\text{Relative density} = \frac{\text{weight of any volume of the substance}}{\text{weight of an equal volume of water}}$$

This will explain why relative density has also been called *specific gravity*: the word gravity implying weight. At the present time, the term relative density is recommended rather than specific gravity. Note that relative density has no units: it is simply a number or ratio. On the other hand, density is expressed in kg/m³ or g/cm³.

Self-assessment questions

1 Name the unit of measurement for a quantity of heat.

2 State two liquids that may be used in thermometers.

3 What are heat-sensitive crayons?

4 Name three good conductors of heat.

5 Which has the greater coefficient of thermal expansion: zinc, iron, steel, or polythene?

6 State the temperature at which the maximum density of water occurs.

7 To which state of matter, i.e. solids, liquids or gases, does ‘Boyle’s Law’ relate?

8 State the SI unit of pressure.

9 How much does 1 litre of water weigh?

10 Define the term ‘density’.

11 Glossary

Abutment An intersection between a roof surface and a wall rising above it.

Ambient Surrounding

Anneal To soften a metal by the application of heat.

Apron flashing A one-piece flashing, such as is used on the lower side of a chimney penetrating a sloping roof.

Architect One who designs and supervises the construction of buildings. His or her main duties are preparing designs, plans and specifications; inspecting sites; obtaining tenders for work and the legal negotiations needed before building commences.

Asbestos A mineral crystal, consisting of thin, tough fibres like textile, which can withstand high temperature.

Asbestos cement Cement mixed with asbestos fibre used in the manufacture of flue pipes, rainwater gutters and downpipes and cold water storage cisterns.

Autogenous A welded joint in which two parts of the same metal are welded together with or without a filler rod of the same metal.

Backflow (cold water) Flow in a direction contrary to the natural or intended direction of flow.

Back gutter A gutter between the upper end of a chimney and a sloping roof.

Backnut A locking nut provided on the screwed shank of a tap, valve or pipe fitting for securing it to some other object. A threaded nut, dished on one face to retain a grommet, used to form a watertight joint on a long threaded connector.

Balanced flue An arrangement of air intake and flue outlet commonly used for domestic gas fired appliances.

Bale tack A lead tack used for securing sheet metal weatherings.

Ball-peen hammer A hammer with a hemispherical peen (also pein or pane).

Base exchange A water softening process in which water is passed through a bed of mineral reagent called zeolite, which absorbs those salts in the water which make it hard.

Bay window A window formed in a projection of the wall beyond its general line, and supported on its own foundations.

Bead Formed on sheet metals for stiffening the edge or fixing it, usually bent round to a circular tube shape.

Benching Sloping surfaces constructed on either side of a channel at the base of a manhole.

Bend A curved length of pipe, tubing or conduit. A 90° bend is called a quarter-bend, a 45° bend is called a one-eighth bend.

Bibtap A water tap which has a horizontal inlet connection which is male threaded.

Birdsbeak A method of terminating an overflow or warning pipe. The pipe end is cut across a centre line and this portion is opened and shaped to form a bird's beak.

Bobbin An egg shaped boxwood tool which is drilled through its centre. It is threaded on to a strong cord and pulled through a bent section of pipework to remove distortions and return the pipe to its correct bore.

Boiler A water heater in which, generally, the water should not boil. It may be heated by gas, oil, electricity or solid fuel.

Bossing The art of shaping malleable materials such as lead and aluminium with hand tools usually made from boxwood or plastics.

Bower-barffing A process for rust proofing mild steel or cast iron in which the metal is raised to red heat and treated with live steam.

Boxwood A hardwood used for making hand tools, kept in good condition by a light application of linseed oil.

Bradawl A short awl with a narrow chisel point, used for making holes for screws or nails.

Bronze An alloy of copper and tin.

Building paper Fibre reinforced bitumen layers of brown paper; may be used as an underfelt between sheet weatherings and the roof structure surface.

Burring reamer A tool used to remove the burr left by a pipe cutter inside the cut end of a pipe.

Butt To meet without overlapping.

By-pass An arrangement of pipes for directing flow of liquid or gas around instead of through a certain pipe or component.

Calcium carbonate The chemical name for chalk, limestone and marble.

Calorifier A cylindrical vessel in which water is heated indirectly, by means of hot water or steam contained in pipe coils, a radiator or a cylinder within a calorifier.

Cap A cover usually with an internal thread or socket joint for sealing the end of a pipe.

Capacity The quantity of liquid contained in a vessel (measured in litres).

Cap and lining A union joint, usually made of brass, generally used for jointing a lead pipe to a BSP thread.

Capillary joint A fine clearance spigot and socket joint into which molten solder is caused to flow by capillary action.

Carbonizing flame The flame produced at the blowpipe tip when there is an excess of acetylene being burned.

Catalyst A substance which increases the speed of a chemical reaction.

Cat ladder A ladder or board with cross cleats fixed to it, laid over a roof slope to protect it and give access to workmen for inspection or repairs (also called a duck board).

Caulking The driving of cold lead into a recess or spigot/socket joint to form a tight mass – usually done with a blunt ended chisel.

Cavitation A phenomenon in the flow of water consisting of the formation and collapse of cavities in the water bringing about pitting and wear on valve and tap seatings and ball valve outlets.

Chain wrench A steel pipe grip which holds the pipe by a chain linked to a bar which is grooved at the end touching the pipe.

Chalk line A length of line well rubbed with chalk, held tight and plucked against a wall, floor or other surface to mark a straight line on it.

Channel pipe An open pipe, semi-circular or three quarter section, used in drainage, particularly at manholes.

Chase wedge A wooden wedge-shaped tool, used for setting in a fold on sheet metal work.

Cheek The vertical side of a dormer.

Chloramine process This involves adding ammonia to water to remove the taste of chlorine.

Chlorination A method of treating water with chloride to sterilize it by destroying harmful bacteria.

Cistern An open topped container for water in which the water is subject to atmospheric pressure only. The water usually enters the cistern via a ball valve.

Cladding The non-loadbearing covering of the walls or roof of a building. The skin used to keep the weather out.

Cleaning eye An access cover.

Close coupled A term used to describe a toilet suite where the cistern bolts or clamps to a projection formed on the top back edge of the pan.

Clout nail A galvanized nail usually between 12 mm and 50 mm long with a large round flat head.

Communication pipe The pipe between the water authority's main and the consumer's stop valve or his or her boundary, whichever is nearer to the main. It is part of the consumer's service pipe which belongs to the water authority.

Compression joint A fitting used to joint copper, stainless steel and polythene tubes.

Contraction The reverse of expansion, i.e. the decrease in size of a solid or liquid substance.

Cornice A projection usually of concrete or stone located in a wall to throw rainwater clear of the wall face below, and improve the appearance of the building.

Corrosion A chemical action which takes place on the surface of the metal, and usually started by atmospheric conditions, attack by acids, or electrolytic action. The 'eating away' of a surface.

Cupro-solvency The ability of some waters to dissolve copper.

Damp proof course A layer of impervious material laid in a wall to prevent the passage of water.

Daywork A method of payment for building work, involving agreement between the clerk of works and the contractor on the hours of work done by each worker and the materials used. Proof of this agreement in shown by the clerk of works' signature on the contractor's daywork sheets. Payment to the contractor consists of his or her expenses in labour and materials, plus an agreed percentage of overheads and profit.

Dead leg (1) A section of hot water draw-off pipe in which the contained water does not circulate, except when it is being drawn off. The water cools down between draw-offs and the dead leg wastes both heat and water. Model water bye-laws specify the maximum permissible length of dead leg allowed.

Dead leg (2) A section of distribution pipe which cannot be emptied via a tap or valve fitted to an appliance. The section of pipework can only be emptied via a drain cock fitted to this section of pipework.

Dead soft temper The softness of copper sheet required for roof weatherings.

Deep seal trap A trap with a 75 mm water seal.

Density A word used to indicate the mass or weight of a body of known or stated volume, thus giving a convenient method of comparison for materials.

Deoxidation The process of separating oxygen from a substance. This process is also called reduction and is a necessary part of the treatment of many metallic oxides and ores.

Development A geometrical method by which the whole of the surface area of a solid may be set out in one plane, as on sheet metal or drawing paper.

Die An internally threaded metal block for cutting male threads on pipes or tubes. The die block is held within a stock.

Distortion A term used in welding to indicate the deformation of the metal being welded, due to unequal expansion and contraction of the heated metal causing bending and buckling.

Distribution pipe A pipe conveying cold water from the storage cistern to fitments below.

Dog ear A corner or angle formed in sheet metal roofworks by folding the material. No cutting takes place.

Dormer A vertical opening formed in a roof slope to give light or ventilation to rooms formed in the roof space.

Drain chute A drain pipe tapered in its upper half at the point where a drain pipe enters or leaves a manhole. The pipe is shaped in such a way so that rodding shall be easy.

Drain cock A cock fitted at the lowest point of a water system or on any section of pipework which is not self draining, to facilitate complete emptying of the system.

Drain plug An expanding stopper, used to seal a drain pipe usually during a test.

Draught The pressure difference at the base of a chimney between the air outside and that inside the chimney. This pressure difference (caused by the air inside being hotter and lighter) draws air up through the burner or fuel bed into the chimney.

Dresser A hand tool made from boxwood or plastics material, used for flattening or dressing sheet materials such as a lead, copper or aluminium.

Drip A step formed in flat or low pitched roof usually at right angles to the direction of fall.

Drip edge The free lower edge of a sheet material covered roof which drips into a gutter or into the open. This edge is often stiffened with a welt or bead.

Ductility A property which allows a metal to be elongated or drawn out without breaking or fracturing.

Eaves The lowest overhanging part of a sloping roof.

Eaves fascia A board or edge nailed along the foot of the rafters. It is used to carry the eaves gutter and may also act as a tilting fillit.

Eaves gutter A rainwater gutter along the eaves.

Elbow A sharp corner of change of direction in a pipe, usually a manufactured fitting.

Element A material composed entirely of atoms of one kind. An element cannot be split into anything simpler than itself.

Enamel Vitreous enamel is a glass like surface attached by firing to cast iron or pressed steel articles such as baths or sink unit tops.

Epoxide resin A synthetic resin used for glueing metal or concrete.

Erosion The wearing away of a surface.

Eutectic A term used in metallurgy in connection with the solidifying or setting of alloys. When alloyed in varying proportions, one particular combination of metals will give the lowest solidifying point. This is known as the eutectic point.

Evaporation The loss of moisture in vapour form from a liquid.

Expansion An increase in the size of a material or substance usually brought about by an increase in temperature.

False ceiling A ceiling which is built with a gap between it and the floor above, to provide space for services.

Feed cistern A cold water storage cistern which supplies hot water apparatus.

Ferrule A type of fitting, normally used for the connection between a water main and the communication pipe.

Fillet A narrow strip of wood fixed to the angle between two surfaces.

Fire cement Refractory cement used for jointing flue pipes.

Flammable Combustible, burns with a flame.

Flashing A strip of impervious material, which excludes water from the junction between a roof covering and another surface. Flashings, at their upper end, are usually wedged tightly into the joints between horizontal bricks, the joint having been raked or cut out to receive the flashing.

Flash point The lowest temperature at which a substance momentarily ignites when a flame is put to it.

Flat A level, or low pitched platform, usually a roof.

Flaunching A cement mortar surface on the top of a chimney stack to shed off the rain – also to secure the chimney pot or terminal.

Flow pipe A pipe which conveys hot water from a boiler to a tank, cylinder or heating system.

Flue A passage for smoke in a chimney.

Flue lining The process of lining an existing flue with a liner. Terracotta, asbestos, fireclay, aluminium or stainless steel are materials which may be used depending on the type of appliance and fuel to be used.

Flue pipe Usually a metal or asbestos cement pipe which conveys smoke or the products of combustion from an appliance.

Flush To discharge a quantity of water down a pipe, into a channel or into an appliance to clean it.

Flushing trough A long water cistern extending above, across or behind a range of WCs, and supplying them with water. It has the advantage that any WC can be flushed at short intervals without the usual waiting period needed for the filling of a cistern.

Flux A fusible substance like borax or tallow which covers a joint and prevents oxidation of the cleaned surfaces.

Footprints An adjustable wrench with serrated jaws.

Force cup A tool used for unblocking wastes or drains. It consists of a rubber cup fixed on the end of a wooden handle.

Fretted lead H-shaped lead strip used for leaded lights (lead cams).

Full-way valve A valve which does not impede the flow of water, used where the water pressure is low and on distribution pipework.

Furred A term to describe pipes, boilers or components which have become encrusted with hard water lime or other salts deposited from the water heated in them.

Fuse Usually a small piece of wire in an electric circuit which melts when the current exceeds a certain value. The fuse is protection against short circuiting.

Fusible link A metal link made from a material with a low melting point and usually incorporated into an anti-fire device.

Gable The triangular part of the end wall of a building with a sloping roof.

Galvanize A process used for protecting metals, usually steel. The process involves dipping the metal to be protected into molten zinc.

Invar A nickel/steel alloy. Its coefficient of expansion is very low and for this reason it is used in bimetallic types of thermostat and steel measuring tapes.

Invert The lowest point of the internal surface of a pipe or channel at any cross-section.

Isometric A drawing method based on the principle that all vertical lines are drawn vertical while all horizontal lines are drawn at an angle of 30°. With this method several surfaces of the object can be exposed to view.

Killed spirits A term applied to zinc chloride which is used as a flux for soldering zinc, copper and brass. The chloride of zinc is made by dissolving small pieces of zinc in hydrochloric acid.

Latent heat The amount of heat required to change the state of a substance from a solid to a liquid or from a liquid to a gas. The heat applied does not bring about a temperature rise.

Leadwelding A process of fusing together pieces of lead sheet or pipe. The basic process involves forming a small molten pool of lead, and adding further lead from a rod in order to reinforce the joint.

Main contractor A contractor who is responsible for the bulk of the work on a site, including the work of the sub-contractors.

Make good To repair as new.

Mallet A tool like a hammer with a wooden, hide, rubber or plastics head.

Mandrel A cylindrical piece of hardwood which is pushed through a lead pipe to remove distortion.

Manipulative joint A compression joint in which the ends of the copper tubes are opened out.

Mansard roof A roof which is often gabled, and has on each side a relatively flat top slope and a steeper lower slope, usually containing dormers.

Margin The exposed surface of a slate or tile.

Masking A form of tape, applied as a protection to sanitary ware.

Mastic A plastic permanently water-proof material, which hardens on its surface so that it can be painted. Used for sealing gaps in expansion joints, gutters and flashings.

Metal coating A thin film of copper, nickel, cadmium, chromium, aluminium or zinc applied to corrodible metal surfaces.

Milled lead Lead rolled into sheets from cast slabs.

Module A unit of length by which the planning of structures can to some extent be standardized.

Mortar A mixture of Portland cement, sand and water.

Multi-point water heater A water heater (usually gas) which supplies hot water to several taps.

Nail A fixing device. Clout nails are large headed. Nails of brass, aluminium alloy or copper are used to provide fixings in certain types of roofing.

Nail punch A short blunt steel rod which tapers at one end. It is struck by a hammer to drive a nail head below its surrounding timber surface.

Neoprene The trade name for an American synthetic rubber which has excellent properties of non-infammability.

Nipple A short section of pipe threaded at each end.

Non-manipulative joint A compression joint in which the ends of the copper tube are cut square and the internal and external burrs removed. Jointing is usually achieved with the aid of a cone, ring or olive which is compressed into the tube wall by the action of the joint nut being screwed on to the joint body.

North-light roof A sloping roof having one steep and one shallow slope. The steeper slope is usually glazed and faces north.

Notch A groove in a timber to receive another timber or pipe.

Offset A double bend in a pipeline, formed so that the pipe continues in its original direction.

Ordinary Portland cement A hydraulic cement made by heating to clinker in a kiln a slurry of clay and limestone.

Orifice A small opening intended for the passage of a fluid.

Overcloak In sheet metal roofing, that part of an upper sheet which laps over a lower sheet or undercloak at a drip or roll joint.

Pantile A single-lap tile shaped like an 'S' laid horizontally.

Parallel thread A thread screwed to a uniform diameter. Used on mechanical connections such as bolts, but not generally on low carbon steel pipe fittings except running nipples and connectors. Compression fittings, taps and ball valves all make use of parallel threads.

Parapet A low wall guarding the edge of a roof.

Patina A thin, protective film of sulphate which forms on metals exposed to air, particularly the green coating on copper or its alloys.

Perspex A transparent acrylic resin.

Pet-cock A small tap or drain cock.

Pilot hole A guiding hole, drilled in a material to form a route for a larger drill or bit.

Pilot light A small gas flame used to ignite the main burners on a gas fired appliance.

Pipe cutter A tool for cutting copper, iron or steel pipes. Cutting is achieved by hard steel discs or wheels which bite into the pipe walls as the tool is revolved around the pipe.

Pipe ring A ring-shaped clamp or bracket, made in halves for screwing or bolting together, which forms part of an assembly for supporting a pipe.

Pipe wrench A heavy wrench with serrated jaws for gripping, screwing or unscrewing low carbon steel pipes and fittings.

Pitch The ratio of the height to the span of a roof, or its angle of inclination to the horizontal.

Pitcher tee A tee on which the branch is swept into the main pipe with a gently curved turn.

Pitch fibre A compound of wood or asbestos fibre and refined coal tar pitch used for making pipes, which are suitable for surface water drainage.

Plastics A name commonly used to describe a group of materials including polyvinyl chloride, polystyrene, polythene, perspex etc. Plastics are either thermosetting (those which harden once and for all time when heated), or thermoplastics (those which soften whenever they are heated).

Pliers A holding or gripping tool, pivoted like a pair of scissors. Some types have blades for cutting thin wire built into the jaws.

Plug A small threaded fitting used to screw into a female connection to seal it off.

Plug cock A simple valve, in which the liquid or gas passes through a hole in a tapered plug. The valve is opened or closed by turning the plug through 90°.

Polystyrene A material used in its expanded form as an insulator for pipes and cisterns.

Poly tetra flour ethylene (PTFE) A plastics material which is used as a thin tape or paste as a jointing medium in pipe threads.

Polythene Abbreviated from polyethylene, a chemically inert synthetic rubber used for making pipes and cisterns for cold water services.

Polyurethane A plastics which in foamed form is used as an insulating material in cavity walls.

Polyvinyl chloride (PVC) A vinyl resin, which is impervious to water, oils and petrol and is particularly incombustible. Used for making gutters, soil and waste pipes, storage cisterns, drainage components.

Portable electric tool A hand-held tool, driven by an electric motor, e.g. electric drill.

Pouring rope Also called a running rope or squirrel's tail, usually made from asbestos fibre rope and secured with a steel spring loaded clip. Used for containing poured molten lead when jointing cast iron pipes in a horizontal position.

Pressed steel A sheet steel which is hot pressed into plumbing components such as baths, sink units and flushing cisterns. The steel is protected by a layer of vitreous enamel.

Primary flow and return The pipes in which water circulates between a boiler and a hot water storage tank or cylinder.

Propane A colourless gas which burns in air to carbon dioxide and water.

Quarter bend A 90° bend in a pipe. Other bends are proportional to this, a one-eighth bend being 45°.

Radiator A sealed container, usually for water with tappings for pipework connections if part of a central heating system.

Radius A straight line running from the centre of a circle to any point on the circumference. Its length is half that of the diameter.

Rainwater head The enlarged entrance at the head of a rainwater pipe, often used as a collection point for other rainwater pipes.

Ramp A short length of pipeline or channel laid at a steeper gradient than the adjoining portions.

Reducer A fitting which enables a pipe diameter or socket connection to be reduced in bore.

Rest bend A drainage fitting in the form of a bend with a support web or shoe, to secure the bend at the base of a discharge pipe.

Ridge The apex or highest point of a roof.

Rodding eye A cover, plate, plug or cap which when removed provides access to the inside of a pipeline.

Rose Head or outlet of a shower fitting.

Saddle A drainage fitting used to connect a branch drain to a larger drain or sewer pipe.

Saddle clip A fixing which passes round the front of a pipe and is screwed to the surface behind the pipe via two lugs which are part of the saddle.

Saddle piece A piece of sheet weathering, formed to cover a joint at a vulnerable position, e.g. where a ridge meets an abutment wall or another roof.

Sal-ammoniac (NH_4Cl, ammonium chloride). A flux used in soldering.

Sarking felt Bituminous flax felt laid over rafters prior to battening and slating or tiling.

Scribe To cut a line on the surface of a material with a sharp pointed tool.

Scribing plate A metal plate used with a pair of dividers to mark out a branch joint on lead pipework.

Seal The depth of water contained in a trap to prevent foul air or gases passing through. The seal depth is measured from the water level down to the crown of the U-shaped part of the trap.

Sealing compound A material used to fill and seal the surface of an expansion joint. It can be applied like a mastic from a pressure cartridge or gun.

Seam A joint, fold or welt formed in sheet metal weatherings. A seamed edge is often formed at the front of an apron flashing to stiffen the edge, provide a more attractive finish, resist the possibility of capillary attraction between the roof and slates or surface and the flashing, or to form a safety edge so that the material is easier to handle and fix.

Seaming pliers A pair of pliers with jaws specially shaped and extended for forming seams or welts.

Secondary circulation A pipework circuit which supplies hot water to appliances which are located some distance from the hot water storage vessel. The circuit may have gravity or pumped circulation.

Secret gutter A nearly hidden gutter, in which the gutter is concealed by the roof covering.

Secret tack A strip of lead, soldered or welded on to the back of a lead sheet. The strip is passed through a slot in the roof boarding and screwed to the inside of it, providing a hidden fixing. Often used for securing vertical lead bays or panels.

Se-duct A flue which is also used as an air intake for gas appliances in multi-storey buildings, enabling many appliances to be supplied from the same flue.

Service pipe A water or gas pipe between the main and the premises receiving the supply.

Shave hook A hand tool used for shaving or cleaning the surface of lead pipes or sheet before soldering or welding.

Sheet A description of aluminium, copper, lead or zinc which is thicker than foil, thinner than 6.35 mm and more than 450 mm wide. Foil is thinner than the size quoted. Strip is narrower than the width quoted.

Sherardizing The coating of small iron or steel components with zinc by heating them with zinc dust in a revolving drum at about 350 °C.

Shoe A rainwater pipe fitting located at the base of a stack, used to direct water away from the structure.

Single point heater A small gas or electric water heater which usually supplies water to one appliance only.

Single stack system A form of one-pipe system which may have waste water and soil discharging into it, all or most of the trap ventilating pipes are omitted.

Sink A waste water fitting usually located in a kitchen or in an area where food preparation occurs.

Skylight A window or light incorporated into the slope of a roof.

Sleeve A pipe built into a wall or floor to allow a smaller diameter service to pass through, leaving it free to expand or contract, without damaging the fabric or structure of the building.

Slop sink A hopper-shaped sink, with a flushing rim and outlet similar to those of a WC pan, for the reception and discharge of human excreta.

Small bore system A system of heating pipework in which the sizes of circuit involved do not exceed 19 mm bore. Circulation is pumped.

Smoke test A method of tracing leaks in sanitation or drainage pipework. The test used is an air (pneumatic) test, and smoke is introduced to find a suspected leak after the air test has failed. Smoke should not be used on PVC pipework systems.

Soaker A small piece of flexible material bent to form a watertight joint at an abutment between a roof and a wall.

Socket The enlarged end of a pipe into which a similar pipe (spigot) is fixed, or, a fitting threaded internally, used for jointing threaded ends of pipe or tube.

Soil The waste products of the human body, as opposed to waste water, i.e. water from bathing or food preparation.

Solder An alloy used for joining other metals.

Soldered dot A method of securing sheet lead to flat, vertical or sloping surfaces.

Solvent A liquid capable of dissolving solids.

Space nipple A short section of threaded pipe with a space between the threads. Nipples with a formed grip surface are called hexagon nipples.

Sparge pipe A perforated pipe used for flushing a urinal stall.

Splash lap That part of an overcloak of a roll or drip which extends on to the flat surface of the next sheet.

Split collar A collar cut lengthwise and secured together with a steel band or clip. Used on flue pipes to assist with disconnection or removal of the appliance.

Spring A steel coil used for bending copper and lead pipes.

Stack A name used to describe a vertical soil, waste or rainwater discharge pipe.

Stainless steel A steel containing chromium and nickel. It is highly corrosion-resistant and is used for waste and sanitary appliances and small diameter plumbing pipework.

Standing seam A raised seam in flexible metal roofing usually running from ridge to eaves.

Step flashing A sheet weathering built into the horizontal joints of brickwork to make a watertight joint between a wall or chimney and the sloping part of a roof. The flashing steps down the thickness of a brick and bed joint with the slope of the roof.

Step turner A hardwood tool used for forming the turned step (usually 25 mm wide through 90°) on sheet metal stepped flashing.

Stopcock Or stop valve, a control valve used for regulating the supply of water in a service pipe or high pressure pipeline.

Storage cistern An open-topped vessel used for storing a quantity of cold water to supply cold water draw-off points at a lower level. The cistern should have a close fitting lid or cover to keep its contents clean and uncontaminated.

Storage water heater A gas or electric water heater which heats a quantity of water and stores it for use at a later time.

Straight edge A long piece of seasoned timber or metal with parallel and straight edges often used in conjunction with a spirit level.

S trap A trap in which the outlet is vertical and in line with the trap inlet.

Strap boss A fitting designed to clip around and on to a pipe, enabling a branch connection to be made to that pipe.

Surface water Rainwater.

Swan neck A name for an offset fitting, particularly that between an eaves gutter and a rainwater discharge pipe.

Sweat To unite or bond metal surfaces together by allowing molten solder to flow between them and adhere to their surfaces.

Sweep tee A pitcher tee, usually for copper or low carbon steel pipes in which the branch connection gently curves into the main run.

Tack A form of fixing or cleat, used mainly for securing sheet weatherings.

Taft joint A small wiped soldered joint contained in the opened out end of a piece of lead pipe and whatever spigot is entering the lead. Also called a finger wiped joint.

Tang The pointed end of a steel tool such as a file, rasp or wood chisel, which is driven into the wooden handle.

Tank A closed straight sided storage vessel generally used for storing hot water or oil.

Tan pin A conical shaped boxwood tool used for opening out the end of a lead pipe. Steel tan pins are available for copper pipes.

Tap A screwed plug, accurately threaded, made of hard steel and used for cutting internal threads.

Taper pipe A drainage fitting which is used as an increaser or reducer in a pipeline.

Taper thread A standard screwed thread used on pipes and fittings to ensure a watertight or gas tight joint.

Tee A fitting, which is a short section of pipe with three openings, one of which is a branch which is usually set at a right angle and located midway between the other two openings.

Temper To toughen steels and non-ferrous metals by the application of controlled heat and cooling.

Template A full size pattern of metal or wood used for forming shapes or testing the accuracy of a manufactured component.

Terminal A cowl or open hood used to finish the end of a gas flue etc.

Thermal movement Movement caused by expansion or contraction due to temperature change.

Thermoplastic Description of a synthetic resin or other material which softens on heating and hardens again on cooling.

Thermostat A device, usually electrical, for sensing temperature rise or fall. They are often used for maintaining a constant temperature, making or breaking a circuit, which turns off or on the gas or electrical supply.

Tilting fillet A small timber of triangular cross-section, used in roofworks to tilt slates or tiles slightly less steeply than the rest of the roof.

Tingle A tab or cleat, usually a strip of aluminium, copper, lead or zinc used for holding down the edge of a metal weathering.

Tinman's solder A fine solder containing more tin than wiping solder (grade D), so that its melting point is lower.

Tinning Coating copper, brass, lead or other metals with a film of tin or tin alloy (solder).

Tin snips Strong scissors used for cutting sheet metals.

Toe board A scaffold board set on its edge at the side of a scaffold to prevent materials or tools being kicked or knocked off the working platform.

Toggle bolt A fixing device which enables a sound fixing to be made to thin board such as hardboard or plasterboard.

Tommy bar A loose bar or rod inserted into a hole in a box spanner or capstan to provide the leverage for turning it.

Torus roll A horizontal wooden roll usually located at the intersection between the two slopes of a mansard roof, weathered with a sheet material.

Trap A U-shaped bend in a pipe or sanitary appliance used to retain a quantity of water to prevent foul air or gases passing through.
Tubular trap A fitting formed from pipe, thus ensuring an unrestricted flow through.
Twist drill A hardened steel bit with helical cutting edges, used in electric or hand drills for cutting circular holes in metal or wood.
Two-pipe system (1) A sanitation system consisting of a soil stack and a waste water stack, each with its own anti-siphonage or vent pipe.
Two-pipe system (2) A heating circuit with a flow and return connected to each radiator or heat emitter.

'U' gauge A glass or plastics 'U' tube half filled with water, one end of the 'U' being connected by flexible tube to a system of pipes under an air pressure test. A manometer.
Undercloak The lower layer or cover of sheet weathering material at a drip or roll which is covered by the overcloak.
Underlay A layer of felt or building paper laid beneath sheet metal weatherings to assist movement and provide a degree of noise and heat insulation.
Union A screwed pipe fitting, usually of brass or low carbon steel. It enables pipes or appliances to be quickly connected or disconnected.
Upstand That part of a sheet weathering or flashing which turns up against a vertical surface without being fitted into it. The upstand is usually covered with a flashing.
'U' value A thermal transmittance value, determined by experiment for a m^2 of a certain floor, wall, roof in a particular situation.

Valley The intersection between two sloping surfaces of a roof.
Valve A device to open or close a flow of liquid, gas or air, or to regulate a flow.
Vent An outlet for air or gases.
Verdigris Green basic acetate of copper formed as a protective patina over copper exposed to the air.
Verge The edge of a sloping roof which overhangs a gable.
Vermiculite A mica which is used as a light insulating aggregate.
Vice A screwed metal or timber clamp, usually fixed to a workbench or tripod and used for holding materials while they are being worked.
Vitreous china A ceramic material used for the manufacture of sanitary appliances.
Vitreous enamel A hard, smooth, glass-like surface attached to cast iron or pressed steel.
Vitrified clay A vitreous form of clay used for the manufacture of drainage pipes and components.
Volt The unit of electrical pressure, related to the units of flow (amperes) and power (watts), watts = volts × amperes. Electromotive force.

Warning pipe An overflow pipe from a cistern, which discharges water in a conspicuous position.
Washer A flat ring made of rubber, leather, plastics or fibrous composition used to form or make a seal between two surfaces. Alternatively, a flat ring made of metal.
Waste discharge pipe A pipe carrying waste water from a waste appliance.
Waste disposal unit An electrically driven rubbish grinder in domestic properties located in the kitchen sink, used for cutting and shredding kitchen waste. The waste is disposed of via a trap and discharge pipe into the drainage system.
Waste water All dirty water discharged from a waste appliance (bath, wash basin, sink, shower) but excluding soil or rainwater discharge.
Water level An instrument for setting out or transferring levels on a building site. It consists of a rubber tube connecting two vertical glass tubes containing water. The level of water in one tube is the same as that in the other if there are no kinks or air locks in the rubber tube.
Water softener A chemical plant for treating water. It removes from the water the calcium and manganese salts which cause hardness. The most used types are base exchange softeners, and the soda-lime process.
Water table The level of water in the ground or sub-soil.

Water test A form of drain testing, used generally for new installations. The drain is filled with water and subjected to hydrostatic pressure for a specified period of time.

Water waste preventer A cistern for flushing a WC or a slop sink (abbreviation WWP).

Watt A unit of power.

Weathering A sheet covering to the roof or part of the roof of a building.

Wedge A tapered piece of metal used for securing a flashing in between two bricks or similar.

Welt A folded joint or seam used in sheet metal roofing or weatherings.

Wing nut A thumb screw nut, which has wings, enabling it to be turned by hand without the use of a spanner.

Wiped joint A joint made with plumber's solder (grade D) between suitable materials, e.g. lead to lead, or lead to brass. The solder is moulded around the joint with a wiping cloth.

Wood roll A piece of timber, round-topped and tapering towards its base, fixed on to a roof surface to enable flexible metal materials to be lapped over it and jointed.

Wrench A form of gripping tool or spanner, usually adjustable.

Zeolites Minerals which are used in the base exchange process of water softening.

Self-assessment questions

1 An intersection between a roof surface and a wall rising above it is called:
(a) a mansard
(b) eaves
(c) an abutment
(d) an open soffit

2 The property which allows a metal to be drawn out or elongated without fracturing is called:
(a) eutectic
(b) malleability
(c) softness
(d) ductility

3 To keep boxwood tools in good condition it is advisable to:
(a) keep them dry
(b) rub frequently with linseed oil
(c) rub with glasspaper
(d) rub gently with grease

4 To prevent the plaster cracking, when a hot water pipe passes through a brick wall it should have:
(a) a reducing joint
(b) a flexible joint
(c) a soldered joint
(d) a metal sleeve

5 The welding of sheet lead can be described as:
(a) autogenous
(b) electrolytic
(c) capillary
(d) brazing

6 A warning pipe is fitted to a:
(a) hot water boiler
(b) hot water cylinder
(c) storage cistern
(d) hot water tank

7 The loss of moisture in vapour form from a liquid is called:
(a) expansion
(b) evaporation
(c) erosion
(d) condensation

8 A scaffold board set on its edge at the side of a scaffold to prevent tools or material falling below is called a:
(a) straight edge
(b) guard rail
(c) toe board
(d) putlog

9 The lowest point of the internal surface of a pipe or channel at any cross section is called:
(a) invert
(b) crown
(c) soffit
(d) benching

10 Pipes and components which have become encrusted with hard lime or other salts deposited from the hot water heater are said to be:
(a) solvent
(b) corroded
(c) furred
(d) eroded

Answers to self-assessment questions

Chapter 1 Safety

1	(a)	**6**	(a)
2	(d)	**7**	(d)
3	(c)	**8**	(c)
4	(b)	**9**	(d)
5	(d)	**10**	(a)

Chapter 2 Materials

1 Aluminium, copper, brass, bronze, lead, zinc, etc.
2 A metallic substance produced by mixing different metals or by mixing metals with non-metallic elements
3 Asbestos – asphalt
4 Brass, bronze, gunmetal, pewter, steel, solder
5 Bauxite, copper pyrites, galena
6 Copper tube to BS 2871 Part 1 Table Y
7 Cast, milled
8 Thermoplastics, thermosetting plastics
9 (a) High rate of linear expansion
(b) Low resistance to damage from fire attack
10 (a) Lighter
(b) Non-corrosive
(c) Poor conductors of heat

Chapter 3 Tools and equipment

1	(c)	**6**	(a)
2	(b)	**7**	(b)
3	(d)	**8**	(d)
4	(a)	**9**	(c)
5	(b)	**10**	(b)

Chapter 4 Communication: sketching, drawing and geometry

1	(c)	**6**	(d)
2	(a)	**7**	(b)
3	(d)	**8**	(a)
4	(c)	**9**	(c)
5	(a)	**10**	(d)

Chapter 5 Cold water supply

1	(b)	**6**	(d)
2	(b)	**7**	(a)
3	(c)	**8**	(b)
4	(b)	**9**	(a)
5	(b)	**10**	(c)

Chapter 6 Hot water supply

1	(c)	**6**	(b)
2	(d)	**7**	(d)
3	(b)	**8**	(a)
4	(b)	**9**	(a)
5	(a)	**10**	(b)

Chapter 7 Sanitary appliances

1	(c)	**6**	(d)
2	(c)	**7**	(c)
3	(b)	**8**	(b)
4	(a)	**9**	(d)
5	(b)	**10**	(c)

Chapter 8 Roofwork and sheet weatherings

1 (c)
2 (b)
3 (a)
4 (d)
5 (c)
6 (c)
7 (b)
8 (a)
9 (d)
10 (c)

Chapter 9 Calculations

1 8.372
2 27.19
3 1304.53
4 8.04
5 $\frac{3}{7}$
6 $\frac{17}{40}$
7 £10.66
8 8.4 m^2
9 0.728 m^3
10 £24.20 and £84.70

Chapter 10 Science

1 Joule
2 Mercury, alcohol
3 A crayon which changes colour when the component or material with which it is marked reaches a predetermined temperature
4 Copper, steel, brass
5 Polythene
6 4 °C
7 Gases
8 Newton
9 1 kg
10 The density of a substance is defined as its mass per unit of volume

Chapter 11 Glossary

1 (c)
2 (d)
3 (b)
4 (d)
5 (a)
6 (c)
7 (b)
8 (c)
9 (a)
10 (c)

Index